Philosophical Studies Series

Volume 144

Philosophical Studies Series aims to provide a forum for the best current research in contemporary philosophy broadly conceived, its methodologies, and applications. Since Wilfrid Sellars and Keith Lehrer founded the series in 1974, the book series has welcomed a wide variety of different approaches, and every effort is made to maintain this pluralism, not for its own sake, but in order to represent the many fruitful and illuminating ways of addressing philosophical questions and investigating related applications and disciplines.

The book series is interested in classical topics of all branches of philosophy including, but not limited to:

- Ethics
- Epistemology
- Logic
- Philosophy of language
- Philosophy of logic
- Philosophy of mind
- Philosophy of religion
- Philosophy of science

Special attention is paid to studies that focus on:

- the interplay of empirical and philosophical viewpoints
- the implications and consequences of conceptual phenomena for research as well as for society
- philosophies of specific sciences, such as philosophy of biology, philosophy of chemistry, philosophy of computer science, philosophy of information, philosophy of neuroscience, philosophy of physics, or philosophy of technology; and
- contributions to the formal (logical, set-theoretical, mathematical, information-theoretical, decision-theoretical, etc.) methodology of sciences.

Likewise, the applications of conceptual and methodological investigations to applied sciences as well as social and technological phenomena are strongly encouraged.

Philosophical Studies Series welcomes historically informed research, but privileges philosophical theories and the discussion of contemporary issues rather than purely scholarly investigations into the history of ideas or authors. Besides monographs, *Philosophical Studies Series* publishes thematically unified anthologies, selected papers from relevant conferences, and edited volumes with a well-defined topical focus inside the aim and scope of the book series. The contributions in the volumes are expected to be focused and structurally organized in accordance with the central theme(s), and are tied together by an editorial introduction. Volumes are completed by extensive bibliographies.

The series discourages the submission of manuscripts that contain reprints of previous published material and/or manuscripts that are below 160 pages/88,000 words.

For inquiries and submission of proposals authors can contact the editor-in-chief Mariarosaria Taddeo via: mariarosaria.taddeo@oii.ox.ac.uk

More information about this series at http://www.springer.com/series/6459

Luciano Floridi
Editor

Ethics, Governance, and Policies in Artificial Intelligence

Springer

Editor
Luciano Floridi
Oxford Internet Institute
University of Oxford
Oxford, UK

ISSN 0921-8599 ISSN 2542-8349 (electronic)
Philosophical Studies Series
ISBN 978-3-030-81906-4 ISBN 978-3-030-81907-1 (eBook)
https://doi.org/10.1007/978-3-030-81907-1

This Springer imprint is published by the registered company Springer Nature Switzerland AG
The registered company address is: Gewerbestrasse 11, 6330 Cham, Switzerland

Acknowledgements

I shall not repeat here the acknowledgements that can be found in each chapter and corresponding article, but rather thank all the people who have made this book possible. First of all, Danuta Farah, my personal assistant. She carefully and patiently edited the original articles and skilfully managed the production process. Without her contribution and organisational support, this book would have been impossible. Next, my colleague and co-director of the Digital Ethics Lab, Mariarosaria Taddeo, for her many ideas and suggestions in the past that led to this book, and her encouragement to pursue the project of a unified anthology of "the best of" in the ethics of AI by the DELab. And finally, all the authors whose brilliant intellectual work is showcased in this volume: Nikita Aggarwal, Monica Beltrametti, Christopher Burr, Raja Chatila, Patrice Chazerand, Josh Cowls, Alexander Denev, Virginia Dignum, Anat Elhalal, Robert Gorwa, Indra Joshi, Thomas C. King, Libby Kinsey, Michelle Seng Ah Lee, Christoph Luetge, Caio C. V. Machado, Tom McCutcheon, Robert Madelin, Jessica Morley, Carl Ohman, Ugo Pagallo, Huw Roberts, Francesca Rossi, Burkhard Schafer, Mariarosaria Taddeo, Andreas Tsamados, Peggy Valcke, Effy Vayena, Vincent Wang, and David S. Watson.

Contents

Contributors

Nikita Aggarwal Faculty of Law, Oxford Internet Institute, University of Oxford, Oxford, UK

Monica Beltrametti Naver Corporation, Grenoble, France

Christopher Burr Alan Turing Institute, London, UK

Raja Chatila French National Center of Scientific Research, Paris, France
Institute of Intelligent Systems and Robotics at Pierre, Marie Curie University, Paris, France

Patrice Chazerand Digital Europe, Brussels, Belgium

Josh Cowls Oxford Internet Institute, University of Oxford, Oxford, UK
Alan Turing Institute, London, UK

Alexander Denev Deloitte, London, UK

Virginia Dignum Department of Computing Science, University of Umeå, Umeå, Sweden
Delft Design for Values Institute, Delft University of Technology, Delft, the Netherlands

Anat Elhalal Digital Catapult, London, UK

Luciano Floridi Oxford Internet Institute, University of Oxford, Oxford, UK

Robert Gorwa Department of Politics and International Relations, Saint Anthony's College, University of Oxford, Oxford, UK

Indra Joshi NHSX, London, UK

Thomas C. King Oxford Internet Institute, University of Oxford, Oxford, UK
Amherst, Cheltenham, UK

Libby Kinsey Digital Catapult, London, UK

Michelle Seng Ah. Lee University of Cambridge, Cambridge, UK

Christoph Luetge TUM School of Governance, Technical University of Munich, Munich, Germany

Caio C. V. Machado Oxford Internet Institute, University of Oxford, Oxford, UK

Robert Madelin Defence Science and Technology Laboratories, Salisbury, UK
Centre for Technology and Global Affairs, University of Oxford, Oxford, UK

Tom McCutcheon Defence Science and Technology Laboratories, Salisbury, UK

Jessica Morley Oxford Internet Institute, University of Oxford, Oxford, UK

Carl Ohman Uppsala University, Uppsala, Sweden

Ugo Pagallo Department of Law, University of Turin, Turin, Italy

Huw Roberts Oxford Internet Institute, University of Oxford, Oxford, UK

Francesca Rossi IBM Research, Albany, NY, USA
University of Padova, Padova, Italy

Burkhard Schafer School of Law, University of Edinburgh Law School, Edinburgh, UK

Mariarosaria Taddeo Oxford Internet Institute, University of Oxford, Oxford, UK
Alan Turing Institute, London, UK

Andreas Tsamados Oxford Internet Institute, University of Oxford, Oxford, UK

Peggy Valcke Centre for IT & IP Law, Catholic University of Leuven, Leuven, Flanders, Belgium
Bocconi University, Milan, Italy

Effy Vayena Bioethics, Health Ethics and Policy Lab, ETH Zurich, Zurich, Switzerland

Vincent Wang Department of Computer Science, University of Oxford, Oxford, UK

David S. Watson Oxford Internet Institute, University of Oxford, Oxford, UK
Department of Statistical Science, University College London, London, UK

Chapter 1
Introduction – The Importance of an Ethics-First Approach to the Development of AI

Luciano Floridi

Abstract This is the introduction to the volume. It highlights the various "seasons" through which the development of AI has gone, and how the failures and successes of AI raise ethical questions, and require an ethical approach.

Keywords Artificial Intelligence (AI) · Ethics of AI · Summer of AI · Winter of AI

The trouble with seasonal metaphors is that they are cyclical. If you say that artificial intelligence (AI) got through a bad winter, you must also remember that winter will return, and you better be ready. An AI winter is that stage when technology, business, and the media get out of their warm and comfortable bubble, cool down, temper their sci-fi speculations and unreasonable hypes, and come to terms with what AI can or cannot really do as a technology (Floridi 2019), without exaggeration. Investments become more discerning, and journalists stop writing about AI, to chase some other fashionable topics and fuel the next fad.

AI has had several winters.[1] Among the most significant, there was one in the late seventies, and another at the turn of the eighties and nineties. Today, we are talking about another predictable winter (Nield 2019; Walch 2019; Schuchmann 2019).[2] AI is subject to these hype cycles because it is a hope or fear that we have entertained since we were thrown out of paradise: some form of agency that does everything for us, instead of us, better than us, with all the dreamy advantages (we shall be on holiday forever) and the nightmarish risks (we are going to be enslaved) that this entails. For some people, speculating about all this is irresistible. It is the wild west of

[1] https://en.wikipedia.org/wiki/AI_winter

[2] Even the BBC, which has contributed to the hype (see for example: https://www.bbc.co.uk/programmes/p031wmt7), now acknowldges it might have been… a hype: https://www.bbc.co.uk/news/technology-51064369

L. Floridi (✉)
Oxford Internet Institute, University of Oxford, Oxford, UK
e-mail: luciano.floridi@oii.ox.ac.uk

L. Floridi (ed.), *Ethics, Governance, and Policies in Artificial Intelligence*, Philosophical Studies Series 144, https://doi.org/10.1007/978-3-030-81907-1_1

"what if" scenarios. But I hope the reader will forgive me for a "I told you so" moment. For some time, I have been warning against commentators and "experts", who were competing to see who could tell the tallest tale (Floridi 2016). A web of myths ensued. They spoke of AI as if it were the ultimate panacea, which would solve everything and overcome everything; or as the final catastrophe, a superintelligence that would destroy millions of jobs, replacing lawyers and doctors, journalists and researchers, truckers and taxi drivers, and ending by dominating human beings as if they were pets at best. Many followed Elon Musk in declaring the development of AI the greatest existential risk run by humanity. As if most of humanity did not live in misery and suffering. As if wars, famine, pollution, global warming, social injustice, and fundamentalism were science fiction, or just negligible nuisances, unworthy of their considerations. They insisted that law and regulations were always going to be too late and never catch up with AI, when in fact laws and norms are not about the speed but about the direction of innovation, for they should steer the proper development of a society (if we like where we are heading, we cannot go there quickly enough). Today, we know that legislation is coming, at least in the EU. They claimed AI was a magic black box, which we could never explain, when in fact it is a matter of the correct level of abstraction (Floridi 2008) at which to interpret the complex interactions engineered – even car traffic downtown becomes a black box if you wish to know why every single individual is there at that moment. Today there is a growing development of adequate tools to monitor and understand how machine learning systems reach their outcomes (Watson and Floridi 2020). They spread scepticism about the possibility of an ethical framework that would synthesise what we mean by socially good AI, when in fact the EU, the OECD, and China have converged on very similar principles that offer a common platform for further agreements (Floridi and Cowls 2019). Sophists in search of headlines. They should be ashamed and apologize. Not only for their untenable comments, but also for the great irresponsibility and alarmism, which have misled public opinion both about a potentially useful technology – that could provide helpful solutions, from medicine to security and monitoring systems (Taddeo and Floridi 2018) – and about the real risks – which we know are concrete but so much less fancy, from everyday manipulation of choices (Milano et al. 2019) to increased pressure on individual and group privacy (Floridi 2014), from cyberconflicts to the use of AI by organised crime for money laundering and identity theft (King et al. 2020).

The risk of every AI summer is that over-inflated expectations turn into a mass distraction. The risk of every AI winter is that the backlash is excessive, the disappointment too negative, and potentially valuable solutions are thrown out with the water of the illusions. Managing the world is an increasingly complex task: megacities and their "smartification" offer a good example. And we have planetary problems – such as global warming, social injustice, and migration – which require ever higher degrees of coordination to be solved. It seems obvious that we need all the good technology that we can design, develop, and deploy to cope with these challenges, and all human intelligence we can exercise to put this technology in the service of a better future. AI can play an important role in all this because we need increasingly smarter ways of processing immense quantities of data, sustainably and efficiently. But AI must be treated as a normal technology, neither as a miracle nor as a plague, and as one of the many solutions that human ingenuity has managed to devise. This is also why the ethical debate is and will always remain an entirely human question, and a very crucial one, as this volume shows.

Now that the new winter is coming, we may try to learn some lessons, and avoid this yo-yo of unreasonable illusions and exaggerated disillusions. Let us not forget that the winter of AI should not be the winter of its opportunities. It certainly won't be the winter of its risks and ethical challenges. We need to ask ourselves whether AI solutions are really going to *replace* previous solutions – as the automobile has done with the carriage – *diversify* them – as did the motorcycle with the bicycle – or *complement* and *expand* them – as the digital smart watch has done with the analog one. What will the level of social acceptability or preferability be in whatever way AI survives the new winter? Are we really going to be wearing some kind of strange glasses to live in a virtual or augmented world created by AI? Consider that today many people are reluctant to wear glasses even when they seriously need them, just for aesthetic reasons. And then, are there feasible AI solutions in everyday life? Are the necessary skills, datasets, infrastructure, and business models in place to make an AI application successful? The futurologists find these questions boring. They like a single, simple idea, which interprets and changes everything, that can be spread thinly across an easy book that makes the reader feel intelligent, a book to be read by everyone today and ignored by all tomorrow. It is the bad diet of junk fast-food for thoughts and the curse of the airport bestseller. We need to resist oversimplification. This time let us think more deeply and extensively on what we are doing and planning with AI. The exercise is called philosophy, not futurology.

This volume is meant to contribute to such an exercise in slower and deeper thinking. It collects some of the most significant outcomes of the research on the ethics of AI conducted by members of the Digital Ethics Lab (DELab), the OII research group that I direct at the University of Oxford, also in collaboration with other colleagues. The chapters have appeared before in a variety of peer-reviewed, international journals, but never together. For the sake of consistency, they have not been modified in content, only in format. The hope is that the reader will find having the whole collection in one place not just convenient, but also intellectually useful, to see the patterns and developments in reasonings and conclusions. As the ethical debate on AI becomes increasingly specialised, mainstream, and practically oriented, the hope is that the chapter in this book may help establish a robust foundation for further studies. Whether this hope is realistic only the reader can judge.

References

Floridi, Luciano. 2008. The method of levels of abstraction. *Minds and Machines* 18 (3): 303–329.

———. 2014. Open data, data protection, and group privacy. *Philosophy & Technology* 27 (1): 1–3.

———. 2016. Should we be afraid of AI. *Aeon Essays*. https://aeon.co/essays/true-ai-is-both-logically-possible-and-utterly-implausible

———. 2019. What the near future of artificial intelligence could be. *Philosophy & Technology* 32 (1): 1–15. https://doi.org/10.1007/s13347-019-00345-y.

Floridi, Luciano, and Josh Cowls. 2019. A unified framework of five principles for AI in society. *Harvard Data Science Review* 1 (1): 99.

King, Thomas C., Nikita Aggarwal, Mariarosaria Taddeo, and Luciano Floridi. 2020. Artificial intelligence crime: An interdisciplinary analysis of foreseeable threats and solutions. *Science and Engineering Ethics* 26 (1): 89–120.

Milano, Silvia, Mariarosaria, Taddeo, and Luciano, Floridi. 2019. *Recommender systems and their ethical challenges*. Available at SSRN 3378581.

Nield, Thomas. 2019. Is deep learning already hitting its limitations? And is another AI winter coming? *Towards Data Science*, January 5. https://towardsdatascience.com/is-deep-learning-already-hitting-its-limitations-c81826082ac3

Schuchmann, Sebastian. 2019. Probability of an approaching AI winter. *Towards Data Science*, August 17. https://towardsdatascience.com/probability-of-an-approaching-ai-winter-c2d818fb338a

Taddeo, Mariarosaria, and Luciano Floridi. 2018. How AI can be a force for good. *Science* 361 (6404): 751–752.

Walch, Kathleen. 2019. Are we heading for another AI winter soon? *Forbes*, October 20. https://www.forbes.com/sites/cognitiveworld/2019/10/20/are-we-heading-for-another-ai-winter-soon/#783bf81256d6

Watson, David S., and Luciano Floridi. 2020. The explanation game: A formal framework for interpretable machine learning. *Synthese*. https://doi.org/10.1007/s11229-020-02629-9.

Chapter 2
A Unified Framework of Five Principles for AI in Society

Luciano Floridi and **Josh Cowls**

Abstract Artificial Intelligence (AI) is already having a major impact on society. As a result, many organizations have launched a wide range of initiatives to establish ethical principles for the adoption of socially beneficial AI. Unfortunately, the sheer volume of proposed principles threatens to overwhelm and confuse. How might this problem of 'principle proliferation' be solved? In this paper, we report the results of a fine-grained analysis of several of the highest-profile sets of ethical principles for AI. We assess whether these principles converge upon a set of agreed-upon principles, or diverge, with significant disagreement over what constitutes 'ethical AI.' Our analysis finds a high degree of overlap among the sets of principles we analyze. We then identify an overarching framework consisting of five core principles for ethical AI. Four of them are core principles commonly used in bioethics: beneficence, non-maleficence, autonomy, and justice. On the basis of our comparative analysis, we argue that a new principle is needed in addition: explicability, understood as incorporating both the epistemological sense of intelligibility (as an answer to the question 'how does it work?') and in the ethical sense of accountability (as an answer to the question: 'who is responsible for the way it works?'). In the ensuing discussion, we note the limitations and assess the implications of this ethical framework for future efforts to create laws, rules, technical standards, and best practices for ethical AI in a wide range of contexts.

Keywords Accountability · Autonomy · Artificial Intelligence · Beneficence · Ethics · Explicability · Fairness · Intelligibility · Justice · Non-maleficence

---

L. Floridi (✉)
Oxford Internet Institute, University of Oxford, Oxford, UK
e-mail: luciano.floridi@oii.ox.ac.uk

J. Cowls
Oxford Internet Institute, University of Oxford, Oxford, UK

Alan Turing Institute, London, UK
e-mail: josh.cowls@oii.ox.ac.uk

L. Floridi (ed.), *Ethics, Governance, and Policies in Artificial Intelligence*,
Philosophical Studies Series 144, https://doi.org/10.1007/978-3-030-81907-1_2

2.1 Introduction

Artificial Intelligence (AI) is already having a major impact on society. The key questions are how, where, when, and by whom the impact of AI will be felt. As a result, many organizations have launched a wide range of initiatives to establish ethical principles for the adoption of socially beneficial AI. Unfortunately, the sheer volume of proposed principles threatens to become overwhelming and confusing, posing two potential problems.[1] Either the various sets of ethical principles for AI are similar, leading to unnecessary repetition and redundancy, or, if they differ significantly, confusion and ambiguity will result instead. The worst outcome would be a 'market for principles' where stakeholders may be tempted to 'shop' for the most appealing ones (Floridi 2019b).

How might this problem of 'principle proliferation' be solved? In this paper, we report the results of a fine-grained analysis of several of the highest-profile sets of ethical principles for AI. We assess whether these principles are convergent, with a set of agreed-upon principles, or divergent, with significant disagreement over what constitutes 'ethical AI.' Our analysis finds a high degree of overlap among the sets of principles we analyze. We then identify an overarching framework consisting of five core principles for ethical AI. In the ensuing discussion, we note the limitations and assess the implications of this ethical framework for future efforts to create laws, rules, standards, and best practices for ethical AI in a wide range of contexts.

2.2 Artificial Intelligence: A Research Area in Search of a Definition

AI has been defined in many ways. Today, it comprises several techno-scientific branches, well summarized in Fig. 2.1 (see also the articles by Dick and Jordan in this issue for enlightening analyses).

Altogether, AI paradigms still satisfy the classic definition provided by John McCarthy, Marvin Minsky, Nathaniel Rochester, and Claude Shannon in their seminal Proposal for the Dartmouth Summer Research Project on Artificial Intelligence, the founding document and later event that established the new field of AI in 1955:

> For the present purpose the artificial intelligence problem is taken to be that of making a machine behave in ways that would be called intelligent if a human were so behaving. (Quotation from the 2006 re-issue in McCarthy et al. 2006 [1955]).

This is a counterfactual: were a human to behave in that way, that behaviour would be called intelligent. It does not mean that the machine is intelligent, or even thinking. The latter scenario is a fallacy, and smacks of superstition. Just because a

---

[1] These are not the only problems, see (Floridi 2019b).

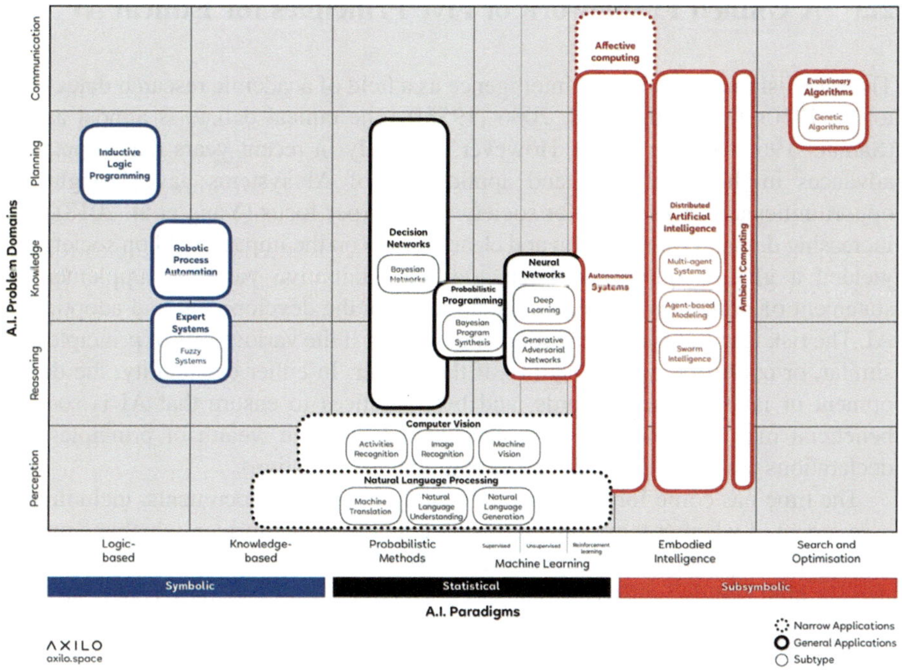

Fig. 2.1 AI Knowledge Map (AIKM). (Source: Corea (2019), reproduced with permission courtesy of F. Corea)

dishwasher cleans the dishes as well as (or even better than) I do does not mean that it cleans them like I do, or needs any intelligence to achieve its task. The same counterfactual understanding of AI underpins the Turing test (Floridi et al. 2009), which, in this case, checks the ability of a machine to perform a task in such a way that the outcome would be indistinguishable from the outcome of a human agent working to achieve the same task (Turing 1950).

The classic definition enables one to conceptualize AI as a growing resource of interactive, autonomous, and often self-learning agency (in the machine learning sense, see Fig. 2.1), that can deal with tasks that would otherwise require human intelligence and intervention to be performed successfully. In short, AI is defined on the basis of engineered outcomes and actions and so, in what follows, we shall treat AI as a reservoir of smart agency on tap (see also Floridi 2019a). This is sufficiently general to capture the many ways in which AI is discussed in the documents we analyse in the rest of this article.

2.3 A Unified Framework of Five Principles for Ethical AI

The establishment of artificial intelligence as a field of academic research dates back to the 1950s (McCarthy et al. 2006 [1955]). The ethical debate is almost as old (Samuel 1960; Wiener 1960). However, it is only in recent years that impressive advances in the capabilities and applications of AI systems have brought the opportunities and risks of AI for society into sharper focus (Yang et al. 2018). The increasing demand for reflection and clear policies on the impact of AI on society has yielded a glut of initiatives. Each additional initiative yields a supplementary statement of principles, values, or tenets to guide the development and adoption of AI. The risk is unnecessary repetition and overlap, if the various sets of principles are similar, or confusion and ambiguity, if they differ. In either eventuality, the development of laws, rules, standards, and best practices to ensure that AI is socially beneficial may be delayed by the need to navigate the wealth of principles and declarations set out by an ever-expanding array of initiatives.

The time has come for a comparative analysis of these documents, including an assessment of whether they converge or diverge and, if the former, whether a unified framework may therefore be synthesised. For this comparative analysis, we identified six high-profile initiatives established in the interest of socially beneficial AI:

1. The Asilomar AI Principles, developed under the auspices of the Future of Life Institute, in collaboration with attendees of the high-level Asilomar conference of January 2017 (hereafter 'Asilomar'; Asilomar AI Principles 2017)
2. The Montreal Declaration for Responsible AI, developed under the auspices of the University of Montreal, following the Forum on the Socially Responsible Development of AI of November 2017 (hereafter 'Montreal'; Montreal Declaration 2017)[2]
3. The General Principles offered in the second version of Ethically Aligned Design: A Vision for Prioritizing Human Well-being with Autonomous and Intelligent Systems. This crowd-sourced global treatise received contributions from 250 global thought leaders to develop principles and recommendations for the ethical development and design of autonomous and intelligent systems, and was published in December 2017 (hereafter 'IEEE'; IEEE 2017, p. 6)[3]
4. The Ethical Principles offered in the Statement on Artificial Intelligence, Robotics and 'Autonomous' Systems, published by the European Commission's European Group on Ethics in Science and New Technologies, in March 2018 (hereafter 'EGE'; EGE 2018, pp. 16–20)

[2]The Montreal Declaration is currently open for comments as part of a redrafting exercise. The principles we refer to here are those which were publicly announced as of May 1, 2018.

[3]The third version of Ethically Aligned Design will be released in 2019 following wider public consultation.

5. The 'five overarching principles for an AI code' offered in UK House of Lords Artificial Intelligence Committee's report, AI in the UK: ready, willing and able?, published in April 2018 (hereafter 'AIUK'; House of Lords 2018, §417)
6. The Tenets of the Partnership on AI, a multi-stakeholder organization consisting of academics, researchers, civil society organisations, companies building and utilising AI technology, and other groups (hereafter 'the Partnership'; Partnership on AI 2018).

Each set of principles meets three basic criteria: they are recent, published within the last 3 years; directly relevant to AI and its impact on society as a whole (thus excluding documents specific to a particular domain, industry, or sector); and highly reputable, published by authoritative, multi-stakeholder organizations with at least national scope.[4] Taken together, they yield 47 principles.[5] Overall, we find a degree of coherence and overlap between the six sets of principles that is impressive and reassuring. This convergence can most clearly be shown by comparing the sets of principles with the four core principles commonly used in bioethics: beneficence, non-maleficence, autonomy, and justice (Beauchamp and Childress 2012). The comparison should not be surprising. Of all areas of applied ethics, bioethics is the one that most closely resembles digital ethics in dealing ecologically with new forms of agents, patients, and environments (Floridi 2013). Yet while the four bioethical principles adapt surprisingly well to the fresh ethical challenges posed by artificial intelligence, they do not offer a perfect translation. As we shall see, the underlying meaning of each of the principles is contested, with similar terms often used to mean different things. Nor are the four principles exhaustive. On the basis of our comparative analysis, we argue that a new principle is needed in addition: explicability, understood as incorporating both intelligibility (for non-experts, e.g., patients or business customers, and for experts, e.g., product designers or engineers) and accountability. However, the convergence that we detect between these different sets of principles also demands caution. We explain the reasons for this caution in the following section, but first, we introduce the five principles.

[4] A similar evaluation of AI ethics guidelines has recently been undertaken by Hagendorff (2019), which adopts different criteria of inclusion and assessment. Note that the evaluation includes in its sample the set of principles we describe here.

[5] Of the six documents, the Asilomar Principles offer the largest number of principles with arguably the broadest scope. The 23 principles are organised under three headings, "research issues", "ethics and values", and "longer-term issues". We have omitted consideration of the five "research issues" here as they are related specifically to the practicalities of AI development in the narrower context of academia and industry. Similarly, the Partnership's eight Tenets consist of both intra-organisational objectives and wider principles for the development and use of AI. We include only the wider principles (the first, sixth, and seventh tenets).

2.3.1 Beneficence: Promoting Well-Being, Preserving Dignity, and Sustaining the Planet

The principle of creating AI technology that is beneficial to humanity is expressed in different ways across the six documents, but is perhaps the easiest of the four traditional bioethics principles to observe. Montreal and IEEE principles both use the term "well-being"; for Montreal, "the development of AI should ultimately promote the well-being of all sentient creatures," while IEEE states the need to "prioritize human well-being as an outcome in all system designs." AIUK and Asilomar both characterise this principle as the "common good": AI should "be developed for the common good and the benefit of humanity," according to AIUK. The Partnership describes the intention to "ensure that AI technologies benefit and empower as many people as possible", while the EGE emphasizes the principle of both "human dignity" and "sustainability." Its principle of "sustainability" articulates perhaps the widest of all interpretations of beneficence, arguing that "AI technology must be in line with ... ensur[ing] the basic preconditions for life on our planet, continued prospering for mankind and the preservation of a good environment for future generations." Taken together, the prominence of beneficence firmly underlines the central importance of promoting the well-being of people and the planet with AI.

2.3.2 Non-maleficence: Privacy, Security and 'Capability Caution'

Though 'do only good' (beneficence) and 'do no harm' (non-maleficence) may seem logically equivalent, they are not, and represent distinct principles. While the six documents all encourage the creation of beneficent AI, each one also cautions against various negative consequences of overusing or misusing AI technologies (Cowls et al. 2018). Of particular concern is the prevention of infringements on personal privacy, which is included as a principle in five of the six sets. Several of the documents emphasize avoiding the misuse of AI technologies in other ways. The Asilomar Principles warn against the threats of an AI arms race and of the recursive self-improvement of AI, while the Partnership similarly asserts the importance of AI operating "within secure constraints." The IEEE document meanwhile cites the need to "avoid misuse," and the Montreal Declaration argues that those developing AI "should assume their responsibility by working against the risks arising from their technological innovations." Yet from these various warnings, it is not entirely clear whether it is the people developing AI, or the technology itself, which should be encouraged not to do harm; in other words, whether it is Frankenstein or his monster against whose maleficence we should be guarding. At the heart of this quandary is the question of autonomy.

2.3.3 Autonomy: The Power to Decide (to Decide)

When we adopt AI and its smart agency, we willingly cede some of our decision-making power to technological artefacts. Thus, affirming the principle of autonomy in the context of AI means striking a balance between the decision-making power we retain for ourselves and that which we delegate to artificial agents. The risk is that the growth in artificial autonomy may undermine the flourishing of human autonomy. It is not therefore surprising that the principle of autonomy is explicitly stated in four of the six documents. The Montreal Declaration articulates the need for a balance between human- and machine-led decision-making, stating that "the development of AI should promote the autonomy [italics added] of all human beings". The EGE argues that autonomous systems "must not impair [the] freedom of human beings to set their own standards and norms," while AIUK adopts the narrower stance that "the autonomous power to hurt, destroy or deceive human beings should never be vested in AI." The Asilomar document similarly supports the principle of autonomy, insofar as "humans should choose how and whether to delegate decisions to AI systems, to accomplish human-chosen objectives." It is therefore clear both that the autonomy of humans should be promoted and that the autonomy of machines should be restricted and made intrinsically reversible, should human autonomy need to be protected or re-established (consider the case of a pilot able to turn off the automatic pilot and regain full control of the airplane). This introduces a notion we might call 'meta-autonomy,' or a 'decide-to-delegate' model: humans should retain the power to decide which decisions to take: exercising the freedom to choose where necessary, and ceding it in cases where overriding reasons, such as efficacy, may outweigh the loss of control over decision-making. Any delegation should also remain overridable in principle (i.e., deciding to decide again).

2.3.4 Justice: Promoting Prosperity, Preserving Solidarity, Avoiding Unfairness

The decision to make or delegate decisions does not take place in a vacuum. Nor is this capacity distributed equally across society. The consequences of this disparity in autonomy are addressed in the principle of justice. The importance of 'justice' is explicitly cited in the Montreal Declaration, which argues that "the development of AI should promote justice and seek to eliminate all types of discrimination," while the Asilomar Principles include the need for both "shared benefit" and "shared prosperity" from AI. Under its principle named "Justice, equity and solidarity," the EGE argues that AI should "contribute to global justice and equal access to the benefits" of AI technologies. It also warns against the risk of bias in datasets used to train AI systems, and—unique among the documents—argues for the need to defend against threats to "solidarity," including "systems of mutual assistance such as in social insurance and healthcare." Elsewhere 'justice' has still other meanings (especially in the sense of fairness), variously relating to the use of AI to correct past wrongs such as eliminating unfair discrimination, promoting diversity, and preventing the rise of new threats to justice. The diverse ways in which justice is characterised hints at a broader lack of clarity over AI as a human-made reservoir of 'smart agency.' Put simply, are we (humans) the patient, receiving the 'treatment' of AI, the doctor prescribing it? Or both? This question can only be resolved with the introduction of a fifth principle which emerges from our analysis.

2.3.5 Explicability: Enabling the Other Principles Through Intelligibility and Accountability

The short answer to the question of whether 'we' are the patient or the doctor is that actually we could be either, depending on the circumstances and on who 'we' are in everyday life. The situation is inherently unequal: a small fraction of humanity is currently engaged in the development of a set of technologies that are already transforming the everyday lives of almost everyone else. This stark reality is not lost on the authors whose documents we analyze. All of them refer to the need to understand and hold to account the decision-making processes of AI. Different terms express this principle: "transparency" in Asilomar and EGE; both "transparency" and "accountability" in IEEE; "intelligibility" in AIUK; and as "understandable and interpretable" by the Partnership. Each of these principles captures something seemingly novel about AI: that its workings are often invisible or unintelligible to all but (at best) the most expert observers.

The addition of the principle of 'explicability,' incorporating both the epistemological sense of 'intelligibility' (as an answer to the question 'how does it work?') and in the ethical sense of 'accountability' (as an answer to the question 'who is responsible for the way it works?'), is the crucial missing piece of the AI ethics jigsaw. It complements the other four principles: for AI to be beneficent and non-maleficent, we must be able to understand the good or harm it is actually doing to society, and in which ways; for AI to promote and not constrain human autonomy, our 'decision about who should decide' must be informed by knowledge of how AI would act instead of us; and for AI to be just, we must know whom to hold accountable in the event of a serious, negative outcome, which would require in turn adequate understanding of why this outcome arose.

2.3.6 A Synoptic View

Taken together, these five principles capture every one of the 47 principles contained in the six high-profile, expert-driven documents we analysed. Moreover, each principle is included in almost every statement of principles we analyzed (see

Table 2.1 The five principles in the six documents analysed and their occurrence in three recent documents

	Beneficence	Nonmaleficence	Autonomy	Justice	Explicability
AIUK	•	•	•	•	•
Asilomar	•	•	•	•	•
EGE	•	•	•	•	•
IEEE	•	•			•
Montreal	•	•	•	•	•
Partnership	•	•		•	•
AI4People	•	•	•	•	•
EC HLEG	•	•	•	•	•
OECD	•	•	•	•	•

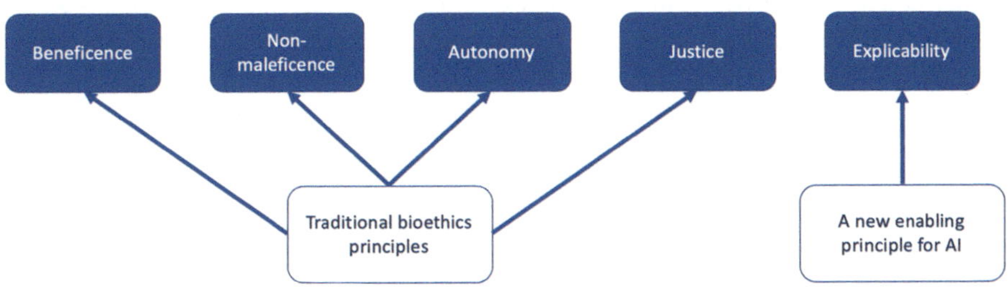

Fig. 2.2 An ethical framework of the five overarching principles for AI which emerged from the analysis

Table 2.1 below). The five principles therefore form an ethical framework within which policies, best practices, and other recommendations may be made. This framework of principles is shown in Fig. 2.2.

2.4 AI Ethics: Whence and for Whom?

It is important to note that each of the six sets of ethical principles for AI that we analyzed emerged either from initiatives with global scope, or from within western liberal democracies. For the framework to be more broadly applicable, it would undoubtedly benefit from the perspectives of regions and cultures presently un- or under-represented in our sample. Of particular interest in this respect is the role of China, which is already home to the world's most valuable AI start-up (Jezard 2018), enjoys various structural advantages in developing AI (Lee and Triolo 2017), and whose government has stated its ambitions to lead the world in state-of-the-art AI technology by 2030 (China State Council 2017). In its State Council Notice on AI and elsewhere, the Chinese government has expressed interest in further consideration of the social and ethical impact of AI (Ding 2018; Webster et al. 2017). Nor is enthusiasm about the use of technologies unique to governments, but it is also shared by general publics—more so those in China and India than in Europe or the USA, as new representative survey research shows (Vodafone Institute 2018).

An executive at the major Chinese technology firm Tencent recently suggested that the European Union should focus on developing AI which has "the maximum benefit for human life, even if that technology isn't competitive to take on [the] American or Chinese market" (Boland 2018). This has been echoed by claims that ethics may be "Europe's silver bullet" in the "global AI battle" (Delcker 2018). We disagree. Ethics is not the preserve of a single continent or culture. Every company, government agency, and academic institution designing, developing or deploying AI has an obligation to do so in line with an ethical framework along the lines of the one we present here, broadened to incorporate a more geographically, culturally, and socially diverse array of perspectives (Cowls et al. n.d.). Similarly, laws, rules, standards and best practices to constrain or control AI—including all those currently under consideration by regulatory bodies, legislatures and industry groups –would also benefit from close engagement with a unified framework of ethical principles.

2.5 Conclusion: From Principles to Practices

If the framework presented in this article provides a coherent and sufficiently comprehensive overview of the central ethical principles for AI (Floridi et al. 2018), then it can serve as the architecture within which laws, rules, technical standards, and best practices are developed for specific sectors, industries, and jurisdictions. In these contexts, the framework may play both an enabling role (consider, for example, the use of AI to help meet the United Nations Sustainable Development Goals), and a constraining one (as in the need to regulate AI technologies in the context of online crime and cyberwarfare: King et al. 2018; Taddeo and Floridi 2018). Indeed, the framework played a valuable role in the work of AI4People, Europe's first global forum on the social impact of AI, which recently adopted it to propose 20 concrete recommendations for a 'Good AI Society' to the European Commission (Floridi et al. 2018). Since then it has been largely adopted by the Ethics Guidelines for Trustworthy AI published by the European Commission's High-Level Expert Group on Artificial Intelligence (HLEGAI 2018, 2019), which in turn has influenced the OECD's Recommendation of the Council on Artificial Intelligence (OECD 2019), reaching 42 countries[6] (see Table 2.1).

The development and use of AI hold the potential for both positive and negative impact on society, to alleviate or to amplify existing inequalities, to cure old problems, or to cause new ones. Charting the course that is socially preferable will depend not only on well-crafted regulation and common standards, but also on the use of a framework of ethical principles, within which concrete actions can be

[6]https://www.oecd.org/science/forty-two-countries-adopt-new-oecd-principles-on-artificial-intelligence.htm

situated. We believe that the framework presented here as emerging from the current debate will serve as valuable architecture for securing positive social outcomes from AI technology and move from good principles to good practices (Cowls et al. 2019; Morley et al. 2019).

Disclosure Floridi chaired the AI4People project and Cowls was the rapporteur. Floridi is also a member of the European Commission's High-Level Expert Group on Artificial Intelligence (HLEGAI).

Funding Floridi's work was supported by (i) Privacy and Trust Stream—Social lead of the PETRAS Internet of Things research hub—PETRAS is funded by the UK Engineering and Physical Sciences Research Council (EPSRC), grant agreement no. EP/N023013/1; (ii) Facebook; and (iii) Google. Cowls is the recipient of a Doctoral Studentship from the Alan Turing Institute.

References

Beauchamp, T.L., and J.F. Childress. 2012. *Principles of biomedical ethics*. Oxford: Oxford University Press.

Boland, H. 2018. Tencent executive urges Europe to focus on ethical uses of artificial intelligence. *The Telegraph*, October 14. https://www.telegraph.co.uk/technology/2018/10/14/tencent-executive-urges-europe-focus-ethical-uses-artificial/

China State Council. 2017. *State Council notice on the issuance of the next generation Artificial Intelligence development plan*, July 8. Retrieved September 18, 2018, from http://www.gov.cn/zhengce/content/2017-07/20/content_5211996.htm. Translation by Creemers, R., G. Webster, P. Triolo, and E. Kania. https://www.newamerica.org/documents/1959/translation-fulltext-8.1.17.pdf

Corea, F. 2019. *AI knowledge map: How to classify AI technologies, a sketch of a new AI technology landscape*. First appeared in Medium—Artificial Intelligence. https://medium.com/@Francesco_AI/ai-knowledge-map-how-to-classify-ai-technologies-6c073b969020. Reproduced in Corea, F. 2019. *An introduction to data*, 26. Springer.

Cowls, J., L. Floridi, and M. Taddeo. 2018. *The challenges and opportunities of ethical AI*. Artificially Intelligent. https://digitransglasgow.github.io/ArtificiallyIntelligent/contributions/04_Alan_Turing_Institute.html

Cowls, J., T. C. King, M. Taddeo, and L. Floridi. 2019. *Designing AI for social good: Seven essential factors*. http://ssrn.com/abstract=3388669

Cowls, J., M.-T. Png, and Y. Au. n.d. *Foundations for geographic representation in algorithmic ethics*. Unpublished.

Delcker, J. 2018. *Europe's silver bullet in global AI battle: Ethics*. Politico, March 3. https://www.politico.eu/article/europe-silver-bullet-global-ai-battle-ethics/

Ding, J. 2018. *Deciphering China's AI dream*, March. https://www.fhi.ox.ac.uk/wp-content/uploads/Deciphering_Chinas_AI-Dream.pdf

European Group on Ethics in Science and New Technologies. 2018. *Statement on Artificial Intelligence, robotics and 'autonomous' systems*, March. https://ec.europa.eu/info/news/ethics-artificial-intelligence-statement-ege-released-2018-apr-24_en

Floridi, L. 2013. *The ethics of information*. Oxford: Oxford University Press.

———. 2019a. What the near future of Artificial Intelligence could be. *Philosophy & Technology* 32 (1): 1–15. https://doi.org/10.1007/s13347-019-00345-y.

———. 2019b. Translating principles into practices of digital ethics: Five risks of being unethical. *Philosophy & Technology* 32 (2): 185–193. https://doi.org/10.1007/s13347-019-00354-x.

Floridi, L., M. Taddeo, and M. Turilli. 2009. Turing's imitation game: Still an impossible challenge for all machines and some judges—An evaluation of the 2008 Loebner contest. *Minds and Machines* 19 (1): 145–150. https://doi.org/10.1007/s11023-008-9130-6.

Floridi, L., J. Cowls, M. Beltrametti, R. Chatila, P. Chazerand, V. Dignum, C. Luetge, R. Madelin, U. Pagallo, F. Rossi, B. Schafer, P. Valcke, and E. Vayena. 2018. AI4People—An ethical framework for a good AI society: Opportunities, risks, principles, and recommendations. *Minds and Machines* 28 (4): 689–707. https://doi.org/10.1007/s11023-018-9482-5.

Hagendorff, T. 2019. *The ethics of AI ethics—An evaluation of guidelines*. https://arxiv.org/abs/1903.03425

HLEGAI [High Level Expert Group on Artificial Intelligence], European Commission. 2018. *Draft ethics guidelines for trustworthy AI*, December 18. https://ec.europa.eu/digital-single-market/en/news/draft-ethics-guidelines-trustworthy-ai

———. 2019. *Ethics guidelines for trustworthy AI*, April 8. https://ec.europa.eu/digital-single-market/en/news/ethics-guidelines-trustworthy-ai

House of Lords Artificial Intelligence Committee. 2018. *AI in the UK: Ready, willing and able*, April 16. https://publications.parliament.uk/pa/ld201719/ldselect/ldai/100/10002.htm

Jezard, A. 2018. *China is now home to the world's most valuable AI start-up*. World Economic Forum, April 11. https://www.weforum.org/agenda/2018/04/chart-of-the-day-china-now-has-the-worlds-most-valuable-ai-startup/

King, T., N. Aggarwal, M. Taddeo, and L. Floridi 2018. *Artificial Intelligence crime: An interdisciplinary analysis of foreseeable threats and solutions*, May 22. https://ssrn.com/abstract=3183238

Lee, K., and P. Triolo 2017. *China's Artificial Intelligence revolution: Understanding Beijing's structural advantages*. Eurasian Group, December. https://www.eurasiagroup.net/live-post/ai-in-china-cutting-through-the-hype

McCarthy, J., M.L. Minsky, N. Rochester, and C.E. Shannon. 2006. A proposal for the Dartmouth summer research project on artificial intelligence, August 31, 1955. *AI Magazine* 27 (4): 12. https://doi.org/10.1609/aimag.v27i4.1904.

Montreal Declaration for a Responsible Development of Artificial Intelligence. 2017. *Announced at the conclusion of the Forum on the Socially Responsible Development of AI*, November 3. https://www.montrealdeclaration-responsibleai.com/the-declaration

Morley, J., L. Floridi, L. Kinsey, and A. Elhalal. 2019. *From what to how. An overview of AI ethics tools, methods and research to translate principles into practices*. ArXiv:1905.06876 [Cs]. Retrieved from http://arxiv.org/abs/1905.06876

OECD. 2019. *Recommendation of the Council on Artificial Intelligence*. https://legalinstruments.oecd.org/en/instruments/OECD-LEGAL-0449

Partnership on AI. 2018. *Tenets*. https://www.partnershiponai.org/tenets/

Samuel, A.L. 1960. Some moral and technical consequences of automation—A refutation. *Science* 132 (3429): 741–742. https://doi.org/10.1126/science.132.3429.741.

Taddeo, M., and L. Floridi. 2018. Regulate artificial intelligence to avert cyber arms race. *Nature* 556 (7701): 296–298.

The IEEE Initiative on Ethics of Autonomous and Intelligent Systems. 2017. *Ethically aligned design, v2*. https://ethicsinaction.ieee.org

Turing, A.M. 1950. Computing machinery and intelligence. *Mind* 5 (236): 433–460. https://doi.org/10.1093/mind/lix.236.433.

Vodafone Institute for Society and Communications. 2018. *New technologies: India and China see enormous potential—Europeans more sceptical*. https://www.vodafone-institut.de/digitising-europe/digitisation-india-and-china-see-enormous-potential/

Webster, G., R. Creemers, P. Triolo, and E. Kania. 2017. *China's plan to 'lead' in AI: Purpose, prospects, and problems*. New America, August, 1. https://www.newamerica.org/cybersecurity-initiative/blog/chinas-plan-lead-ai-purpose-prospects-and-problems/

Wiener, N. 1960. Some moral and technical consequences of automation. *Science* 131 (3410): 1355–1358. https://doi.org/10.1126/science.131.3410.1355.

Yang, G.Z., J. Bellingham, P.E. Dupont, P. Fischer, L. Floridi, R. Full, N. Jacobstein, V. Kumar, M. McNutt, R. Merrifield, B.J. Nelson, B. Scassellati, M. Taddeo, R. Taylor, M. Veloso, Z.L. Wang, and R. Wood. 2018. The grand challenges of science robotics. *Science robotics* 3 (14): eaar7650. https://doi.org/10.1126/scirobotics.aar7650.

Chapter 3
An Ethical Framework for a Good AI Society: Opportunities, Risks, Principles, and Recommendations

Luciano Floridi, Josh Cowls, Monica Beltrametti, Raja Chatila, Patrice Chazerand, Virginia Dignum, Christoph Luetge, Robert Madelin, Ugo Pagallo, Francesca Rossi, Burkhard Schafer, Peggy Valcke, and Effy Vayena

Abstract This article reports the findings of AI4People, a year-long initiative designed to lay the foundations for a "Good AI Society". We introduce the core opportunities and risks of AI for society; present a synthesis of five ethical principles that should undergird its development and adoption; and offer 20 concrete

---

L. Floridi (✉)
Oxford Internet Institute, University of Oxford, Oxford, UK
e-mail: luciano.floridi@oii.ox.ac.uk

J. Cowls
Oxford Internet Institute, University of Oxford, Oxford, UK

Alan Turing Institute, London, UK
e-mail: Josh.cowls@oii.ox.ac.uk

M. Beltrametti
Naver Corporation, Grenoble, France
e-mail: Monica.beltrametti@naverlabs.com

R. Chatila
French National Center of Scientific Research, Paris, France

Institute of Intelligent Systems and Robotics at Pierre, Marie Curie University, Paris, France
e-mail: Raja.chatila@sorbonne-universite.fr; chatila@isir.upmc.fr

P. Chazerand
Digital Europe, Brussels, Belgium
e-mail: patrice.chazerand@digitaleurope.org

V. Dignum
Department of Computing Science, University of Umeå, Umeå, Sweden

Delft Design for Values Institute, Delft University of Technology, Delft, the Netherlands
e-mail: virginia@cs.umu.se

C. Luetge
TUM School of Governance, Technical University of Munich, Munich, Germany
e-mail: luetge@tum.de

L. Floridi (ed.), *Ethics, Governance, and Policies in Artificial Intelligence*,
Philosophical Studies Series 144, https://doi.org/10.1007/978-3-030-81907-1_3

recommendations – to assess, to develop, to incentivise, and to support good AI – which in some cases may be undertaken directly by national or supranational policy makers, while in others may be led by other stakeholders. If adopted, these recommendations would serve as a firm foundation for the establishment of a Good AI Society.

Keywords Artificial intelligence · AI4People · Data Governance · Digital Ethics · Governance · Ethics of AI

3.1 Introduction

AI is not another utility that needs to be regulated once it is mature. It is a powerful force, a new form of smart agency, which is already reshaping our lives, our interactions, and our environments. AI4People was set up to help steer this powerful force towards the good of society, everyone in it, and the environments we share. This White Paper is the outcome of the collaborative effort by the AI4People Scientific Committee – comprising 12 experts and chaired by Luciano Floridi[1] – to propose a series of recommendations for the development of a Good AI Society.

[1] Besides Luciano Floridi, the members of the Scientific Committee are: Monica Beltrametti, Raja Chatila, Patrice Chazerand, Virginia Dignum, Christoph Luetge, Robert Madelin, Ugo Pagallo, Francesca Rossi, Burkhard Schafer, Peggy Valcke, and Effy Vayena. Josh Cowls is the rapporteur. Thomas Burri contributed to an earlier draft.

R. Madelin
Defence Science and Technology Laboratories, Salisbury, UK

Centre for Technology and Global Affairs, University of Oxford, Oxford, UK
e-mail: Robert.Madelin@ec.europa.eu; robert.madelin@fipra.com

U. Pagallo
Department of Law, University of Turin, Turin, Italy
e-mail: ugo.pagallo@unito.it

F. Rossi
IBM Research, Albany, NY, USA

University of Padova, Padova, Italy
e-mail: Francesca.Rossi2@ibm.com

B. Schafer
School of Law, University of Edinburgh Law School, Edinburgh, UK
e-mail: B.Schafer@ed.ac.uk

P. Valcke
Centre for IT & IP Law, Catholic University of Leuven, Leuven, Flanders, Belgium

Bocconi University, Milan, Italy
e-mail: peggy.valcke@kuleuven.be

E. Vayena
Bioethics, Health Ethics and Policy Lab, ETH Zurich, Zurich, Switzerland
e-mail: effy.vayena@hest.ethz.ch

The White Paper synthesises three things: the *opportunities* and associated *risks* that AI technologies offer for fostering human dignity and promoting human flourishing; the *principles* that should undergird the adoption of AI; and twenty specific *recommendations* that, if adopted, will enable all stakeholders to seize the opportunities, to avoid or at least minimise and counterbalance the risks, to respect the principles, and hence to develop a Good AI Society.

The White Paper is structured around four more sections after this introduction. Section 3.2 states the core opportunities for promoting human dignity and human flourishing offered by AI, together with their corresponding risks.[2] Section 3.3 offers a brief, high-level view of the advantages for organisations of taking an ethical approach to the development and use of AI. Section 3.4 formulates 5 ethical principles for AI, building on existing analyses, which should undergird the ethical adoption of AI in society at large. Finally, Sect. 3.5 offers 20 recommendations for the purpose of developing a Good AI Society in Europe.

Since the launch of AI4People in February 2018, the Scientific Committee has acted collaboratively to develop the recommendations in the final section of this paper. Through this work, we hope to have contributed to the foundation of a Good AI Society we can all share.

3.2 The Opportunities and Risks of AI for Society

That AI will have a major impact on society is no longer in question. Current debate turns instead on how far this impact will be positive or negative, for whom, in which ways, in which places, and on what timescale. Put another way, we can safely dispense with the question of *whether* AI will have an impact; the pertinent questions now are *by whom, how, where*, and *when* this positive or negative impact will be felt.

In order to frame these questions in a more substantive and practical way, we introduce here what we consider the four chief opportunities for society that AI offers. They are four because they address the four fundamental points in the understanding of human dignity and flourishing: *who we can become* (autonomous self-realisation); *what we can do* (human agency); *what we can achieve* (individual and societal capabilities); and *how we can interact with each other and the world* (societal cohesion). In each case, AI can be *used* to foster human nature and its potentialities, thus creating opportunities; *underused*, thus creating opportunity costs; or *overused* and *misused*, thus creating risks. As the terminology indicates, the assumption is that the *use* of AI is synonymous with good innovation and positive applications of this technology. However, fear, ignorance, misplaced concerns or excessive reaction may lead a society to *underuse* AI technologies below

[2]The analysis in this and the following two sections is also available in Cowls and Floridi (2018). Further analysis and more information on the methodology employed will be presented in Cowls and Floridi (Forthcoming).

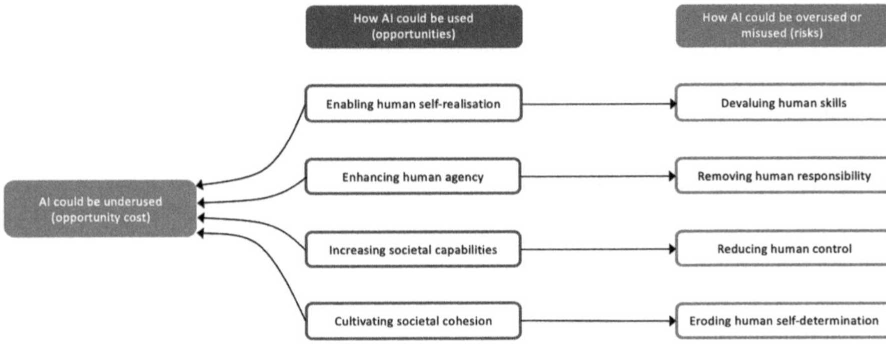

Fig. 3.1 Overview of the four core opportunities offered by AI, four corresponding risks, and the opportunity cost of underusing AI

their full potential, for what might be broadly described as the wrong reasons. This may cause significant opportunity costs. It might include, for example, heavy-handed or misconceived regulation, under-investment, or a public backlash akin to that faced by genetically modified crops (Imperial College 2017). As a result, the benefits offered by AI technologies may not be fully realised by society. These dangers arise largely from unintended consequences and relate typically to good intentions gone awry. However, we must also consider the risks associated with inadvertent *overuse* or wilful *misuse* of AI technologies, grounded, for example, in misaligned incentives, greed, adversarial geopolitics, or malicious intent. Everything from email scams to full-scale cyber-warfare may be accelerated or intensified by the malicious use of AI technologies (Taddeo 2017). And new evils may be made possible (King et al. 2018). The possibility of social progress represented by the aforementioned opportunities above must be weighed against the risk that malicious manipulation will be enabled or enhanced by AI. Yet a broad risk is that AI may be underused out of fear of overuse or misuse. We summarise these risks in Fig. 3.1 below, and offer a more detailed explanation in the text that follows.

3.2.1 *Who We Can Become: Enabling Human Self-Realisation, Without Devaluing Human Abilities*

AI may enable self-realisation, by which we mean the ability for people to flourish in terms of their own characteristics, interests, potential abilities or skills, aspirations, and life projects. Much as inventions, such as the washing machine, liberated people – particularly women – from the drudgery of domestic work, the "smart" automation of other mundane aspects of life may free up yet more time for cultural, intellectual and social pursuits, and more interesting and rewarding work. More AI may easily mean more human life spent more intelligently. The risk in this case is not the obsolescence of some old skills and the emergence of new ones *per se*, but the pace at which this is happening and the unequal distributions of the costs and benefits that result. A very fast devaluation of old skills and hence a quick disruption of the job market and the nature of employment can be seen at the level of both the individual and society. At the level of the individual, jobs are often intimately linked to personal identity, self-esteem, and social role or standing, all factors that may be adversely affected by redundancy, even putting to one side the potential for severe economic harm. Furthermore, at the level of society, the deskilling in sensitive, skill-intensive domains, such as health care diagnosis or aviation, may create dangerous vulnerabilities in the event of AI malfunction or an adversarial attack. Fostering the development of AI in support of new abilities and skills, while anticipating and mitigating its impact on old ones will require both close study and potentially radical ideas, such as the proposal for some form of "universal basic income", which is growing in popularity and experimental use. In the end, we need some intergenerational solidarity between those disadvantaged today and those advantaged tomorrow, to ensure that the disruptive transition between the present and the future will be as fair as possible, for everyone.

3.2.2 What We Can Do: Enhancing Human Agency, Without Removing Human Responsibility

AI is providing a growing reservoir of "smart agency". Put at the service of human intelligence, such a resource can hugely enhance human agency. We can do more, better, and faster, thanks to the support provided by AI. In this sense of "Augmented Intelligence", AI could be compared to the impact that engines have had on our lives. The larger the number of people who will enjoy the opportunities and benefits of such a reservoir of smart agency "on tap", the better our societies will be. Responsibility is therefore essential, in view of what sort of AI we develop, how we use it, and whether we share with everyone its advantages and benefits. Obviously, the corresponding risk is the absence of such responsibility. This may happen not just because we have the wrong socio-political framework, but also because of a "black box" mentality, according to which AI systems for decision-making are seen as being beyond human understanding, and hence control. These concerns apply not only to high-profile cases, such as deaths caused by autonomous vehicles, but also to more commonplace but still significant uses, such as in automated decisions about parole or creditworthiness.

Yet the relationship between the degree and quality of agency that people enjoy and how much agency we delegate to autonomous systems is not zero-sum, either pragmatically or ethically. In fact, if developed thoughtfully, AI offers the opportunity of *improving and multiplying* the possibilities for human agency. Consider examples of "distributed morality" in human-to-human systems such as peer-to-peer lending (Floridi 2013). Human agency may be ultimately supported, refined and expanded by the embedding of "facilitating frameworks", designed to improve the likelihood of morally good outcomes, in the set of functions that we delegate to AI systems. AI systems could, if designed effectively, amplify and strengthen shared moral systems.

3.2.3 What We Can Achieve: Increasing Societal Capabilities, Without Reducing Human Control

Artificial intelligence offers myriad opportunities for improving and augmenting the capabilities of individuals and society at large. Whether by preventing and curing diseases or optimising transportation and logistics, the use of AI technologies presents countless possibilities for reinventing society by radically enhancing what humans are collectively capable of. More AI may support better coordination, and hence more ambitious goals. Human intelligence augmented by AI could find new solutions to old and new problems, from a fairer or more efficient distribution of resources to a more sustainable approach to consumption. Precisely because such technologies have the potential to be so powerful and disruptive, they also introduce proportionate risks. Increasingly, we may not need to be either 'in or on the loop' (that is, as part of the process or at least in control of it), if we can delegate our tasks to AI. However, if we rely on the use of AI technologies to augment our own abilities in the wrong way, we may delegate important tasks and above all decisions to autonomous systems that should remain at least partly subject to human supervision and choice. This in turn may reduce our ability to monitor the performance of these systems (by no longer being 'on the loop' either) or preventing or redressing errors or harms that arise ('post loop'). It is also possible that these potential harms may accumulate and become entrenched, as more and more functions are delegated to artificial systems. It is therefore imperative to strike a balance between pursuing the ambitious opportunities offered by AI to improve human life and what we can achieve, on the one hand, and, on the other hand, ensuring that we remain in control of these major developments and their effects.

3.2.4 How We Can Interact: Cultivating Societal Cohesion, Without Eroding Human Self-Determination

From climate change and antimicrobial resistance to nuclear proliferation and fundamentalism, global problems increasingly have high degrees of coordination complexity, meaning that they can be tackled successfully only if all stakeholders co-design and co-own the solutions and cooperate to bring them about. AI, with its data-intensive, algorithmic-driven solutions, can hugely help to deal with such coordination complexity, supporting more societal cohesion and collaboration. For example, efforts to tackle climate change have exposed the challenge of creating a cohesive response, both within societies and between them. The scale of this challenge is such that we may soon need to decide between engineering the climate directly and designing societal frameworks to encourage a drastic cut in harmful emissions. This latter option might be undergirded by an algorithmic system to cultivate societal cohesion. Such a system would not be imposed from the outside; it would be the result of a self-imposed choice, not unlike our choice of not buying chocolate if we had earlier chosen to be on a diet, or setting up an alarm clock to wake up. "Self-nudging" to behave in socially preferable ways is the best form of nudging, and the only one that preserves autonomy. It is the outcome of human decisions and choices, but it can rely on AI solutions to be implemented and facilitated. Yet the risk is that AI systems may erode human self-determination, as they may lead to unplanned and unwelcome changes in human behaviours to accommodate the routines that make automation work and people's lives easier. AI's predictive power and relentless nudging, even if unintentional, should be at the service of human self-determination and foster societal cohesion, not undermining of human dignity or human flourishing.

Taken together, these four opportunities, and their corresponding challenges, paint a mixed picture about the impact of AI on society and the people in it. Accepting the presence of trade-offs, seizing the opportunities while working to anticipate, avoid, or minimise the risks head-on will improve the prospect for AI technologies to promote human dignity and flourishing. Having outlined the potential benefits to individuals and society at large of an ethically engaged approach to AI, in the next section we highlight the "dual advantage" to organisations of taking such an approach.

3.3 The Dual Advantage of an Ethical Approach to AI

Ensuring socially preferable outcomes of AI relies on resolving the tension between incorporating the benefits and mitigating the potential harms of AI, in short, simultaneously avoiding the misuse and underuse of these technologies. In this context, the value of an ethical approach to AI technologies comes into starker relief. Compliance with the law is merely necessary (the leas that is required), but significantly insufficient (not the most than can be done) (Floridi 2018). With an analogy, it is the difference between playing according to the rules, and playing well, so that one may win the game. Adopting an ethical approach to AI confers what we define here as a "dual advantage". On one side, ethics enables organisations to take advantage of the social value that AI enables. This is the advantage of being able to identify and leverage new opportunities that are socially acceptable or preferable. On the other side, ethics enables organisations to anticipate and avoid or at least minimise costly mistakes. This is the advantage of prevention and mitigation of courses of action that turn out to be socially unacceptable and hence rejected, even when legally unquestionable. This also lowers the opportunity costs of choices not made or options not grabbed for fear of mistakes.

Ethics' dual advantage can only function in an environment of public trust and clear responsibilities more broadly. Public acceptance and adoption of AI technologies will occur only if the benefits are seen as meaningful and risks as potential, yet preventable, minimisable, or at least something against which one can be protected, through risk management (e.g. insurance) or redressing. These attitudes will depend in turn on public engagement with the development of AI technologies, openness about how they operate, and understandable, widely accessible mechanisms of regulation and redress. In this way, an ethical approach to AI can also be seen as an early warning system against risks which might endanger entire organisations. The clear value to any organisation of the dual advantage of an ethical approach to AI amply justifies the expense of engagement, openness, and contestability that such an approach requires.

3.4 A Unified Framework of Principles for AI in Society

AI4People is not the first initiative to consider the ethical implications of AI. Many organisations have already produced statements of the values or principles that should guide the development and deployment of AI in society. Rather than conduct a similar, potentially redundant exercise here, we strive to move the dialogue forward, constructively, from principles to proposed policies, best practices, and concrete recommendations for new strategies. Such recommendations are not offered in a vacuum. But rather than generating yet another series of principles to serve as an ethical foundation for our recommendations, we offer a synthesis of existing sets of principles produced by various reputable, multi-stakeholder organisations and initiatives. A fuller explanation of the scope, selection and method of assessing these sets of principles is available in Cowls and Floridi (Forthcoming). Here, we focus on the commonalities and noteworthy differences observable across these sets of principles, in view of the 20 recommendations offered in the rest of the paper. The documents we assessed are:

1. the Asilomar AI Principles, developed under the auspices of the Future of Life Institute, in collaboration with attendees of the high-level Asilomar conference of January 2017 (hereafter "Asilomar"; Asilomar AI Principles 2017);
2. the Montreal Declaration for Responsible AI, developed under the auspices of the University of Montreal, following the Forum on the Socially Responsible Development of AI of November 2017 (hereafter "Montreal"; Montreal Declaration 2017);[3]
3. the General Principles offered in the second version of *Ethically Aligned Design: A Vision for Prioritizing Human Well-being with Autonomous and Intelligent Systems*. This crowd-sourced global treatise received contributions from

[3]The Montreal Declaration is currently open for comments as part of a redrafting exercise. The principles we refer to here are those which were publicly announced as of 1st May, 2018.

250 global thought leaders to develop principles and recommendations for the ethical development and design of autonomous and intelligent systems, and was published in December 2017 (hereafter "IEEE"; IEEE 2017);[4]
4. the Ethical Principles offered in the *Statement on Artificial Intelligence, Robotics and 'Autonomous' Systems,* published by the European Commission's European Group on Ethics in Science and New Technologies, in March 2018 (hereafter "EGE"; EGE 2018);
5. the "five overarching principles for an AI code" offered in paragraph 417 of the UK House of Lords Artificial Intelligence Committee's report, *AI in the UK: ready, willing and able?*, published in April 2018 (hereafter "AIUK"; House of Lords 2018); and
6. the Tenets of the Partnership on AI, a multistakeholder organisation consisting of academics, researchers, civil society organisations, companies building and utilising AI technology, and other groups (hereafter "the Partnership"; Partnership on AI 2018).

Taken together, they yield 47 principles.[5] Overall, we find an impressive and reassuring degree of coherence and overlap between the six sets of principles. This can most clearly be shown by comparing the sets of principles with the set of four core principles commonly used in bioethics: beneficence, non-maleficence, autonomy, and justice. The comparison should not be surprising. Of all areas of applied ethics, bioethics is the one that most closely resembles digital ethics in dealing ecologically with new forms of agents, patients, and environments (Floridi 2013). The four bioethical principles adapt surprisingly well to the fresh ethical challenges posed by artificial intelligence. But they are not exhaustive. On the basis of the following comparative analysis, we argue that one more, new principle is needed in addition: *explicability*, understood as incorporating both intelligibility and accountability.

[4]The third version of *Ethically Aligned Design* will be released in 2019 following wider public consultation.

[5]Of the six documents, the Asilomar Principles offer the largest number of principles with arguably the broadest scope. The 23 principles are organised under three headings, "research issues", "ethics and values", and "longer-term issues". We have omitted consideration of the five "research issues" here as they are related specifically to the practicalities of AI development, particularly in the narrower context of academia and industry. Similarly, the Partnership's eight Tenets consist of both intra-organisational objectives and wider principles for the development and use of AI. We include only the wider principles (the first, sixth, and seventh tenets).

3.4.1 Beneficence: Promoting Well-Being, Preserving Dignity, and Sustaining the Planet

Of the four core bioethics principles, beneficence is perhaps the easiest to observe across the six sets of principles we synthesise here. The principle of creating AI technology that is beneficial to humanity is expressed in different ways, but it typically features at the top of each list of principles. Montreal and IEEE principles both use the term "well-being": for Montreal, "the development of AI should ultimately promote the well-being of all sentient creatures"; while IEEE states the need to "prioritize human well-being as an outcome in all system designs". AIUK and Asilomar both characterise this principle as the "common good": AI should "be developed for the common good and the benefit of humanity", according to AIUK. The Partnership describes the intention to "ensure that AI technologies benefit and empower as many people as possible"; while the EGE emphasises the principle of both "human dignity" and "sustainability". Its principle of "sustainability" represents perhaps the widest of all interpretations of beneficence, arguing that "AI technology must be in line with ... ensur[ing] the basic preconditions for life on our planet, continued prospering for mankind and the preservation of a good environment for future generations". Taken together, the prominence of these principles of beneficence firmly underlines the central importance of promoting the well-being of people and the planet.

3.4.2 Non-maleficence: Privacy, Security and "Capability Caution"

Though "do only good" (beneficence) and "do no harm" (non-maleficence) seem logically equivalent, in both the context of bioethics and of the ethics of AI they represent distinct principles, each requiring explication. While they encourage well-being, the sharing of benefits and the advancement of the public good, each of the six sets of principles also cautions against the many potentially negative consequences of overusing or misusing AI technologies. Of particular concern is the prevention of infringements on personal privacy, which is listed as a principle in five of the six sets, and as part of the "human rights" principles in the IEEE document. In each case, privacy is characterised as being intimately linked to individuals' access to, and control over, how personal data is used.

Yet the infringement of privacy is not the only danger to be avoided in the adoption of AI. Several of the documents also emphasise the importance of avoiding the misuse of AI technologies in other ways. The Asilomar Principles are quite specific on this point, citing the threats of an AI arms race and of the recursive self-improvement of AI, as well as the need for "caution" around "upper limits on future AI capabilities". The Partnership similarly asserts the importance of AI operating "within secure constraints". The IEEE document meanwhile cites the need to "avoid misuse", while the Montreal Declaration argues that those developing AI "should assume their responsibility by working against the risks arising from their technological innovations", echoed by the EGE's similar need for responsibility.

From these various warnings, it is not entirely clear whether it is the people developing AI, or the technology itself, which should be encouraged not to do harm – in other words, whether it is Frankenstein or his monster against whose maleficence we should be guarding. Confused also is the question of intent: promoting non-maleficence can be seen to incorporate the prevention of both accidental (what we above call "overuse") and deliberate (what we call "misuse") harms arising. In terms of the principle of non-maleficence, this need not be an either/or question: the point is simply to prevent harms arising, whether from the intent of humans or the unpredicted behaviour of machines (including the unintentional nudging of human behaviour in undesirable ways). Yet these underlying questions of agency, intent and control become knottier when we consider the next principle.

3.4.3 Autonomy: The Power to Decide (Whether to Decide)

Another classic tenet of bioethics is the principle of autonomy: the idea that individuals have a right to make decisions for themselves about the treatment they do or not receive. In a medical context, this principle of autonomy is most often impaired when patients lack the mental capacity to make decisions in their own best interests; autonomy is thus surrendered involuntarily. With AI, the situation becomes rather more complex: when we adopt AI and its smart agency, we *willingly* cede some of our decision-making power to machines. Thus, affirming the principle of autonomy in the context of AI means striking a balance between the decision-making power we retain for ourselves and that which we delegate to artificial agents.

The principle of autonomy is explicitly stated in four of the six documents. The Montreal Declaration articulates the need for a balance between human- and machine-led decision-making, stating that "the development of AI should *promote* the autonomy of all human beings *and control* ... the autonomy of computer systems" (italics added). The EGE argues that autonomous systems "must not impair [the] freedom of human beings to set their own standards and norms and be able to live according to them", while AIUK adopts the narrower stance that "the autonomous power to hurt, destroy or deceive human beings should never be vested in AI". The Asilomar document similarly supports the principle of autonomy, insofar as "humans should choose how and whether to delegate decisions to AI systems, to accomplish human-chosen objectives".

These documents express a similar sentiment in slightly different ways, echoing the distinction drawn above between beneficence and non-maleficence: not only should the autonomy of humans be promoted, but also the autonomy of machines should be restricted and made intrinsically reversible, should human autonomy need to be re-established (consider the case of a pilot able to turn off the automatic pilot and regain full control of the airplane). Taken together, the central point is to protect the intrinsic value of human choice – at least for significant decisions – and, as a corollary, to contain the risk of delegating too much to machines. Therefore, what seems most important here is what we might call "meta-autonomy", or a "decide-to-delegate" model: humans should always retain the power to *decide which decisions to take*, exercising the freedom to choose where necessary, and ceding it in cases where overriding reasons, such as efficacy, may outweigh the loss of control over decision-making. As anticipated, any delegation should remain overridable in principle (deciding to decide again).

The decision to make or delegate decisions does not take place in a vacuum. Nor is this capacity to decide (to decide, and to decide again) distributed equally across society. The consequences of this potential disparity in autonomy are addressed in the final of the four principles inspired by bioethics.

3.4.4 Justice: Promoting Prosperity and Preserving Solidarity

The last of the four classic bioethics principles is justice, which is typically invoked in relation to the distribution of resources, such as new and experimental treatment options or simply the general availability of conventional healthcare. Again, this bioethics principle finds clear echoes across the principles for AI that we analyse. The importance of "justice" is explicitly cited in the Montreal Declaration, which argues that "the development of AI should promote justice and seek to eliminate all types of discrimination", while the Asilomar Principles include the need for both "shared benefit" and "shared prosperity" from AI. Under its principle named "Justice, equity and solidarity", the EGE argues that AI should "contribute to global justice and equal access to the benefits" of AI technologies. It also warns against the risk of bias in datasets used to train AI systems, and – unique among the documents – argues for the need to defend against threats to "solidarity", including "systems of mutual assistance such as in social insurance and healthcare". The emphasis on the protection of social support systems may reflect geopolitics, insofar as the EGE is a European body. The AIUK report argues that citizens should be able to "flourish mentally, emotionally and economically alongside artificial intelligence". The Partnership, meanwhile, adopts a more cautious framing, pledging to "respect the interests of all parties that may be impacted by AI advances".

As with the other principles already discussed, these interpretations of what justice means as an ethical principle in the context of AI are broadly similar, yet contain subtle distinctions. Across the documents, justice variously relates to

(a) using AI to correct past wrongs such as eliminating unfair discrimination;
(b) ensuring that the use of AI creates benefits that are shared (or at least shareable); and
(c) preventing the creation of *new* harms, such as the undermining of existing social structures.

Notable also are the different ways in which the position of AI, *vis-à-vis* people, is characterised in relation to justice. In Asilomar and EGE respectively, it is AI technologies themselves that "should benefit and empower as many people as possible" and "contribute to global justice", whereas in Montreal, it is "the *development* of AI" that "should promote justice" (italics added). In AIUK, meanwhile, people should flourish merely "alongside" AI. Our purpose here is not to split semantic hairs. The diverse ways in which the relationship between people and AI is described in these documents hints at broader confusion over AI as a man-made reservoir of "smart agency". Put simply, and to resume our bioethics analogy, are we (humans) the patient, receiving the "treatment" of AI, the doctor prescribing it? Or both? It seems that we must resolve this question before seeking to answer the next question of whether the treatment will even work. This is the core justification for our identification within these documents of a new principle, one that is not drawn from bioethics.

3.4.5 Explicability: Enabling the Other Principles Through Intelligibility and Accountability

The short answer to the question of whether "we" are the patient or the doctor is that actually we could be either – depending on the circumstances and on who "we" are in our everyday life. The situation is inherently unequal: a small fraction of humanity is currently engaged in the design and development of a set of technologies that are already transforming the everyday lives of just about everyone else. This stark reality is not lost on the authors whose documents we analyse. In all, reference is made to the need to *understand* and *hold to account* the decision-making processes of AI. This principle is expressed using different terms: "transparency" in Asilomar; "accountability" in EGE; both "transparency" and "accountability" in IEEE; "intelligibility" in AIUK; and as "understandable and interpretable" for the Partnership. Though described in different ways, each of these principles captures something seemingly novel about AI: that its workings are often invisible or unintelligible to all but (at best) the most expert observers.

The addition of this principle, which we synthesise as "explicability" both in the epistemological sense of "intelligibility" (as an answer to the question "how does it work?") and in the ethical sense of "accountability" (as an answer to the question: "who is responsible for the way it works?"), is therefore the crucial missing piece of the jigsaw when we seek to apply the framework of bioethics to the ethics of AI. It complements the other four principles: for AI to be beneficent and non-maleficent, we must be able to understand the good or harm it is actually doing to society, and in which ways; for AI to promote and not constrain human autonomy, our "decision about who should decide" must be informed by knowledge of how AI would act instead of us; and for AI to be just, we must ensure that the technology – or, more accurately, the people and organisations developing and deploying it – are held

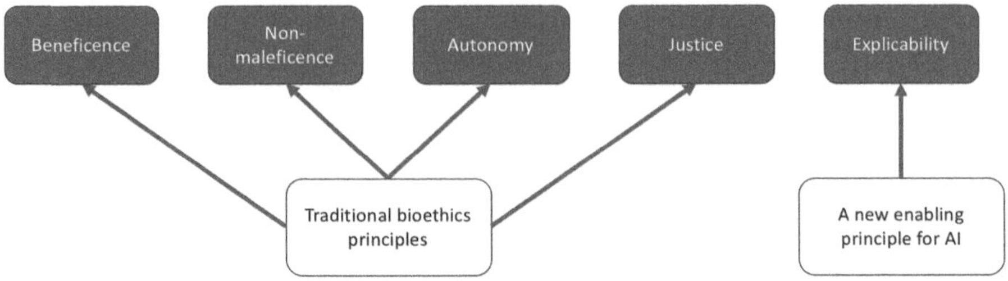

Fig. 3.2 An ethical framework for AI, formed of four traditional principles and a new one

accountable in the event of a negative outcome, which would require in turn some understanding of why this outcome arose. More broadly, we must negotiate the terms of the relationship between ourselves and this transformative technology, on grounds that are readily understandable to the proverbial person "on the street".

Taken together, we argue that these five principles capture the meaning of each of the 47 principles contained in the six high-profile, expert-driven documents, forming an ethical framework within which we offer our recommendations below. This framework of principles is shown in Fig. 3.2.

3.5 Recommendations for a Good AI Society

This section introduces the Recommendations for a Good AI Society. It consists of two parts: a Preamble, and 20 Action Points.

There are four kinds of Action Points: to *assess,* to *develop,* to *incentivise* and to *support.* Some recommendations may be undertaken directly, by national or European policy makers, in collaboration with stakeholders where appropriate. For others, policy makers may play an enabling role for efforts undertaken or led by third parties.

3.5.1 Preamble

We believe that, in order to create a Good AI Society, the ethical principles identified in the previous section should be embedded in the default practices of AI. In particular, AI should be designed and developed in ways that decrease inequality and further social empowerment, with respect for human autonomy, and increase benefits that are shared by all, equitably. It is especially important that AI be explicable, as explicability is a critical tool to build public trust in, and understanding of, the technology.

We also believe that creating a Good AI Society requires a multistakeholder approach, which is the most effective way to ensure that AI will serve the needs of society, by enabling developers, users and rule-makers to all be on board and collaborating from the outset.

Different cultural frameworks inform attitudes to new technology. This document represents a European approach, which is meant to be complementary to other approaches. We are committed to the development of AI technology in a way that *secures people's trust*, *serves the public interest*, and *strengthens shared social responsibility*.

Finally, this set of recommendations should be seen as a "living document". The Action Points are designed to be dynamic, requiring not simply single policies or one-off investments, but rather, continuous, ongoing efforts for their effects to be sustained.

3.5.2 Action Points

3.5.2.1 Assessment

1. ***Assess* the capacity of existing institutions, such as national civil courts, to redress the mistakes made or harms inflicted by AI systems**. This assessment should evaluate the presence of sustainable, majority-agreed foundations for liability from the design stage onwards in order to reduce negligence and conflicts (see also **Recommendation 5**).[6]
2. ***Assess* which tasks and decision-making functionalities should *not* be delegated to AI systems,** through the use of participatory mechanisms to ensure alignment with societal values and understanding of public opinion. This assessment should take into account existing legislation and be supported by ongoing dialogue between all stakeholders (including government, industry, and civil society) to debate how AI will impact society opinion (**in concert with Recommendation 17**).
3. ***Assess* whether current regulations are sufficiently grounded in ethics to provide a legislative framework that can keep pace with technological developments.** This may include a framework of key principles that would be applicable to urgent and/or unanticipated problems.

[6]Determining accountability and responsibility may usefully borrow from lawyers in Ancient Rome who would go by this formula 'cuius commoda eius et incommoda' ('the person who derives an advantage from a situation must also bear the inconvenience'). A good 2,200 years old principle that has a well-established tradition and elaboration could properly set the starting level of abstraction in this field.

3.5.2.2 Development

4. ***Develop* a framework to enhance the explicability of AI systems which make socially significant decisions.** Central to this framework is the ability for individuals to obtain a factual, direct, and clear explanation of the decision-making process, especially in the event of unwanted consequences. This is likely to require the development of frameworks specific to different industries, and professional associations should be involved in this process, alongside experts in science, business, law, and ethics.
5. ***Develop* appropriate legal procedures and improve the IT infrastructure of the justice system to permit the scrutiny of algorithmic decisions in court.** This is likely to include the creation of a framework for AI explainability as indicated in **Recommendation 4**, specific to the legal system. Examples of appropriate procedures may include the applicable disclosure of sensitive commercial information in IP litigation, and – where disclosure poses unacceptable risks, for instance to national security – the configuration of AI systems to adopt technical solutions by default, such as zero-knowledge proofs in order to evaluate their trustworthiness.
6. ***Develop* auditing mechanisms for AI systems to identify unwanted consequences, such as unfair bias, and (for instance, in cooperation with the insurance sector) a solidarity mechanism to deal with severe risks in AI-intensive sectors.** Those risks could be mitigated by multistakeholder mechanisms upstream. Pre-digital experience indicates that, in some cases, it may take a couple of decades before society catches up with technology by way of rebalancing rights and protection adequately to restore trust. The earlier that users and governments become involved – as made possible by ICT – the shorter this lag will be.
7. ***Develop* a redress process or mechanism to remedy or compensate for a wrong or grievance caused by AI.** To foster public trust in AI, society needs a widely accessible and reliable mechanism of redress for harms inflicted, costs incurred, or other grievances caused by the technology. Such a mechanism will necessarily involve a clear and comprehensive allocation of accountability to humans and/or organisations. Lessons could be learnt from the aerospace industry, for example, which has a proven system of handling unwanted consequences thoroughly and seriously. The development of this process must follow from the assessment of existing capacity outlined in **Recommendation 1**. If a lack of capacity is identified, additional institutional solutions should be developed at national and/or EU levels, to enable people to seek redress. Such solutions may include:

   - an "AI ombudsperson" to ensure the auditing of allegedly unfair or inequitable uses of AI;
   - a guided process for registering a complaint akin to making a Freedom of Information request; and

- the development of liability insurance mechanisms, which would be required as an obligatory accompaniment of specific classes of AI offerings in EU and other markets. This would ensure that the relative reliability of AI-powered artefacts, especially in robotics, is mirrored in insurance pricing and therefore in the market prices of competing products.[7]

Whichever solutions are developed, these are likely to rely on the framework for intelligibility proposed in **Recommendation 4.**

8. ***Develop* agreed-upon metrics for the trustworthiness of AI products and services, to be undertaken either by a new organisation, or by a suitable existing organisation. These metrics would serve as the basis for a system that enables the user-driven benchmarking of all marketed AI offerings.** In this way, an index for trustworthy AI can be developed and signalled, in addition to a product's price. This "trust comparison index" for AI would improve public understanding and engender competitiveness around the development of safer, more socially beneficial AI (e.g., "IwantgreatAI.org"). In the longer term, such a system could form the basis for a broader system of certification for deserving products and services, administered by the organisation noted here, and/or by the oversight agency proposed in **Recommendation 9**. The organisation could also support the development of codes of conduct (see **Recommendation 18**). Furthermore, those who own or operate inputs to AI systems and profit from it could be tasked with funding and/or helping to develop AI literacy programs for consumers, in their own best interest.
9. ***Develop* a new EU oversight agency responsible for the protection of public welfare through the scientific evaluation and supervision of AI products, software, systems or services.** This may be similar, for example, to the European Medicines Agency. Relatedly, a "post-release" monitoring system for AIs similar to, for example, the one available for drugs should be developed, with reporting duties for some stakeholders and easy reporting mechanisms for other users.
10. ***Develop* a European observatory for AI**. The mission of the observatory would be to watch developments, provide a forum to nurture debate and consensus, provide a repository for AI literature and software (including concepts and links to available literature), and issue step-by-step recommendation and guidelines for action.
11. ***Develop* legal instruments and contractual templates to lay the foundation for a smooth and rewarding human-machine collaboration in the work environment.** Shaping the narrative on the 'Future of Work' is instrumental to winning "hearts and minds". In keeping with 'A Europe that protects', the idea of "inclusive innovation" and to smooth the transition to new kinds of jobs, a

[7] Of course, to the extent that AI systems are 'products', general tort law still applies in the same way to AI as it applies in any instance involving defective products or services that injure users or do not perform as claimed or expected.

European AI Adjustment Fund could be set up along the lines of the European Globalisation Adjustment Fund.

3.5.2.3 Incentivisation

12. ***Incentivise* financially, at the EU level, the development and use of AI technologies within the EU that are socially preferable (not merely acceptable) and environmentally friendly (not merely sustainable but favourable to the environment)**. This will include the elaboration of methodologies that can help assess whether AI projects are socially preferable and environmentally friendly. In this vein, adopting a 'challenge approach' (see DARPA challenges) may encourage creativity and promote competition in the development of specific AI solutions that are ethically sound and in the interest of the common good.
13. ***Incentivise* financially a sustained, increased and coherent European research effort**, tailored to the specific features of AI as a scientific field of investigation. This should involve a clear mission to advance AI for social good, to serve as a unique counterbalance to AI trends with less focus on social opportunities.
14. ***Incentivise* financially cross-disciplinary and cross-sectoral cooperation and debate concerning the intersections between technology, social issues, legal studies, and ethics.** Debates about technological challenges may lag behind the actual technical progress, but if they are strategically informed by a diverse, multistakeholder group, they may steer and support technological innovation in the right direction. Ethics should help seize opportunities and cope with challenges, not only describe them. It is essential in this respect that diversity infuses the design and development of AI, in terms of gender, class, ethnicity, discipline and other pertinent dimensions, in order to increase inclusivity, toleration, and the richness of ideas and perspectives.
15. ***Incentivise* financially the inclusion of ethical, legal and social considerations in AI research projects.** In parallel, **incentivise regular reviews of legislation to test the extent to which it fosters socially positive innovation.** Taken together, these two measures will help ensure that AI technology has ethics at its heart and that policy is oriented towards innovation.
16. ***Incentivise* financially the development and use of lawfully de-regulated special zones within the EU for the empirical testing and development of AI systems.** These zones may take the form of a "living lab" (or *Tokku*), building on the experience of existing "test highways" (or *Teststrecken*). In addition to aligning innovation more closely with society's preferred level of risk, sandbox experiments such as these contribute to hands-on education and the promotion of accountability and acceptability at an early stage. "Protection by design" is intrinsic to this kind of framework.
17. ***Incentivise* financially research about public perception and understanding of AI and its applications, and the implementation of structured public**

consultation mechanisms to design policies and rules related to AI. This may include the direct elicitation of public opinion via traditional research methods, such as opinion polls and focus groups, as well as more experimental approaches, such as providing simulated examples of the ethical dilemmas introduced by AI systems, or experiments in social science labs. This research agenda should not serve merely to measure public opinion, but should also lead to the co-creation of policies, standards, best practices, and rules as a result.

3.5.2.4 Support

18. ***Support* the development of self-regulatory codes of conduct for data and AI related professions, with specific ethical duties.** This would be along the lines of other socially sensitive professions, such as medical doctors or lawyers, i.e., with the attendant certification of 'ethical AI' through trust-labels to make sure that people understand the merits of ethical AI and will therefore demand it from providers. Current attention manipulation techniques may be constrained through these self-regulating instruments.
19. ***Support* the capacity of corporate boards of directors to take responsibility for the ethical implications of companies' AI technologies**. For example, this may include improved training for existing boards and the potential development of an ethics committee with internal auditing powers. This could be developed within the existing structure of both one-tier and two-tier board systems, and/or in conjunction with the development of a mandatory form of "corporate ethical review board" to be adopted by organisations developing or using AI systems, to evaluate initial projects and their deployment with respect to fundamental principles.
20. ***Support* the creation of educational curricula and public awareness activities around the societal, legal, and ethical impact of Artificial Intelligence.** This may include:

    - curricula for schools, supporting the inclusion of computer science among the basic disciplines to be taught;
    - initiatives and qualification programmes in businesses dealing with AI technology, to educate employees on the societal, legal, and ethical impact of working alongside AI;
    - a European-level recommendation to include ethics and human rights in the degrees of data and AI scientists and other scientific and engineering curricula dealing with computational and AI systems;
    - the development of similar programmes for the public at large, with a special focus on those involved at each stage of management of the technology, including civil servants, politicians and journalists;
    - engagement with wider initiatives such as the ITU AI for Good events and NGOs working on the UN Sustainable Development Goals.

3.6 Conclusion

Europe, and the world at large, face the emergence of a technology that holds much exciting promise for many aspects of human life, and yet seems to pose major threats as well. This White Paper – and especially the Recommendations in the previous section – seek to nudge the tiller in the direction of ethically and socially preferable outcomes from the development, design and deployment of AI technologies. Building on our identification of both the core opportunities and the risks of AI for society as well as the set of five ethical principles we synthesised to guide its adoption, we formulated 20 Action Points in the spirit of collaboration and in the interest of creating *concrete* and *constructive* responses to the most pressing social challenges posed by AI.

With the rapid pace of technological change, it can be tempting to view the political process in the liberal democracies of today as old-fashioned, out-of-step, and no longer up to the task of preserving the values and promoting the interests of society and everyone in it. We disagree. With the Recommendations we offer here, including the creation of centres, agencies, curricula, and other infrastructure, we have made the case for an ambitious, inclusive, equitable programme of policy making and technological innovation, which we believe will contribute to securing the benefits and mitigating the risks of AI, for all people, and for the world we share.

Acknowledgements This publication would not have been possible without the generous support of Atomium – European Institute for Science, Media and Democracy. We are particularly grateful to Michelangelo Baracchi Bonvicini, Atomium's President, to Guido Romeo, its Editor in Chief, the staff of Atomium for their help, and to all the partners of the AI4People project and members of its Forum (http://www.eismd.eu/ai4people) for their feedback. The authors of this article are the only persons responsible for its contents and any remaining mistakes.

References

Asilomar AI Principles. 2017. *Principles developed in conjunction with the 2017 Asilomar conference* [Benevolent AI 2017]. Retrieved September 18, 2018, from https://futureoflife.org/ai-principles.

Cowls, J., and L. Floridi. 2018. *Prolegomena to a White Paper on Recommendations for the Ethics of AI* (June 19, 2018). Available at SSRN: https://ssrn.com/abstract=3198732.

———. Forthcoming. The utility of a principled approach to AI ethics.

European Group on Ethics in Science and New Technologies. 2018, March. *Statement on artificial intelligence, robotics and 'autonomous' systems*. Retrieved September 18, 2018, from https://ec.europa.eu/info/news/ethics-artificial-intelligence-statement-ege-released-2018-apr-24_en.

Imperial College London. 2017, October 11. *Written Submission to House of Lords Select Committee on Artificial Intelligence* [AIC0214]. Retrieved September 18, 2018, from http://bit.ly/2yleuET.

Floridi, L. 2018. Soft ethics and the governance of the digital. *Philosophy & Technology* 2018: 1–8.

———. 2013. *The ethics of information*. Oxford: Oxford University Press.

House of Lords Artificial Intelligence Committee. 2018, April 16. *AI in the UK: Ready, willing and able?* Retrieved September 18, 2018, from https://publications.parliament.uk/pa/ld201719/ldselect/ldai/100/10002.htm.

King, T., N. Aggarwal, M. Taddeo, and L. Floridi. 2018, May 22. *Artificial intelligence crime: An interdisciplinary analysis of foreseeable threats and solutions*. Available at SSRN: https://ssrn.com/abstract=3183238.

Montreal Declaration for a Responsible Development of Artificial Intelligence. 2017, November 3. *Announced at the conclusion of the Forum on the Socially Responsible Development of AI.* Retrieved September 18, 2018, from https://www.montrealdeclaration-responsibleai.com/the-declaration.

Partnership on AI. 2018. *Tenets*. Retrieved September 18, 2018, from https://www.partnershiponai.org/tenets/.

Taddeo, M. 2017. The limits of deterrence theory in cyberspace. *Philosophy & Technology* 2017: 1–17.

The IEEE Initiative on Ethics of Autonomous and Intelligent Systems. 2017. *Ethically aligned design, v2*. Retrieved September 18, 2018, from https://ethicsinaction.ieee.org.

Chapter 4
Establishing the Rules for Building Trustworthy AI

Luciano Floridi

Abstract In this chapter, I argue that the European commission's report, 'Ethics guidelines for trustworthy AI', provides a clear benchmark to evaluate the responsible development of AI systems, and to facilitate international support for AI solutions that are good for humanity and the environment.

Keywords Artificial Intelligence · AI Ethical principles · Ethics · European Commission · Trustworthy AI

AI is revolutionizing everyone's life, and it is crucial that it does so in the right way. AI's profound and far-reaching potential for transformation concerns the engineering of systems that have some degree of autonomous agency. This is epochal and requires establishing a new, ethical balance between human and artificial autonomy.

4.1 Careful Planning Rather Than Beta Testing

As a new kind of autonomous, smart agency, AI could bring enormous benefits—individually, socially and environmentally. It could represent a force for good in a world that is increasingly complex and requires sophisticated solutions to deal with large-scale and interrelated issues. The 17 UN Sustainable Development Goals show that humanity is struggling with many challenges, on many vital fronts, and it would be unwise not to make use of AI solutions. However, what processes and decisions are going to be delegated to AI systems, what kinds of effects the trade-offs between human and artificial agency are going to have, and what forms of assessment, control, revision and redressing must be put in place, are crucial questions that should not be answered through trial and error. AI should never be beta-tested on

L. Floridi (✉)
Oxford Internet Institute, University of Oxford, Oxford, UK
e-mail: luciano.floridi@oii.ox.ac.uk

L. Floridi (ed.), *Ethics, Governance, and Policies in Artificial Intelligence*, Philosophical Studies Series 144, https://doi.org/10.1007/978-3-030-81907-1_4

humans or the environment. The development of AI requires socio-political deliberation and consensus, in view of a long-term strategy about what kind of AI should be developed, for what purpose, for whom, and according to which ethical priorities. This is a main aim of the ethics guidelines report from the European Commission (EC).

The report, published on 8 April 2019 after several versions and more than 500 public consultations, is put together by an independent, High-Level Expert Group (HLEG) (European Commission 2019a). The HLEG was appointed by the EC in June 2018 and consists of 52 experts (disclosure: I am one of them), with relevant expertise from academia, civil society and industry. The work of the HLEG is expected to inform the European Union's (EU) policies and legislation about AI, to support the implementation of the EU strategy on AI, and to serve as the steering group for the European AI Alliance's work. The guidelines support a responsible approach to the development of AI, which should be (1) lawful, respecting all applicable laws and regulations; (2) ethical, respecting ethical principles and values (Fig. 1 summarizes the principles grounding the guidelines, which were informed (European Commission 2018) by the AI4People's research (Floridi et al. 2018)); and (3) robust, both technically and in terms of its social environment.

Since AI will become increasingly important and pervasive, it must work reliably, in ways that anyone can trust will be for the benefit of humanity and the whole environment. The alternative is that AI may be misused, overused or underused. Ethical uncertainty breeds both reckless risk-taking and excessive caution. This is why the guidelines are so important. They represent a good step in the right direction of a clear, shared and socially preferable framework for ethical AI.

4.2 Ethics First to Inform Legislation

The guidelines have been praised and welcomed by many, but have also been criticized (Metzinger 2019; Meyer 2019) for being weak, because they are part of a mere self-regulatory strategy, which is not legally enforced, and unhelpful, because they are too general, and join so many other initiatives that have so far had little impact. These and similar criticisms can be countered. First, the guidelines contain principles and clarifications that are robust, in terms of social expectations, and consistent with the current state of the debate on the ethics of AI. Of course, both law and ethics about AI are needed. The guidelines presuppose and are aligned with the EU legislation. The EU is at the forefront of the international debate on AI, also thanks to the General Data Protection Regulation (GDPR). Ethics can contribute to the shaping of new legislation (for example, about facial recognition systems) or act as a guide in its absence.

Sometimes, ethics is needed to interpret existing legislation (for example, the GDPR). Other times, ethics may recommend not to do something that legislation does not prohibit (for example, leaving a medical decision entirely to an algorithm without supervision or explanation), or recommend to do something that legislation does not require (for example, designing an algorithm that minimizes the environmental impact of domestic central heating). In all these cases, compliance with the law is necessary but insufficient, and, as the guidelines acknowledge, it must be complemented by a post-compliance 'soft ethics' approach (Floridi 2018), because the law provides the rules of the game, but does not indicate how to play well according to the rules. Second, granted: the guidelines are not very original or innovative, but that would have been astonishing and perhaps a bit concerning, after more than a half a century of discussion on the topic (Wiener 1960; Samuel 1960). There are in fact currently more than 70 frameworks and lists of principles about the ethics of AI (AlgorithmWatch 2019; Winfield 2019). This mushrooming of declarations is generating inconsistency and confusion, among stakeholders, regarding which document may be preferable. It also puts pressure on private and public actors that develop or deploy AI solutions to produce their own declarations for fear to be seen to be left behind, thus further contributing to the noise. And it risks creating a supermarket of principles and values, where private and public actors may shop for the kind of ethics that is best retrofitted to justify their behaviours, rather than revising their behaviours to make them consistent with a socially accepted ethical framework. However, the guidelines resolve these challenges because they are the closest thing available in the EU to a comprehensive and authoritative standard, offering a clear frame of reference and a common, conceptual vocabulary. They have been designed to establish a benchmark for what may or may not qualify, from now on, as trustworthy AI.

4.3 Further Steps for a Global Stage

In some cases, a regulative approach may be premature, too prescriptive or stifle valuable innovation. An ethical approach leads to more flexible and still demanding expectations. It is important to remember that the publication of the guidelines is also just the first step. They will contribute to inform EU legislation and policies, but they also represent a roadmap for the rapid transformations enabled by AI technology (Floridi and Lord Clement-Jones 2019). In June 2019, the HLEG will issue its recommendations for the EU's AI research agenda, and on how the EU may strengthen its competitiveness in the development and deployment of AI, in line with the guidelines. And this summer, the EC will launch a pilot project to test the guidelines in collaboration with stakeholders to identify potential improvements and promote practical applications. The HLEG will review the outcome in early 2020 and further refine its output. In the long run, the EC "wants to bring this approach to AI ethics to the global stage ... [and] strengthen cooperation with like-minded partners such as Japan, Canada or Singapore ... [as well as] the G7 and G20" (European Commission 2019b). Some critics concede all this but still object that one cannot become a leader in ethical AI without becoming a leader in AI first (Delcker 2019; Vincent 2019). Yet 'innovate first, fix later' is a mistake that, in the case of AI, could also be very costly and may cause a public backlash against AI, similar to the one against genetically modified crops in the past (Cookson 2018). The climate change disaster and the trouble with social media platforms interfering in democracy should have taught us to plan innovation more carefully. This is why the EU wants to determine a long-term strategy in which ethics is an innovation enabler that offers a competitive advantage, and which ensures that fundamental rights and values are fostered, the public interest is served, and the natural environment thrives. Ethics-first is the right approach to set global standards for AI. The era of 'move fast and break things' is over. It is time to 'make haste slowly' (festina lente) in the development of AI.

	Seven essentials for achieving trustworthy AI
	Trustworthy AI should respect all applicable laws and regulations, as well as a series of requirements; specific assessment lists aim to help verify the application of each of the key requirements:
1	**Human agency and oversight**: AI systems should enable equitable societies by supporting human agency and fundamental rights, and not decrease, limit or misguide human autonomy
2	**Robustness and safety**: Trustworthy AI requires algorithms to be secure, reliable and robust enough to deal with errors or inconsistencies during all life cycle phases of AI systems
3	**Privacy and data governance**: Citizens should have full control over their own data, while data concerning them will not be used to harm or discriminate against them
4	**Transparency**: The traceability of AI systems should be ensured
5	**Diversity, non-discrimination and fairness**: AI systems should consider the whole range of human abilities, skills and requirements, and ensure accessibility
6	**Societal and environmental well-being**: AI systems should be used to enhance positive social change and enhance sustainability and ecological responsibility
7	**Accountability**: Mechanisms should be put in place to ensure responsibility and accountability for AI systems and their outcomes
	European Commission, ref. IP/19/1893

Acknowledgements L.F. is a member of the High-Level Expert Group (HLEG). Some of his research on the ethical impact of automation and algorithms has been funded by academic grants from the UK Research Council, the EU and Google Europe (also a member of the HLEG).

References

AlgorithmWatch. 2019. https://algorithmwatch.org/en/project/ai-ethics-guidelines-global-inventory/

Cookson, C. 2018. *Financial Times*. https://www.ft.com/content/0b301152-b0f8-11e8-99ca-68cf89602132

Delcker, J. 2019. *Politico*. https://www.politico.eu/article/europe-silver-bullet-global-ai-battle-ethics/

European Commission. 2018. *Draft ethics guidelines for trustworthy AI*. https://ec.europa.eu/digital-single-market/en/news/draft-ethics-guidelines-trustworthy-ai

———. 2019a. *Ethics guidelines for trustworthy AI*. https://ec.europa.eu/digital-single-market/en/news/ethics-guidelines-trustworthy-ai

———. 2019b. *Artificial Intelligence: Commission takes forward its work on ethics guidelines*. http://europa.eu/rapid/press-release_IP-19-1893_en.htm

Floridi, L. 2018. Soft ethics, the governance of the digital and the general data protection regulation. *Philosophical Transactions of the Royal Society A* 376: 20180081.

Floridi, L., and T. Lord Clement-Jones. 2019. *New statesman*. https://tech.newstatesman.com/policy/ai-ethics-framework

Floridi, L., et al. 2018. AI4People—An ethical framework for a good AI society: Opportunities, risks, principles, and recommendations. *Minds and Machines* 28: 689–707.

Metzinger, T. Der. 2019. *Tagesspiegel*. https://www.tagesspiegel.de/politik/eu-guidelines-ethics-washing-made-in-europe/24195496.html

Meyer, D. 2019. *Fortune*. http://fortune.com/2019/04/08/eu-ai-ethics-principles/

Samuel, A.L. 1960. Some moral and technical consequences of automation—A refutation. *Science* 132: 741–742.

Vincent, J. 2019. *The verge*. https://www.theverge.com/2019/4/8/18300149/eu-artificial-intelligence-ai-ethical-guidelines-recommendations

Wiener, N. 1960. Some moral and technical consequences of automation. *Science* 131: 1355–1358.

Winfield, A.. 2019. *Alan Winfield's web log*. http://alanwinfield.blogspot.com/2019/04/an-updated-round-up-of-ethical.html

Chapter 5
The Chinese Approach to Artificial Intelligence: An Analysis of Policy, Ethics, and Regulation

Huw Roberts, Josh Cowls, Jessica Morley, Mariarosaria Taddeo, Vincent Wang, and Luciano Floridi

Abstract In July 2017, China's State Council released the country's strategy for developing artificial intelligence (AI), entitled 'New Generation Artificial Intelligence Development Plan' (新一代人工智能发展规划). This strategy outlined China's aims to become a world leader in AI by 2030, to monetise AI into a trillion-yuan ($150 billion) industry, and to emerge as the driving force in defining ethical norms and standards for AI. Several reports have analysed specific aspects of China's AI policies or have assessed the country's technical capabilities. Instead, in this article, we focus on the socio-political background and policy debates that are shaping China's AI strategy. In particular, we analyse the main strategic areas in which China is investing in AI and the concurrent ethical debates that are delimiting its use. By focusing on the policy backdrop, we seek to provide a more comprehensive and critical understanding of China's AI policy by bringing together debates and analyses of a wide array of policy documents.

Keywords Artificial Intelligence · China · Cyber warfare · Digital ethics · Economic growth · Governance · Innovation · International competition · New generation artificial intelligence development plan · Policy · Privacy · Social governance

---

H. Roberts · J. Morley · L. Floridi (✉)
Oxford Internet Institute, University of Oxford, Oxford, UK
e-mail: luciano.floridi@oii.ox.ac.uk

J. Cowls · M. Taddeo
Oxford Internet Institute, University of Oxford, Oxford, UK

Alan Turing Institute, London, UK

V. Wang
Department of Computer Science, University of Oxford, Oxford, UK

L. Floridi (ed.), *Ethics, Governance, and Policies in Artificial Intelligence*, Philosophical Studies Series 144, https://doi.org/10.1007/978-3-030-81907-1_5

5.1 Introduction

In March 2016, a Google DeepMind AI designed for playing the board game Go (AlphaGo) defeated Lee Sedol, a South Korean professional Go player. At the time, Sedol had the second-highest number of Go international championship victories, yet lost against AlphaGo by four games to one (Borowiec 2016). While the match received some coverage in the West, it was a major event in China, where over 280 million people watched it live. Two government insiders described this match as a 'Sputnik moment' for the development of Artificial Intelligence (AI) within China (Lee 2018, p. 3). Although there had been AI policy initiatives in the country previously, the victory for AlphaGo contributed to an increase in focus, as indicated by the 2017 'New Generation Artificial Intelligence Development Plan' (AIDP). The AIDP set out strategic aims and delineated the overarching goal of making China a world leader in AI by 2030.[1]

A limited number of reports have attempted to assess the plausibility of China's AI strategy given China's current technical capabilities (Ding 2018; "China AI Development Report" 2018). Others have sought to understand specific areas of development, for instance, security or economic growth (Allen 2019; Barton et al. 2017; China AI Development Report 2018; "Net Impact of AI on Jobs in China" 2018). However, in order to grasp the ramified implications and direction of the AIDP, it is insufficient to analyse specific elements in isolation or to consider only technical capabilities. Instead, a more comprehensive and critical analysis of the driving forces behind China's AI strategy, its political economy, cultural specificities, and the current relevant policy debates, is required in order to understand China's AI strategy. This is the task we undertake in this Chapter.

In order to provide this contextualised understanding, Sect. 5.2 maps relevant AI governance in China. We argue that, although previous policy initiatives have stated an intent to develop AI, these efforts have been fractious and viewed AI as one of many tools in achieving a different set goal. In contrast, the AIDP is the first national-level governance effort that focuses explicitly on the development of AI as a unified strategy. Following this, Sect. 5.3 analyses the interventions and impact of the AIDP on three strategic areas identified in the document, namely: *international competition*, *economic growth*, and *social governance*. Section 5.4 focuses on China's aim to develop *ethical norms and standards for AI*. There we argue that, although the debate is in its early stages, the desire to define normative boundaries for acceptable uses of AI is present and pressing. Altogether, this article seeks to provide a detailed and critical understanding of the reasons behind, and the current trajectory of, China's AI strategy. It emphasises that the Chinese government is aware of the potential benefits, practical risks, and the ethical challenges that AI presents, and that the direction of China's AI strategy will largely be determined by

[1] In the rest of this article, we shall use 'China' or 'Chinese' to refer to the political, regulatory, and governance approach decided by the Chinese national government concerning the development and use of AI capabilities.

the interplay of these factors and by the extent to which government's interests may outweigh ethical concerns. Section 5.5 concludes the paper by summarising the key findings of our analysis.

5.2 AI Governance in China

Since 2013, China has published several national-level policy documents, which reflect the intention to develop and deploy AI in a variety of sectors. For example, in 2015, the State Council released guidelines on China's 'Internet+' action. It sought to integrate the internet into all elements of the economy and society. The document clearly stated the importance of cultivating emerging AI industries and investing in research and development ("Internet Plus" 2015) In the same year, the ten-year plan 'Made in China 2025' was released, with the aim to transform China into the dominant player in global high-tech manufacturing, including AI (McBride and Chatzky 2019). Another notable example is the Central Committee of the Communist Party of China's 13th Five-Year Plan,[2] published in March 2016. The document mentioned AI as one of the six critical areas for developing the country's emerging industries ("The 13th Five Year Plan" 2016), and as an important factor in stimulating economic growth. When read together, these documents indicate that there has been a conscious effort to develop and use AI in China for some time, even before 'the Sputnik moment'. However, prior to 2016, AI was presented merely as one technology among many others, which could be useful in achieving a range of policy goals. This changed with the release of the AIDP.

5.2.1 The New Generation Artificial Intelligence Development Plan (AIDP)

Released in July 2017 by the State Council (which is the chief administrative body within China), the 'New Generation Artificial Intelligence Development Plan' (AIDP) acts as a unified document that outlines China's AI policy objectives. Chinese media have referred to it as 'year one of China's AI development strategy' ("China AI Development Report" 2018, p. 63). The overarching aim of the policy, as articulated by the AIDP, is to make China the world centre of AI innovation by 2030, and make AI 'the main driving force for China's industrial upgrading and economic transformation' (Webster et al. 2017b). The AIDP also indicates the importance of using AI in a broader range of sectors, including defence and social welfare, and focuses on the need to develop standards and ethical norms for the use of

[2]The five-year plans are a central pillar in China's economic growth policy (Heilmann and Melton 2013; Hu 2013).

AI. Altogether, the Plan provides a comprehensive AI strategy, and challenges other leading powers in many key areas.

The AIDP delineates three key steps, each of which contains a series of goals, some of which are tightly defined, while others are vaguer. They are summarised as follows and in Fig. 5.1 below:

1. By 2020, China aims to maintain competitiveness with other major powers and optimise its AI development environment. In monetary terms, China intends to create an AI industry worth more than 150 billion yuan (ca. 21 billion dollars). Lastly, it seeks to establish initial ethical norms, policies and regulations for vital areas of AI.
2. By 2025, China aims to have achieved a 'major breakthrough' (as stated in the document) in basic AI theory and to be world-leading in some applications ('some technologies and applications achieve a world-leading level'). Overall, China targets an increase in the worth of its core AI industry to over 400 billion yuan (ca. 58 billion dollars), and plans to expand upon, and codify in law, ethical standards for AI.
3. By 2030, China seeks to become the world's innovation centre for AI. By then, growth in the core AI industry is expected to more than double again and be valued at 1 trillion yuan (ca 147 billion dollars), and further upgrades in the laws and standards are also to be expected, in order to deal with newly emerging challenges.

5.2.2 *Implementing the AIDP*

The Plan will be guided by a new AI Strategy Advisory Committee, established in November 2017, and will be coordinated by the Ministry of Science and Technology

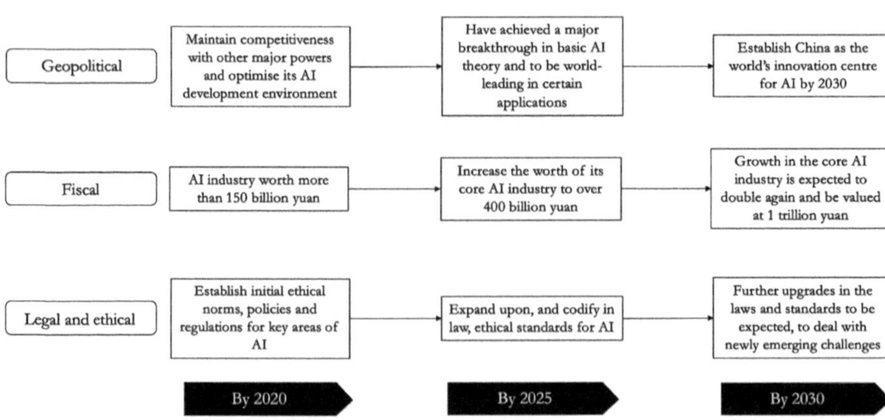

Fig. 5.1 Visualising China's AIDP

(MIST), alongside the AI Plan Promotion Office, and other relevant bodies ("AI Policy—China" n.d.).[3] Although these bodies will provide central guidance, the Plan is not meant to act as a centrally enacted initiative. The AIDP instead functions as a stamp of approval for de-risking and actively incentivising local projects that make use of AI. Recognising this point is important: the AIDP is an ambitious strategy set by central government, but the actual innovation and transformation is expected to be driven by the private sector and local governments. In other words, it is more appropriate to view the AIDP as a highly incentivised 'wish list', to nudge and coordinate other relevant stakeholders, rather than a central directive (Sheehan 2018). This is why the three-year plan promoting the AIDP (2018–2020) emphasises coordination between provinces and with local governments.

With regard to the private sector, China has selected 'AI national champions': businesses endorsed by the government to focus on developing specific sectors of AI. For example, Baidu has been tasked with the development of autonomous driving, Alibaba with the development of smart cities, and Tencent with computer vision for medical diagnoses (Jing and Dai 2017). Being endorsed as a national champion involves a deal whereby private companies agree to focus on the government's strategic aims. In return, these companies receive preferential contract bidding, easier access to finance, and sometimes market share protection. Although other companies can compete in these fields, historically the status of 'national champion' has helped larger companies dominate their respective sectors (Graceffo 2017).

With this said, the new AI 'national team' differs from previous state-sponsored national champions in that they are already internationally successful in their respective fields, independently of this preferential treatment. Furthermore, there is extensive domestic competition in the areas where national champions have been selected. This suggests that competition may not be stymied in the traditional manner. For instance, all the companies selected as AI national champions are developing technologies in Alibaba's designated area of smart cities (Ding 2019). In parallel with this, patronage does not prohibit smaller companies benefiting from the financial incentive structure. Technology start-ups within China often receive government support and subsidies for developing AI technologies. As an example, Zhongguancun Innovation Town is a purpose-built, government subsidised, incubator workspace that provides a suite of services to help Chinese technology start-ups succeed, often in the sectors where national champions have been selected. Finally, there are also cases where there is no specific endorsement. For example, while the AIDP promotes smart courts, with a stated desire to develop AI for evidence collection, case analysis, and legal document reading, as of April 2020 there is no national champion selected for developing AI applications for the administration of justice.

[3] It should be noted that, although MIST has been tasked with coordinating the AIDP, it was the Ministry of Industry and Information Technology (MIIT) that released the guidance for the implementing the first step of the AIDP.

Concerning local governments, the political structure within China creates a system of incentives for fulfilling national government policy aims. Short term limits for provincial politicians, and promotions based on economic performance provide strong incentives for following centrally-defined government initiatives (Li and Zhou 2005; Persson and Zhuravskaya 2016). Thus, local governments become hotbeds for testing and developing central government policy. The strength of this incentive system can be seen in the decision made by the administration of the city of Tianjin to establish a $5 billion fund for the development of AI, around the same time as the publication of the AIDP (Mozur 2017). At the same time, it is important to recognise how the absence of an effective accountability review of local government spending creates problems within this system. Notably, it has facilitated a mindset in which local politicians know that central government will bail them out for failed projects, leading to poor budget management (Ji 2014). A clear example of this are the large-scale port building initiatives developed by provincial governments in East coast provinces that were based more on prestige than any economic rationale, and which led to overcapacity and disorderly competition (Zhu 2019).

These incentive structures contain a subtle distinction. A national team has been selected to lead the research and development in a handful of designated strategic areas. Beyond these selected companies, there are few specific guidelines provided to industry and local state agents as to which items to pursue on the AIDP's 'wish list'. This enables companies to cherry-pick the technologies they want to develop, and provides local governments with a choice of private sector partners for integrating AI into city infrastructure or governance (Sheehan 2018). Subsequent documentation has emphasised the importance of strengthening organisation and implementation,[4] including between provinces and ministries, yet it is unclear how this coordination would function in practice. Thus, the AIDP may work as a 'wish list', but the exact guidance, incentivisation and risk differs depending on the type of stakeholder.

The AIDP should not be read in isolation when considering China's AI strategy (Ding 2018), but it does provide the most transparent and influential indication of the driving forces behind China's AI strategy. Because of the AIDP's significance (in terms of policy) and importance (in terms of strategy), in the rest of this article, we shall use it as the organisational skeleton for explaining the drivers and ethical boundaries shaping China's approach to AI.

[4]To accompany the three steps outlined earlier, the Ministry of Industry and Information Technology (MIIT) provides documents to flesh out these aims. The first of these, 'Three-Year Action Plan for Promoting Development of a New Generation Artificial Intelligence Industry (2018–2020)', has already been released.

5.3 China's AI Strategic Focus

The AIDP provides a longitudinal perspective on China's strategic situation regarding AI, including its *comparative capabilities*, the *opportunities* offered, and the potential *risks*. Following a technology-first approach, it may be tempting to concentrate one's attention on the stated capabilities of AI, in order to gain an insight into the types of technologies in which China is investing. However, this would likely offer only a short-term perspective and would soon be out of date as technological innovation advances rapidly. Furthermore, it would do little to explain why China is seeking to develop a strong AI sector in the decades to come. To this end, it is more useful to try to understand China's strategic focus from a policy-first approach, by analysing the areas where China considers that AI presents opportunities. In this section, we focus on these areas of particular importance to China, on how and what China expects to gain from developing AI in each of them, and on some of the perceived risks present in each of these areas. The AIDP highlights three areas where AI can make a substantial difference within China: *international competition*, *economic development*, and *social governance*. They are strictly interrelated but, for the sake of clarity, we shall analyse them separately, and contextualise each of them by discussing the relevant literature surrounding the broader political backdrop and contemporary policy debates.

5.3.1 International Competition

The AIDP states that AI has become a new focus of international competition and that 'the development of AI [is] [...] a major strategy to enhance national competitiveness and protect national security' (Webster et al. 2017b). It emphasises that China should take the strategic opportunity afforded by AI to make 'leapfrog developments'[5] in military capabilities. Although China and the US are regularly portrayed as geopolitical rivals (Mearsheimer 2010; Zhao 2015), the military budgets of the two powers remain significantly different. China has the world's second-largest military budget, with $175 billion allocated in 2019 (Chan and Zhen 2019), but its spending is still only a third of the US budget (Martina and Blanchard 2019). Rather than outspending the US in conventional weaponry, China considers investing in AI as an opportunity to make radical breakthroughs in military technologies and thus overtake the US.

Attempts to use technologies to challenge US hegemony are nothing new within China's military strategy. Since the late 1990s, the country has been following a policy of '*shashoujian*' (杀手锏), which roughly translates as 'trump-card' (Bruzdzinski 2004). Rather than directly competing with the US, China has sought

[5] This term refers to 'an actor, which lags behind its competitors in terms of development, coming up with a radical innovation that will allow it to overtake its rivals' (Brezis et al. 1993)

to develop asymmetric capabilities, which could provide a critical advantage in warfare and credible deterrence in peacetime (Blasko 2011). This trump-card strategy seeks to use unorthodox technologies against enemies' weaknesses to gain the initiative in war (Peng and Yao 2005). The trump-card approach was echoed by the former Party Chairman, Jiang Zemin, who emphasised that technology should be the foremost focus of the military, especially the technology that the 'enemy fears [the] most' (Cheung et al. 2016).

One area in which China has been developing these asymmetric tactics is cyber warfare, where capabilities have been developed for targeting the US military's battle-critical networks, if needed (Kania 2017b). Alongside this, evidence points to the persistent use of cyberattacks to collect scientific, technological and commercial intelligence (Inkster 2010). The Chinese position on these capabilities is ambivalent. On the one hand, China has officially promoted international initiatives for regulating hostile state-run activities in cyberspace, and to fill the existing regulatory gap for state behaviour in this domain (Austin 2016; Ku 2017; Taddeo 2012, 2016). For example, China co-sponsored the *International Code of Conduct for Information Security* at the UN General Assembly in September 2011, which sought a commitment against using information technologies in acts of aggression and has provided continued support for dialogue by the UN Group of Government Experts in preventing cyberconflicts (Meyer 2020). On the other hand, China has also run cyber operations targeting US infrastructure and aiming at extracting commercial and scientific information as well as acquiring relevant intelligence against several countries, including Australia, Philippines, Hong Kong, and the US.[6]

The desire to leapfrog the US is echoed in statements from China's political and military leadership. For instance, President Xi Jinping stated in 2017 that 'under a situation of increasingly fierce international military competition, only the innovators win' (Kania 2020, p. 2). This sentiment is shared by Lieutenant General Liu Guozhi, deputy of the 19th National Congress and director of the Science and Technology Committee of the Central Military Commission, who stated in an interview that AI presented a rare opportunity for taking shortcuts to achieve innovation and surpass rivals ("Accelerate the Process of Military Reform" 2017). In parallel, academics affiliated with the People's Liberation Army (PLA) highlight that AI will be used to predict battlefield situations and identify optimal approaches, facilitating 'winning before the war' (Li 2019). Some members of the PLA go further than this in anticipating a battlefield 'singularity', where AI outpaces human decision-making (Kania 2017a). These statements emphasise the belief, which is widespread throughout China's military and defence circles, in the importance of utilising emergent technologies including AI to achieve a competitive military advantage.

As China has developed economically and militarily, the focus of the country's military strategy has also matured. Over the past few years, China's strategy has coalesced around efforts to develop 'new concept weapons' to surpass the US's

[6]https://www.csis.org/programs/technology-policy-program/significant-cyber-incidents

military capabilities. These are not limited to AI alone, and are applicable to China's investments in other fields of emerging military technologies, like hypersonic weaponry (Kania 2017b). Therefore, China's efforts to use technology to gain an advantage in military affairs should not be seen as something new, but instead understood within a broader historical context of finding innovative ways to challenge the hegemony of the US.

Although the push for leapfrog developments marks a continuation of previous policy, there are strong concurrent indications that Chinese officials are also concerned about AI causing an arms race and potential military escalation. Statements of senior officials seem to suggest a belief in cooperation and arms control in order to mitigate the risks that AI's military development poses. In particular, three major risks are central to the debate:

(i) human involvement and control once AI-based weapons are deployed;
(ii) the absence of well-defined norms for state behaviour and use of AI weapons; which in turn increases
(iii) the likelihood of misperceptions or unintentional conflict escalation (Allen 2019; Taddeo and Floridi 2018).

These concerns underpin China's support to restrict the use of autonomous weapons, as expressed at the 5th Convention on Certain Conventional Weapons ("Chinese Position Paper" 2016) and, more recently, the desire to ban autonomous lethal weapons (Kania 2018a). Despite concerns (i)–(iii), it is crucial to stress that China is the actor pursuing the most aggressive strategy for developing AI for military uses among the major military powers (Pecotic 2019).

Digging more deeply into China's actions on the international stage is revealing. The ban that China advocated encompassed only *usage* and not *development* or *production* of autonomous lethal weapon systems. Thus, it would not prevent the existence of autonomous lethal weapons serving as a deterrent, in much the same way that China has a putative 'no first use' (NFU) doctrine for nuclear weapons. Furthermore, the definition of autonomy embraced by China is extremely narrow, including only fully autonomous weapons ("UN Seeks to Retain Human Control over Force" 2018). Some commentators argue that this juxtaposition of cautious concerns about deployment, on the one hand, and an aggressive approach to development, on the other, can be explained by the Chinese efforts to exert pressure on other militaries whose democratic societies are more sensitive to the controversies of using automated weapons (Kania 2018a). This is a reasonable claim: a continuation of propaganda may be part of the explanation. For instance, China was the first nuclear power to pledge 'no first use' of nuclear weapons (so far only India has a similar pledge; other countries, including the US and the UK, have pledged to use nuclear weapons only defensively). But rather than offering a genuine commitment to NFU, this pledge was meant as internal and external propaganda tool, which would be circumvented by semantics if needed (Schneider 2009).

Taken together, China's focus on military AI can be considered as a continuation of a longer-term strategy, which privileges developing (with the threat of deploying) technology to gain a military advantage. There remains a conscious recognition, by several actors in China, that developing AI presents an especially fraught risk of igniting an arms race or causing unintentional escalation due to the autonomy of these technologies (Allen 2019; Taddeo and Floridi 2018). But at the political level, efforts to curtail the use of military AI internationally may also be seen as part of a propaganda strategy.

5.3.2 Economic Development

Economic development is the second strategic opportunity explicitly mentioned in the AIDP. It is stated that AI will be the driving force behind a new round of industrial transformation, which will 'inject new kinetic energy into China's economic growth' (Webster et al. 2017b). The reconstruction of economic activity is targeted in all sectors, with manufacturing, agriculture, logistics, and finance being the examples promoted in the AIDP.

China's rapid growth has frequently been referred to as an 'economic miracle', due to the country's shift from having a slow-growth economy to enjoying some of the world's highest growth rates for over two decades (Naughton and Tsai 2015; Ray 2002). A number of factors facilitated this economic growth, of which the demographic dividend is one. A large workforce, in combination with a small dependent population, fostered high levels of savings and heavy investment (Cai and Lu 2013). Structural changes, including a conscious shift from a predominantly agricultural to a manufacturing economy, and the opening up of markets, are additional, critical factors. By 2012, China's labour force growth dropped to around zero, and its shift from an agricultural to manufacturing economy had largely matured. These trends have led Chinese policymakers to the realisation that an alternative development model is necessary for maintaining high rates of growth. This model rests on the shift from heavy investment in industry to growth stimulated by an innovative society (Naughton and Tsai 2015). Recently, science and technology have been put forward as a crucial means for achieving this type of innovative growth (Zhang 2018).

Some commentators have argued that maintaining these high levels of growth is particularly important for China due to the implicit trading by citizens of political freedoms for economic growth and embourgeoisement (Balding 2019). Research has highlighted that support for the party and a relatively lacklustre desire for democracy stems from satisfaction with employment and material aspects of life, particularly within the middle classes (Chen 2013). Slowing economic growth would likely sow dissatisfaction within the populace and make inherent features within the Chinese political system, such as corruption, less tolerable (Diamond 2003; Pei 2015). A lack of a democratic outlet for this frustration could lower the overall support that the government currently receives. Some maintain that this creates a 'democratise or die' dynamic (Huang 2013), however this may be unfeasible, given China's political control (Chin 2018; Dickson 2003).

Against this backdrop, a report by PwC suggested that China is the country that has the most to gain from AI, with a boost in GDP of up to 26% by 2030 ("Sizing the

Prize" 2017). Estimates also suggest that AI could facilitate an increase in employment by 12% over the next two decades ("Net Impact of AI on Jobs in China" 2018). Because of these potential benefits, President Xi has frequently spoken of the centrality of AI to the country's overall economic development (Hickert and Ding 2018; Kania 2018b). China has been pursuing the potential economic benefits of AI concretely and proactively for some time. For example, there has been a 500% increase in annual installation of robotic upgrades since 2012. This rate is staggering, especially when compared to a rate of just over 100% in Europe (Shoham et al. 2018), equating to over double the number of robot installations in China than Europe.

AI can be a double-edged sword, because the benefits and improvements brought about by AI come with the risk, amongst others, of labour market disruptions. This is a concern explicitly stated in the AIDP. Although the aforementioned PwC report predicts that automation will increase the net number of jobs in China, disruption will likely be unevenly spread ("Net Impact of AI on Jobs in China" 2018). Smarter automation will most immediately affect low- and medium-skilled jobs, while creating opportunities for higher-skilled technical roles (Barton et al. 2017). China has been active in its efforts to adapt to such AI-related risks, especially with an education overhaul promoted by the 'National Medium- and Long-term Education Reform and Development Plan (2010–2020)'. This plan has the goal of supporting the skilled labour required in the information age ("Is China Ready for Intelligent Automation?" 2018). In the same vein, China is addressing the shortage in AI skills specifically by offering higher education courses on the subject (Fang 2019). Accordingly, China seems to be preparing better than other middle-income countries to deal with the longer-term challenges of automation ("The Automation Readiness Index 2018" 2018).

Although these efforts will help to develop the skillset required in the medium and long term, they do little to ease the short-term structural changes. Estimates show that, by 2030, automation in manufacturing might have displaced a fifth of all jobs in the sector, equating to 100 million workers ("Is China Ready for Intelligent Automation" 2018). These changes are already underway, with robots having replaced up to 40% of workers in several companies in China's export-manufacturing provinces of Zhejiang, Jiangsu and Guangdong (Yang and Liu 2018). In the southern city of Dongguan alone, reports suggest that 200,000 workers have been replaced with robots ("Is China Ready for Intelligent Automation" 2018). When this is combined with China's low international ranking in workforce transition programmes for vocational training ("Is China Ready for Intelligent Automation" 2018), it can be suggested that the short-term consequence of an AI-led transformation is likely to be significant disruptions to the workforce, potentially exacerbating China's growing inequality (Barton et al. 2017).

5.3.3 Social Governance

Social governance, or more literally in Chinese 'social construction',[7] is the third area in which AI is promoted as a strategic opportunity for China. Alongside an economic slowdown, China is facing emerging social challenges, hindering its pursuit of becoming a 'moderately prosperous society' (Webster et al. 2017b). An ageing population and constraints on the environment and other resources are explicit examples provided in the AIDP of the societal problems that China is facing. Thus, the AIDP outlines the goal of using AI within a variety of public services to make the governance of social services more precise and, in doing so, mitigate these challenges and improve people's lives.

China has experienced some of the most rapid structural changes of any country in the past 40 years. It has been shifting from a planned to a market economy and from a rural to an urban society (Naughton 2007). These changes have helped facilitate economic development, but also introduced a number of social issues. One of the most pressing social challenges China is facing is the absence of a well-established welfare system (Wong 2005). Under the planned economy, workers were guaranteed cradle-to-grave benefits, including employment security and welfare benefits, which were provided through local state enterprises or rural collectives (Selden and You 1997). China's move towards a socialist market economy since the 1990s has accelerated a shift of these provisions from enterprises and local collectives to state and societal agencies (Ringen and Ngok 2017). In practice, China has struggled to develop mature pension and health insurance programmes, creating gaps in the social safety net (Naughton 2007). Although several initiatives have been introduced to alleviate these issues (Li et al. 2013), the country has found it difficult to implement them (Ringen and Ngok 2017).

The serious environmental degradation that has taken place in the course of China's rapid development is another element of concern. For most of China's development period, the focus has been on economic growth, with little or no incentive provided for environmental protection (Rozelle et al. 1997). As a result, significant, negative externalities and several human-induced natural disasters have occurred that have proven detrimental for society. One of the most notable is very poor air quality, which has been linked to an increased chance of illness and is now the fourth leading cause of death in China (Delang 2016). In parallel, 40% of China's rivers are polluted by industry, causing 60,000 premature deaths per year (Economy 2013). Environmental degradation of this magnitude damages the health of the population, lowers the quality of life, and places further strain on existing welfare infrastructure.

The centrality of these concerns could be seen at the 19th National Party Congress in 2017, where President Xi declared that the 'principal contradiction' in China had changed. Although the previous 'contradiction' focused on 'the ever-growing material and cultural needs of the people and backward social production,' Xi stated

---

[7]The Chinese text (社会建设) directly translates to 'society/community' and 'build/construction'.

'what we now face is the contradiction between unbalanced and inadequate development and the people's ever-growing needs for a better life' ("Principal Contradiction has Evolved" 2017). After years of focusing on untempered economic growth, President Xi's remarks emphasise a broader shift in China's approach to dealing with the consequences of economic liberalisation.

These statements are mirrored in several government plans, including the State Council Initiative, 'Healthy China 2030', which seeks to overhaul the healthcare system. Similar trends can be seen in China's efforts to clean up its environment, with a new three-year plan building on previous relevant initiatives (Leng 2018). China has recently focused on AI as a way of overcoming these problems and improving the welfare of citizens. It has been pointed out that China's major development strategies rely on solutions driven by big data (Heilmann 2017). For example, 'Healthy China 2030' explicitly stresses the importance of technology in achieving China's healthcare reform strategy, and emphasises a switch from treatment to prevention, with AI development as a means to achieve the goal (Ho 2018). This approach also shapes environmental protection, where President Xi has been promoting 'digital environmental protection' (数字环保) (Kostka and Zhang 2018). Within this, AI is being used to predict and mitigate air pollution levels (Knight 2015), and to improve waste management and sorting ("AI-Powered Waste Management Underway in China" 2019).

Administration of justice is another area where the Chinese government has been advancing using AI to improve social governance. Under Xi Jinping, there has been an explicit aim to professionalise the legal system, which suffers from a lack of transparency, issues of local protectionism, and interference in court cases by local officials (Finder 2015). A variety of reforms have been introduced in an attempt to curtail these practices including, transferring responsibility for the management of local courts from local to provincial governments, the creation of a database where judges can report attempts at interference by local politicians, and a case registration system that makes it more difficult for courts to reject complex or contentious cases (A. Li 2016).

Of particular interest, when focusing on AI, is the *Several Opinions of the Supreme People's Court on Improving the Judicial Accountability System* (2015), that requires judges to reference similar cases in their judicial reasoning. Furthermore, it stipulates that decisions conflicting with previous similar cases should trigger a supervision mechanism with more senior judges. To help judges minimise inconsistencies, an effort has been made to introduce AI technologies that facilitate making 'similar judgements in similar cases' (Yu and Du 2019). In terms of the technology, two overarching types of system have emerged. The first is a 'similar cases pushing system', where AI is used to identify judgements from similar cases and provide judges these for reference. This type of system has been introduced by, amongst others, Hainan's High People's Court who have also encouraged the use of AI systems in lower-level courts across the province (Yuan 2019). The second technology is 'abnormal judgement warning' that would detect if a judgement made differs from similar cases and if so, it will alert the judge's superiors, prompting an intervention (Yu and Du 2019). The reception of the use of technology has been mixed, with people receptive of the prospect of lessened corruption and judges appreciating the reduced workloads. However, some legal theorists criticised the inhumane effects of using technology in sentencing and the detriment that it could cause for 'legal hermeneutics, legal reasoning techniques, professional training and the ethical personality of the adjudicator' (Ji 2013, p. 205).

Looking forward, the focus on China's use of AI in governance seems most likely to centre on the widely reported 'Social Credit' System, which is premised upon developing the tools required to address China's pressing social problems (Chorzempa et al. 2018). To do this, the system broadly aims at increasing the state's governance capacity, promoting the credibility of state institutions, and building a viable financial credit base (Chai 2018). Currently, the Social Credit System is not one unified nationwide system but rather comprises national blacklists that collate data from different government agencies, individual social credit systems run by local governments, and private company initiatives (Liu 2019). These systems are fractious and, in many cases, the local trials lack technical sophistication, with some versions relying on little more than paper and pen (Gan 2019). Nonetheless, the ambitious targets of the Social Credit System provide a compelling example of the government's intent to rely on digital technology, for social governance and *also* for more fine-grained regulation of the behaviour of its citizens.

5.3.4 Moral Governance

Social governance/construction in China does not just encompass material and environmental features, but also the behaviour of citizens. Scholars have argued that the disruption of the Maoist period followed by an 'opening up' has created a moral vacuum within China (Lazarus 2016; Yan 2009). These concerns are echoed by the Chinese public, with Ipsos Mori finding that concerns over 'moral decline' in China were twice as high as the global average (Atkinson and Skinner 2019).[8] This is something that has been recognised by the Chinese government, with high-level officials, including President Xi, forwarding the idea of a 'minimum moral standard' within society (He 2015). This goal is not limited to ensuring 'good' governance in the traditional sense; it extends to the regulating the behaviour of citizens and enhancing their moral integrity, which is considered a task within the government's remit ("Xi Jinping's Report at 19th CPC National Congress" 2017). In the view of the government, AI can be used to this end.

The AIDP highlights AI's potential for understanding group cognition and psychology (2017). The intention to rely on AI for moral governance can be seen in further legislation, with perhaps the clearest example being the State Council's

[8]It is worth highlighting, however, that the Chinese are more than double the world average, and ranked first, when it comes to answering the question "whether the country is going in the right direction", with 94% of the respondents in agreement.

'Outline for the Establishment of a Social Credit System', released in 2014. This document underscored that the Social Credit System did not just aim to regulate financial and corporate actions of business and citizens, but also the social behaviour of individuals. This document outlines several social challenges that the plan seeks to alleviate, including tax evasion, food safety scares, and academic dishonesty (Chorzempa et al. 2018). As highlighted, current efforts to implement these systems have been fractious, yet a number have already included moral elements, such as publicly shaming bad debtors (Hornby 2019).

Further concrete examples of how China has been utilising AI in social governance can be seen in the sphere of internal security and policing. China has been at the forefront of the development of smart cities, with approximately half of the world's smart cities located within China. The majority of resources that have gone into developing these cities have focused on surveillance technologies, such as facial recognition and cloud computing for ordinary policing (Anderlini 2019). The use of advanced 'counterterrorism'[9] surveillance programmes in the autonomous region of Xinjiang offers clearer and more problematic evidence of governmental efforts to use AI for internal surveillance. This technology is not limited to facial recognition, but also includes mobile phone applications to track the local Uyghur population, who are portrayed by the government as potential dissidents or terrorists (Wang 2019). When government statements are read in parallel with these developments, it seems likely that some form of the Social Credit System(s) will play a central role in the future of China's AI-enabled governance (Ding 2018), putting the rights of citizens under a sharp devaluative pressure. For example, most citizens generate large data footprints, and nearly all day-to-day transactions in cities are cashless and done with mobile apps (Morris 2019), internet providers enact 'real-name registration', linking all online activity to the individual (Sonnad 2017), enabling the government to identify and have access to the digital profile of all citizens using mobile-internet services.

The significant and likely risks related to implementing AI for governance stem from the intertwining of the material aspects of social governance with surveillance and moral control. Articles in the Western media often emphasise the problematic nature of 'the' Social Credit System, due to the authoritarian undertones of this pervasive control (Botsman 2017; Clover 2016). Examples of public dissatisfaction with specific features of locally run social credit systems appear to support this viewpoint (Zhang and Han 2019). In some cases, there have even been cases of public backlash leading to revisions in the rating criteria for local social credit systems. In contrast, some commentators have emphasised that, domestically, a national social credit system may be positively received as a response to the perception of moral decline in China, and a concomitant desire to build greater trust; indeed, it has been suggested that the system may be better named the 'Social Trust' system (Kobie 2019; Song 2018). When looking at the punishments

[9]The word 'counterterrorism' started to be used after 9/11, with the phrase 'cultural integration' favoured before this ("Devastating Blows" 2005).

distributed by the social credit systems, some measures, including blacklisting citizens from travelling due to poor behaviour on trains, have received a positive response on Chinese social media. Government censorship and a chilling effect could account for this support, but there is currently no evidence of censors specifically targeting posts concerning social credit systems (Koetse 2018).

Efforts have also been made to understand public opinion on the systems as a whole, rather than just specific controversies or cases of blacklisting. A nationwide survey by a Western academic on China's social credit systems found high levels of approval within the population (Kostka 2019). With this said, problems with the methodology of the paper, in particular with the translation of 'Social Credit System', indicate that it may be more appropriate to consider a general lack of awareness, rather than a widespread sentiment of support ("Beyond Black Mirror" 2019). These points suggest that it is too early to measure public sentiment in China surrounding the development of the Social Credit System(s).

It is important to recognise that despite the relative mundanity of current applications of the Social Credit System (Daum 2019; Lewis 2019), looking forward, substantial ethical risks and challenges remain in relation to the criteria for inclusion on a blacklist or receiving a low score, and the exclusion that this could cause. In terms of the former, national blacklists are comprised of those who have broken existing laws and regulations, with a clear rationale for inclusion provided (Engelmann et al. 2019). However, the legal documents on which these lists are built are often ill-defined and function within a legal system that is subordinate to the Chinese Communist Party (Whiting 2017). As a result, legislation, like the one prohibiting the spread of information that seriously disturbs the social order, could be used to punish individuals for politically undesirable actions, including free speech (Arsène 2019). Still, it is more appropriate to consider this a problem of the political-legal structure and not a social credit system *per se*. The fundamental, ethical issue of an unacceptable approach to surveillance remains unaddressed.

In relation to local score-based systems that do not solely rely on illegality, assessment criteria can be even vaguer. For instance, social credit scores in Fuzhou account for 'employment strength', which is based on the loosely defined 'hard-working/conscientious and meticulous' (Lewis 2019). This is ethically problematic because of the opaque and arbitrary inclusion standards that are introduced for providing people certain benefits. In tandem with inclusion is the exclusion that these systems can cause. At present, most social credit systems are controlled by separate entities and do not connect with each other (Liu 2019), limiting excessive punishment. Nonetheless, memorandums of understanding are emerging between social credit systems and private companies for excluding those blacklisted from activities such as flying (Arsène 2019). As a result, it is important to emphasise that whilst the Social Credit System(s) is still evolving, the inclusion criteria and potential exclusion caused raise serious ethical questions.

5.4 The Debate on Digital Ethics and AI in China

Alongside establishing material goals, the AIDP outlines a specific desire for China to become the world leader in defining ethical norms and standards for AI. Following the release of the AIDP, government, public bodies, and industry within China were relatively slow to develop AI ethics frameworks (Hickert and Ding 2018; Lee 2018). However, there has been a recent surge in attempts to define ethical principles. In March 2019, China's Ministry of Science and Technology established The National New Generation Artificial Intelligence Governance Expert Committee. In June 2019, this body released eight principles for the governance of AI. The principles emphasise that, above all else, AI development should begin from enhancing the common well-being of humanity. Respect for human rights, privacy and fairness were also underscored within the principles. Finally, they highlighted the importance of transparency, responsibility, collaboration, and agility to deal with new and emerging risks (Laskai and Webster 2019).

In line with this publication, the Standardization Administration of the People's Republic of China, the national-level body responsible for developing technical standards, released a white paper on AI standards. The paper contains a discussion of the safety and ethical issues related to the technology (Ding and Triolo 2018). Three key principles for setting the ethical requirements of AI technologies are outlined. First, the principle of *human interest* states that the ultimate goal of AI is to benefit human welfare. Second, the principle of *liability* emphasises the need to establish accountability as a requirement for both the development and the deployment of AI systems and solutions. Subsumed within this principle is *transparency*, which supports the requirement of understanding what the operating principles of an AI system are. Third, the principle of *consistency of* [sic] *rights and responsibilities* emphasised that, on the one hand, data should be properly recorded and oversight present but, on the other hand, that commercial entities should be able to protect their intellectual property (Ding and Triolo 2018).

Government affiliated bodies and private companies have also developed their own AI ethics principles. For example, the Beijing Academy of Artificial Intelligence, a research and development body including China's leading companies and Beijing universities, was established in November 2018 (Knight 2019). This body then released the 'Beijing AI Principles' to be followed for the research and development, use, and governance of AI ("Beijing AI Principles" 2019). Similar to the principles forwarded by the AIDP Expert Committee's, the Beijing Principles focus on doing good for humanity, using AI 'properly', and having the foresight to predict and adapt to future threats. In the private sector, the most high-profile ethical framework has come from the CEO of Tencent, Pony Ma. This framework emphasises the importance of AI being available, reliable, comprehensible, and controllable (Si 2019). Finally, the Chinese Association for Artificial Intelligence (CAII)[10]

[10]The Chinese Association for Artificial Intelligence (CAAI) is the only state-level science and technology organization in the field of artificial intelligence under the Ministry of Civil Affairs.

has yet to establish ethical principles, but it formed an AI ethics committee in mid-2018 with this purpose in mind ("AI Association to Draft Ethics Guidelines" 2019).

The aforementioned principles bear some similarity to those supported in the Global North (Floridi and Cowls 2019), yet institutional and cultural differences mean that the outcome is likely to be significantly different. China's AI ethics needs to be understood in terms of the country's culture, ideology, and public opinion (Webster et al. 2017a). Although a full comparative analysis is beyond the scope of this article, it might be anticipated, for example, that the principles which emerge from China place a greater emphasis on social responsibility and group and community relations, with relatively less focus on individualistic rights, thus echoing earlier discussions about Confucian ethics on social media (Wong 2013).

In the following sections, we shall focus on the debate about AI ethics as it is emerging in connection with privacy and medical ethics, because these are two of the most mature areas where one may grasp a more general sense of the current 'Chinese approach' to digital ethics. The analysis of the two areas is not meant to provide an exhaustive map of all the debates about ethical concerns over AI in China. Instead, it may serve to highlight some of the contentious issues that are emerging, and inform a wider understanding of the type of boundaries which may be drawn in China when a normative agenda in the country is set.

5.4.1 Privacy

All of the sets of principles for ethical AI outlined above mention the importance of protecting privacy. However, there is a contentious debate within China over exactly what types of data should be protected. China has historically had weak data protection regulations—which has allowed for the collection and sharing of enormous amounts of personal information by public and private actors—and little protection for individual privacy. In 2018, Robin Li, co-founder of Baidu, stated that 'the Chinese people are more open or less sensitive about the privacy issue. If they are able to trade privacy for convenience, safety and efficiency, in a lot of cases, they are willing to do that' ("Baidu Chief under Fire" 2018). This viewpoint—which is compatible with the apparently not too negative responses to the Social Credit System—has led some Western commentators to misconstrue public perceptions of privacy in their evaluations of China's AI strategy (Webb 2019). However, Li's understanding of privacy is not one that is widely shared, and his remarks sparked fierce backlash on Chinese social media ("Baidu Chief under Fire" 2018). This concern for privacy is reflective of survey data from the Internet Society of China, with 54% stating that they considered the problem of personal data breaches as 'severe' (Sun 2018). When considering some cases of data misuse, this number is unsurprising. For example, a China Consumers Association survey revealed that 85% of people had experienced a data leak of some kind (Yang 2018). Thus, contrary to what may be inferred from some high-profile statements, there is a general sentiment of concern within the Chinese public over the misuse of personal information.

As a response to these serious concerns, China has been implementing privacy protection measures, leading one commentator to refer to the country as 'Asia's surprise leader on data protection' (Lucas 2018). At the heart of this effort has been the Personal Information Security Specification (the Specification), a privacy standard released in May 2018. This standard was meant to elaborate on the broader privacy rules, which were established in the 2017 Cybersecurity Law. In particular, it focused on both protecting personal data and ensuring that people are empowered to control their own information (Hong 2018). A number of the provisions within the standard were particularly all-encompassing, including a broad definition of sensitive personal information, which includes features such as reputational damage. The language used in the standard led one commentator to argue that some of the provisions were more onerous than that those of the GDPR (Sacks 2018).

Despite the previous evidence, the nature of the standard means that it is not really comparable to the GDPR. On the one hand, rather than being a piece of formally enforceable legislation, the Specification is merely a 'voluntary' national standard created by the China National Information Security Standardization Technical Committee (TC260). It is on this basis that one of the drafters stated that this standard was not comparable to the GDPR, as it is only meant as a guiding accompaniment to previous data protection legislation, such as the 2017 Cyber Security law (Hong 2018). On the other hand, there remains a tension that is difficult to resolve because, although it is true that standards are only voluntary, standards in China hold substantive clout for enforcing government policy aims, also through certification schemes (Sacks and Li 2018). Thus, in June 2018, a certification standard for privacy measures was established, with companies such as Alipay and Tencent Cloud receiving certification (Zhang and Yin 2019). Further, the Specification stipulates the specificities of the enforceable Cybersecurity Law, with Baidu and AliPay both forced to overhaul their data policies due to not 'complying with the spirit of the Personal Information Security Standard' (Yang 2019).

In reality, the weakness in China's privacy legislation is due less to its 'non-legally binding' status and more to the many loopholes in it, the weakness of China's judicial system, and the influential power of the government, which is often the last authority, not held accountable through democratic mechanisms. In particular, significant and problematic exemptions are present for the collection and use of data, including when related to security, health, or the vague and flexibly interpretable 'significant public interests'. It is these large loopholes that are most revealing of China's data policy. It may be argued that some broad consumer protections are present, but actually this is not extended to the government (Sacks and Laskai 2019). Thus, the strength of privacy protection is likely to be determined by the government's decisions surrounding data collection and usage, rather than legal and practical constraints. This is alarming.

It is important to recognise that the EU's GDPR contains a similar 'public interest' basis for lawfully processing personal data where consent or anonymisation are impractical, and that these conditions are poorly defined in legislation and often neglected in practice (Stevens 2017). But the crucial and stark difference between the Chinese and EU examples concerns the legal systems underpinning the two approaches. The EU's judicial branch has substantive influence, including the capacity to interpret legislation and to use judicial review mechanisms to determine the permissibility of legislation more broadly.[11] In contrast, according to the Chinese legal system, the judiciary is subject to supervision and interference from the legislature, which has *de jure* legislative supremacy (Ji 2014); this give *de facto* control to the Party (Horsley 2019). Thus, the strength of privacy protections in China may be and often is determined by the government's decisions surrounding data collection and usage rather than legal and practical constraints. As it has been remarked, 'The function of law in governing society has been acknowledged since 2002, but it has not been regarded as essential for the CCP. Rather, morality and public opinion concurrently serve as two alternatives to law for the purpose of governance. As a result, administrative agencies may ignore the law on the basis of party policy, morality, public opinion, or other political considerations' (Wang and Liu 2019, p. 6).

When relating this back to AI policy, China has benefited from the abundance of data that historically lax privacy protections have facilitated (Ding 2018). On the surface, China's privacy legislation seems to contradict other development commitments, such as the Social Credit System, which requires extensive personal data. This situation creates a dual ecosystem whereby the government is increasingly willing to collect masses of data, respecting no privacy, while simultaneously admonishing tech companies for the measures they employ (Sacks and Laskai 2019). Recall that private companies, such as the AI National Team, are relied upon for governance at both a national and local level, and therefore may receive tacit endorsement rather than admonishment in cases where the government's interests are directly served. As a result, the 'privacy strategy' within China appears to aim to protect the privacy of a specific type of *consumer*, rather than that of *citizens* as a whole, allowing the government to collect personal data wherever and whenever it may be merely useful (not even strictly necessary) for its policies. From an internal perspective, one may remark that, when viewed against a backdrop of high levels of trust in the government and frequent private sector leaks and misuses, this trade-off seems more intelligible to the Chinese population. The Specification has substantial scope for revising this duality, with a number of loopholes being closed since the initial release (Zhang and Yin 2019), but it seems unlikely that privacy protections from government intrusion will be codified in the near future. The ethical problem remains unresolved.

[11] As a practical example of this, the Court of Justice of the European Union gave judgment in Rīgas Case (2017) that has been used in defining what is meant by 'legitimate interest.'

5.4.2 *Medical Ethics*

Medical ethics is another significant area impacted by the Chinese approach to AI ethics. China's National Health Guiding Principles have been central to the strategic development and governance of its national healthcare system for the past 60 years (Zhang and Liang 2018). They have been re-written several times, as the healthcare system has transitioned from being a single-tier system, prior to 1978, to a two-tier system that was reinforced by healthcare reform in 2009 (Wu and Mao 2017). The last re-write of the Guiding Principles was in 1996 and the following principles still stand (Zhang and Liang 2018):

(a) People in rural areas are the top priority
(b) Disease prevention must be placed first
(c) Chinese traditional medicine and Western medicine must work together
(d) Health affairs must depend on science and education
(e) Society as a whole should be mobilised to participate in health affairs, thus contributing to the people's health and the country's overall development.

All five principles are relevant for understanding China's healthcare system as a whole but, from the perspective of analysing the ethics of China's use of AI in the medical domain, principles (a), (b), and (e) are the most important. They highlight that—in contrast to the West, where electronic healthcare data are predominantly focused on individual health, and thus AI techniques are considered crucial to unlock 'personalised medicine' (Nittas et al. 2018)—in China, healthcare is predominantly focused on the health of the population. In this context, the ultimate ambition of AI is to liberate data for public health purposes[12] (Li et al. 2019a). This is evident from the AIDP, which outlines the ambition to use AI to 'strengthen epidemic intelligence monitoring, prevention and control,' and to 'achieve breakthroughs in big data analysis, Internet of Things, and other key technologies' for the purpose of strengthening community intelligent health management. The same aspect is even clearer in the State Council's 2016 official notice on the development and use of big data in the healthcare sector, which explicitly states that health and medical big data sets are a national resource, and that their development should be seen as a national priority to improve the nation's health (Zhang et al. 2018).[13]

From an ethical analysis perspective, the promotion of healthcare data as a public good throughout public policy—including documents such as *Measures on*

[12] This is not to imply that the West is not interested in using AI for population health management purposes, or that China is not interested in using AI for personalised health purposes. China is, for example, also developing an integrated data platform for research into precision medicine (Zhang et al. 2018). We simply mean to highlight that the order of priority between these two goals seems to differ.

[13] The challenges section outlines some concrete benefits of implementing AI, illustrating some perceived gains to China. A separate (though more technological than ethical) point substantiated by the article is there is a lot of medical data which could *potentially* be beneficial, but the data are spread out among hospitals, not used for research, and largely unstructured.

Population Health Information and the *Guiding Opinions on Promoting and Regulating the Application of Big Medical and Health Data* (Chen and Song 2018)—is crucial. This approach, combined with lax rules about data sharing *within* China (Liao 2019; Simonite 2019), and the encouragement of the open sharing of public data between government bodies ("Outline for the Promotion of Big Data Development" 2015), promotes the collection and aggregation of health data without the need for individual consent, by positioning group *beneficence* above individual *autonomy*. This is best illustrated with an example. As part of China's 'Made in 2025' plan, 130 companies, including 'WeDoctor' (backed by Tencent, one of China's AI national champions) signed co-operation agreements with local governments to provide medical check-ups comprised of blood pressure, electrocardiogram (ECG), urine and blood tests, free of charge to rural citizens (Hawkins 2019). The data generated by these tests were automatically (i.e. with no consent from the individual) linked to a personal identification number and then uploaded to the WeDoctor cloud, where they were used to train WeDoctor's AI products. These products include the 'auxiliary treatment system for general practice', which is used by village doctors to provide suggested diagnosis and treatments from a database of over 5000 symptoms and 2000 diseases. Arguably, the sensitive nature of the data can make 'companies—and regulators—wary of overseas listings, which would entail greater disclosure and scrutiny' (Lucas 2019). Although this, and other similar practices, do involve anonymisation, they are in stark contrast with the European and US approaches to the use of medical data, which prioritise individual autonomy and privacy, rather than social welfare. A fair balance between individual and societal needs is essential for an ethical approach to personal data, but there is an asymmetry whereby an excessive emphasis on an individualistic approach may be easily rectified with the consensus of the individuals, whereas a purely societal approach remains unethical insofar as it overrides too easily individual rights and cannot be rectified easily.

Societal welfare may end up justifying the sacrifice of individual rights as a means. This remains unethical. However, how this is perceived within China remains a more open question. One needs to recall that China has very poor primary care provision (Wu and Mao 2017), that it achieved 95% health coverage (via a national insurance scheme) only in 2015 (Zhang et al. 2018), it has approximately 1.8 doctors per 1000 citizens compared to the OECD average of 3.4 (Liao 2019), and is founded on Confucian values that promote group-level equality. It is within this context that the ethical principle of the 'duty of easy rescue' may be interpreted more insightfully. This principle prescribes that, if an action can benefit others and poses little threat to the individual, then the ethical option is to complete the action (Porsdam Mann et al. 2016). In this case, from a Chinese perspective one may argue that sharing of the healthcare data may pose little immediate threat to the individual, especially as *Article 6 of the Regulations on the Management of Medical Records of Medical Institutions*, *Article 8 of the Management Regulations on Application of Electronic Medical Records*, *Article 6 of the Measures for the Management of Health Information*, the *Cybersecurity Law of the People's Republic of China,* and the new *Personal Information Security Specification* all provide specific and detailed instructions to ensure data security and confidentiality (Wang 2019). However, it could potentially deliver significant benefit to the wider population.

The previous 'interpretation from within' does not imply that China's approach to the use of AI in healthcare is acceptable or raises no ethical concerns. The opposite is actually true. In particular, the Chinese approach is undermined by at least three main risks.

First, there is a risk of creating a market for human care. China's two-tiered medical system provides state-insured care for all, and the option for individuals to pay privately for quicker or higher quality treatment. This is in keeping with Confucian thought, which encourages the use of private resources to benefit oneself and one's family (Wu and Mao 2017). With the introduction of Ping An [sic] Good Doctor's unmanned 'one-minute clinics' across China (of which there now may be up to 1000 in place), patients can walk in, provide symptoms and medical history, and receive an automated diagnosis and treatment plans (which are only followed up by human clinical advice for new customers), it is entirely possible to foresee a scenario in which only those who are able to pay will be able to access human clinicians. In a field where emotional care, and involvement in decision making, are often as important as the logical deduction of a 'diagnosis,' this could have a significantly negative impact on the level and quality of care accessed across the population and on the integrity of the self (Pasquale 2015),[14] at least for those who are unable to afford human care.

Second, in the context of a population that is still rapidly expanding yet also ageing, China is investing significantly in the social informatisation of healthcare and has, since at least 2015, been linking emotional and behavioural data extrapolated from social media and daily healthcare data (generated from ingestibles, implantables, wearables, carebots, and Internet of Things devices) to Electronic Health Records (Li et al. 2019b), with the goal of enabling community care of the elderly. This further adds to China's culture of State-run, mass-surveillance and, in the age of the Social Credit System, suggests that the same technologies designed to enable people to remain independent in the community as they age may one day be used as a means of social control ("China Is Building The Ultimate Digital Health Paradise. Or Is It?" 2019), to reduce the incidence of 'social diseases'—such as obesity and type II diabetes (Hawkins 2019)—under the guise of 'improving peoples lives' through the use of AI to improve the governance of social services (as stated in the AIDP).

The third ethical risk is associated with CRISPR gene modification and AI. CRISPR is a controversial gene modification technique that can be used to alter the presentation of genes in living organisms, for example for the purpose of curing or preventing genetic diseases. It is closely related to AI, as Machine Learning techniques can be used to identify which gene or genes need to be altered with the CRISPR method. The controversies, and potential significant ethical issues,

[14] Note that the emphasis on individual wellbeing must also be contextualised culturally.

associated with research in this area are related to the fact that it is not always possible to tell where the line is between unmet clinical need and human enhancement or genetic control (Cohen 2019). This became clear when, in November 2018, biophysics researcher He Jiankui revealed that he had successfully genetically modified babies using the CRISPR method to limit their chances of ever contracting HIV (Cohen 2019). The announcement was met by international outcry and He's experiment was condemned by the Chinese government at the time (Belluz 2018). However, the drive to be seen as a world leader in medical care (Cheng 2018), combined with the promise gene editing offers for the treatment of diseases, suggest that a different response may be possible in the future ("China Opens a Pandora's Box" 2018; Cyranoski 2019). Such a change in government policy is especially likely as global competition in this field heats up. The US has announced that it is enrolling patients in a trial to cure an inherited form of blindness (Ledford 2020); and the UK has launched the Accelerating Detection of Disease challenge to create a five-million patient cohort whose data will be used to develop new AI approaches to early diagnosis and biomarker discovery ("Accelerating Detection of Disease" 2019). These announcements create strong incentives for researchers in China to push regulatory boundaries to achieve quick successes (Lei et al. 2019; Tatlow 2015). Notably, China has also filed the largest number of patents for gene-editing on animals in the world (Martin-Laffon et al. 2019). Close monitoring will be essential if further ethical misdemeanours are to be avoided.

5.5 Conclusion

In this article, we analysed the nature of AI policy within China and the context within which it has emerged, by mapping the major national-level policy initiatives that express the intention to utilise AI. We identified three areas of particular relevance: *international competitiveness*, *economic growth*, and *social governance (construction)*. The development and deployment of AI in each of these areas have implications for China and for the international community. For example, although the 'trump-card' policy to gain a military advantage may not be something new, its application to AI technologies risks igniting an arms race and undermining international stability (Taddeo and Floridi 2018). Efforts to counteract this trend seem largely hollow. Our analysis indicates that China has some of the greatest opportunities for economic benefit in areas like automation, and that the country is pushing forward in AI-related areas substantially. Nonetheless, efforts to cushion the disruptions that emerge from using AI in industry are currently lacking. Thus, AI can help foster increased productivity and high levels of growth, but its use is likely to intensify the inequalities present within society and even decrease support for the government and its policies. The AIDP also promotes AI as a way to help deal with some of the major social problems, ranging from pollution to standards of living. However, positive impact in this area seem to come with increased control over individuals' behaviour, with governance extending into the realm of moral behaviour and further erosion of privacy.

Ethics also plays a central role in the Chinese policy effort on AI. The AIDP outlines a clear intention to define ethical norms and standards, yet efforts to do so are at a fledgling stage, being broadly limited to high-level principles, lacking implementation. Analyses of existing Chinese approaches and emerging debates in the areas of privacy and medical ethics provide an insight into the types of frameworks that may emerge. With respect to privacy, on the surface, recently introduced protections may seem robust, with definitions of personal information even broader than that used within the GDPR. However, a closer look exposes the many loopholes and exceptions that enable the government (and companies implicitly endorsed by the government) to bypass privacy protection and fundamental issues concerning lack of accountability and government's unrestrained decisional power about mass-surveillance.

In the same vein, when focusing on medical ethics, it is clear that, although China may agree with the West on the bioethical principles, its focus on the health of the population, in contrast to the West's focus on the health of the individual, may easily lead to unethical outcomes (the sacrifice imposed on one for the benefit of many) and is creating a number of risks, as AI encroaches on the medical space. These are likely to evolve over time, but the risks of unequal care between those who can afford a human clinician and those who cannot, control of social diseases, and of unethical medical research are currently the most significant.

China is a central actor in the international debate on the development and governance of AI. It is important to *understand* China's internal needs, ambitions in the international arena, and ethical concerns, all of which are shaping the development of China's AI policies. It is also important to understand all this not just externally, from a Western perspective, but also internally, from a Chinese perspective. However, some ethical safeguards, constraints and desiderata are universal and are universally accepted and cherished, such as the nature and scope of human rights.[15] They enable one to *evaluate*, after having *understood*, China's approach to the development of AI. This is why in this article we have sought to contribute to a more comprehensive and nuanced analysis of the structural, cultural and political factors that ground China's stance on AI, as well as an indication of its possible trajectory, while also highlighting where ethical problems remain, arise, or are likely to be exacerbated. They should be addressed as early as it is contextually possible.

[15]For arguments on the universality of human rights coming from *within* cultural perspectives, see J. Chan (1999) on Confucianism and human rights.

References

Accelerating Detection of Disease. 2019. *UK research and innovation*. https://www.ukri.org/innovation/industrial-strategy-challenge-fund/accelerating-detection-of-disease/

AI association to draft ethics guidelines. 2019. *Xinhua*, January 9. http://www.xinhuanet.com/english/2019-01/09/c_137731216.htm

AI Policy—China. n.d. *Future of life institute*. Retrieved 12 October 2020, from https://futureoflife.org/ai-policy-china/

AI-powered waste management underway in China. 2019. *People's Daily Online*, February. http://en.people.cn/n3/2019/0226/c98649-9549956.html

Allen, G.C. 2019. *Understanding China's AI strategy*. Center for New American Security. https://www.cnas.org/publications/reports/understanding-chinas-ai-strategy.

Anderlini, J. 2019. How China's smart-city tech focuses on its own citizens. *Financial Times*, June 5. https://www.ft.com/content/46bc137a-5d27-11e9-840c-530737425559

Arsène, S. 2019. *China's social credit system: A chimera with real claws* (No. 11; Asie. Visions, 28). Centre for Asian Studies.

Atkinson, S., and G. Skinner. 2019. *What worries the world—September 2019*. Ipsos. https://www.ipsos.com/en/what-worries-world-september-2019.

Austin, G. 2016. International legal norms in cyberspace: Evolution of China's National Security Motivations. In *International cyber norms: Legal, policy and industry perspectives*, ed. A.-M. Osula and H. Rõigas. NATO Cooperative Cyber Defence Centre of Excellence.

Baidu chief under fire for privacy comments. 2018. *People's Daily Online*, March. http://en.people.cn/n3/2018/0328/c90000-9442509.html

Balding, C. 2019. What's causing China's economic slowdown. *Foreign Affairs*, March 20. https://www.foreignaffairs.com/articles/china/2019-03-11/whats-causing-chinas-economic-slowdown

Barton, D., J. Woetzel, J. Seong, and Q. Tian. 2017. *Artificial intelligence: Implications for China*. McKinsey Global Institute. https://www.mckinsey.com/featured-insights/china/artificial-intelligence-implications-for-china.

Beijing AI Principles. 2019. *Beijing academy of artificial intelligence*. https://www.baai.ac.cn/news/beijing-ai-principles-en.html

Belluz, J. 2018. *Is the CRISPR baby controversy the start of a terrifying new chapter in gene editing?* Vox, November 30. https://www.vox.com/science-and-health/2018/11/30/18119589/crispr-gene-editing-he-jiankui.

Beyond Black Mirror—China's Social Credit System. 2019. *Re:publica 2019*. https://19.re-publica.com/de/session/beyond-black-mirror-chinas-social-credit-system

Blasko, D.J. 2011. 'Technology determines tactics': The relationship between technology and doctrine in Chinese military thinking. *Journal of Strategic Studies* 34 (3): 355–381. https://doi.org/10.1080/01402390.2011.574979.

Borowiec, S. 2016. Google's AI machine v world champion of 'Go': Everything you need to know. *The Guardian*, March 8. https://www.theguardian.com/technology/2016/mar/09/googles-ai-machine-v-world-champion-of-go-everything-you-need-to-know

Botsman, R. 2017. Big data meets Big Brother as China moves to rate its citizens. *Wired UK*, October 21. https://www.wired.co.uk/article/chinese-government-social-credit-score-privacy-invasion

Brezis, E.S., P.R. Krugman, and D. Tsiddon. 1993. Leapfrogging in international competition: A theory of cycles in national technological leadership. *American Economic Review* 83 (5): 1211–1219.

Bruzdzinski, J.E. 2004. Demystifying Shashoujian: "China's Assassin's Mace" concept. In *Civil-military change in China elites, institutes, and ideas after the 16th party congress*, ed. A. Scobell and L. Wortzel. Darby: DIANE Publishing.

Cai, F., and Y. Lu. 2013. Population change and resulting slowdown in potential GDP growth in China. *China & World Economy* 21 (2): 1–14. https://doi.org/10.1111/j.1749-124X.2013.12012.x.

Chai, S. 2018. *The social credit system: Is technology helping China legitimate and augment authoritarian rule?* IPP. http://oxis.oii.ox.ac.uk/wp-content/uploads/sites/77/2018/08/IPP2018-Chai.pdf.

Chan, J. 1999. A confucian perspective on human rights for contemporary China. In *The east Asian challenge for human rights*, ed. J.R. Bauer and D.A. Bell. New York: Cambridge University Press.

Chan, M., and L. Zhen. 2019. Modern military remains top priority as China boosts defence spending. *South China Morning Post*, March 5. https://www.scmp.com/news/china/military/article/2188771/modernising-military-remains-top-priority-china-boosts-defence

Chen, J. 2013. *A middle class without democracy: Economic growth and the prospects for democratization in China*. Oxford: OUP USA.

Chen, Y., and L. Song. 2018. China: Concurring regulation of cross-border genomic data sharing for statist control and individual protection. *Human Genetics* 137 (8): 605–615. https://doi.org/10.1007/s00439-018-1903-2.

Cheng, Y. 2018. China will always be bad at bioethics. *Foreign Policy*, April. https://foreignpolicy.com/2018/04/13/china-will-always-be-bad-at-bioethics/

Cheung, T.M., T. Mahnken, D. Seligsohn, K. Pollpeter, E. Anderson, and F. Yang. 2016. *Planning for innovation: Understanding China's plans for technological, energy, industrial, and defense development*. US-China Economic and Security Review Commission. https://www.uscc.gov/research/planning-innovation-understanding-chinas-plans-technological-energy-industrial-and-defense.

Chin, J.J. 2018. The longest march: Why China's democratization is not imminent. *Journal of Chinese Political Science* 23 (1): 63–82. https://doi.org/10.1007/s11366-017-9474-y.

China AI development report 2018. 2018. *Tsinghua University*. http://www.sppm.tsinghua.edu.cn/eWebEditor/UploadFile/China_AI_development_report_2018.pdf

China Is Building The Ultimate Digital Health Paradise. Or Is It? 2019. *The medical futurist*, February 19. https://medicalfuturist.com/china-digital-health

China Opens a 'Pandora's Box' of Human Genetic Engineering. 2018. *Bloomberg.Com*, November 27. https://www.bloomberg.com/news/articles/2018-11-27/china-opens-a-pandora-s-box-of-human-genetic-engineering

Chorzempa, M., P. Triolo, and S. Sacks. 2018. *China's social credit system: A mark of progress or a threat to privacy?* PIIE, June 25. https://www.piie.com/publications/policy-briefs/chinas-social-credit-system-mark-progress-or-threat-privacy.

Clover, C. 2016. China: When big data meets big brother. *Financial Times*, January 19. https://www.ft.com/content/b5b13a5e-b847-11e5-b151-8e15c9a029fb

Cohen, J. 2019. *The untold story of the 'circle of trust' behind the world's first gene-edited babies*. Science | AAAS, August 1. https://www.sciencemag.org/news/2019/08/untold-story-circle-trust-behind-world-s-first-gene-edited-babies.

Cyranoski, D. 2019. The CRISPR-baby scandal: What's next for human gene-editing. *Nature* 566 (7745): 440–442. https://doi.org/10.1038/d41586-019-00673-1.

Daum, J. 2019. Keeping track of social credit. *China Law Translate*, September 24. https://www.chinalawtranslate.com/keeping-track-of-social-credit/

Delang, C.O. 2016. *China's air pollution problems*. London: Routledge.

Devastating Blows. 2005. *Human Rights Watch*, April 11. https://www.hrw.org/report/2005/04/11/devastating-blows/religious-repression-uighurs-xinjiang

Diamond, L. 2003. The rule of law as transition to democracy in China. *Journal of Contemporary China* 12 (35): 319–331. https://doi.org/10.1080/1067056022000054632.

Dickson, B.J. 2003. *Red capitalists in China: The party, private entrepreneurs, and prospects for political change*. Cambridge: Cambridge University Press.

Ding, J. 2018. *Deciphering Chinas AI dream*. Future of Humanity Institute. https://www.fhi.ox.ac.uk/wp-content/uploads/Deciphering_Chinas_AI-Dream.pdf.

———. 2019. *ChinAI #51: China's AI 'National Team'*, May. https://chinai.substack.com/p/chinai-51-chinas-ai-national-team

Ding, J., and P. Triolo. 2018. *Translation: Excerpts from China's 'White Paper on Artificial Intelligence Standardization'*. New America, June. http://newamerica.org/cybersecurity-initiative/digichina/blog/translation-excerpts-chinas-white-paper-artificial-intelligence-standardization/.

Economy, E.C. 2013. China's water pollution crisis. *The Diplomat*, January. https://thediplomat.com/2013/01/forget-air-pollution-chinas-has-a-water-problem/

Engelmann, S., M. Chen, F. Fischer, C. Kao, and J. Grossklags. 2019. Clear sanctions, vague rewards: How China's social credit system currently defines 'good' and 'bad' behavior. In *Proceedings of the conference on fairness, accountability, and transparency*, 69–78. https://doi.org/10.1145/3287560.3287585.

Fang, A. 2019. *Chinese colleges to offer AI major in challenge to US*. Nikkei Asia, April. https://asia.nikkei.com/Business/China-tech/Chinese-colleges-to-offer-AI-major-in-challenge-to-US.

Finder, S. 2015. China's master plan for remaking its courts. *The Diplomat*, March. https://thediplomat.com/2015/03/chinas-master-plan-for-remaking-its-courts/

Floridi, L., and J. Cowls. 2019. A unified framework of five principles for AI in society. *Harvard Data Science Review* 1 (1). https://doi.org/10.1162/99608f92.8cd550d1.

Full text of Xi Jinping's report at 19th CPC National Congress. 2017. *Xinhua*, November. http://www.xinhuanet.com/english/special/2017-11/03/c_136725942.htm

Gan, N. 2019. China's social credit horrifies the West. But do the Chinese even notice? *South China Morning Post, February 7*. https://www.scmp.com/news/china/politics/article/2185303/hi-tech-dystopia-or-low-key-incentive-scheme-complex-reality

Graceffo, A. 2017. China's national champions: State support makes Chinese companies dominant. *Foreign Policy Journal*, May 15. https://www.foreignpolicyjournal.com/2017/05/15/chinas-national-champions-state-support-makes-chinese-companies-dominant/

Guidance of the State Council on the Active Promotion of the "Internet Plus" 国务院关于积极推进"互联网+"行动的指导意见(国发〔2015〕40号)政府信息公开专栏. 2015. State Council. http://www.gov.cn/zhengce/content/2015-07/04/content_10002.htm

Hawkins, A. 2019. How elderly, sickly farmers are quenching China's thirst for data. *Wired UK*, April 12. https://www.wired.co.uk/article/china-ai-healthcare

He, H. 2015. *Social ethics in a changing China: Moral decay or ethical awakening?* Brookings Institution Press; JSTOR. https://www.jstor.org/stable/10.7864/j.ctt7zsw42.

Heilmann, S. 2017. Big data reshapes China's approach to governance. *Financial Times*, September 28. https://www.ft.com/content/43170fd2-a46d-11e7-b797-b61809486fe2

Heilmann, S., and O. Melton. 2013. *The reinvention of development planning in China, 1993–2012: Modern China*. https://doi.org/10.1177/0097700413497551.

Hickert, C., and J. Ding. 2018. *Read what top Chinese officials are hearing about AI competition and policy*. New America. http://newamerica.org/cybersecurity-initiative/digichina/blog/read-what-top-chinese-officials-are-hearing-about-ai-competition-and-policy/.

Ho, A. 2018. *AI can solve China's doctor shortage. Here's how*. World Economic Forum, September. https://www.weforum.org/agenda/2018/09/ai-can-solve-china-s-doctor-shortage-here-s-how/.

Hong, Y. 2018. *Responses and explanations to the five major concerns of the Personal Information Security Code.*个人信息安全规范》五大重点关切的回应和解释. Network. http://mp.weixin.qq.com/s?__biz=MzIxODM0NDU4MQ==&mid=2247484830&idx=1&sn=6cbd44a98ff48c62db16f31458d1104f&chksm=97eab874a09d3162aa1942b97f26ed01582c3f71f82272bf581b94c6d6eed8e8c14e20d22b6a#rd

Hornby, L. 2019. Chinese app names and shames bad debtors. *Financial Times*, February 3. https://www.ft.com/content/2ad7feea-278e-11e9-a5ab-ff8ef2b976c7

Horsley, J.P. 2019. Party leadership and rule of law in the Xi Jinping era: What does an ascendant Chinese Communist Party mean for China's legal development. *Global China* 20.

Hu, A. 2013. *The distinctive transition of China's five-year plans: Modern China*. https://doi.org/10.1177/0097700413499129.

Huang, Y. 2013. Democratize or die. *Foreign Affairs*. https://www.foreignaffairs.com/articles/china/2012-12-03/democratize-or-die

Inkster, N. 2010. China in cyberspace. *Survival* 52 (4): 55–66. https://doi.org/10.1080/00396338.2010.506820.

Is China Ready for Intelligent Automation?. 2018. *China power project*, October 19. http://chinapower.csis.org/china-intelligent-automation/

Ji, W. 2013. The judicial reform in China: The status quo and future directions. *Indiana Journal of Global Legal Studies* 20 (1). https://core.ac.uk/reader/232664450.

———. 2014. The rule of law in a Chinese way: Social diversification and reconstructing the system of authority. *Asian Journal of Law and Society* 1 (2): 305–338. https://doi.org/10.1017/als.2014.9.

Jing, M., and S. Dai 2017. China recruits Baidu, Alibaba and Tencent to AI 'national team'. *South China Morning Post*, November 21. https://www.scmp.com/tech/china-tech/article/2120913/china-recruits-baidu-alibaba-and-tencent-ai-national-team

Kania, E.B. 2017a. *Battlefield singularity: Artificial intelligence, military revolution, and China's future military power*. Center for a New American Security.

———. 2017b. *杀手锏 and 跨越发展: Trump Cards and Leapfrogging*. The Strategy Bridge, September. https://thestrategybridge.org/the-bridge/2017/9/5/-and-trump-cards-and-leapfrogging

———. 2018a. *China's strategic ambiguity and shifting approach to lethal autonomous weapons systems*. Lawfare, April 17. https://www.lawfareblog.com/chinas-strategic-ambiguity-and-shifting-approach-lethal-autonomous-weapons-systems.

———. 2018b. *China's embrace of AI: Enthusiasm and challenges*. ECFR, November. https://www.ecfr.eu/article/commentary_chinas_embrace_of_ai_enthusiasm_and_challenges.

———. 2020. "AI weapons" in China's military innovation. *Global China* 23.

Knight, W. 2015. *How artificial intelligence can fight air pollution in China*. MIT Technology Review, August. https://www.technologyreview.com/2015/08/31/10611/how-artificial-intelligence-can-fight-air-pollution-in-china/.

———. 2019. *Why does Beijing suddenly care about AI ethics?* MIT Technology Review, May. https://www.technologyreview.com/2019/05/31/135129/why-does-china-suddenly-care-about-ai-ethics-and-privacy/.

Kobie, N. 2019. The complicated truth about China's social credit system. *Wired UK*, June 7. https://www.wired.co.uk/article/china-social-credit-system-explained

Koetse, M. 2018. *Insights into the social credit system on Chinese online media vs its portrayal in Western media*, October. https://www.whatsonweibo.com/insights-into-the-social-credit-system-on-chinese-online-media-and-stark-contrasts-to-western-media-approaches/

Kostka, G. 2019. *China's social credit systems and public opinion: Explaining high levels of approval: New Media & Society*. https://doi.org/10.1177/1461444819826402.

Kostka, G., and C. Zhang. 2018. Tightening the grip: Environmental governance under Xi Jinping. *Environmental Politics* 27 (5): 769–781. https://doi.org/10.1080/09644016.2018.1491116.

Ku, J. 2017. *Tentative observations on China's views on international law and cyber warfare*. Lawfare, August 26. https://www.lawfareblog.com/tentative-observations-chinas-views-international-law-and-cyber-warfare.

Laskai, L., and G. Webster. 2019. *Translation: Chinese expert group offers 'governance principles' for 'responsible AI'*. New America, June. http://newamerica.org/cybersecurity-initiative/digichina/blog/translation-chinese-expert-group-offers-governance-principles-responsible-ai/.

Lazarus, L.M. 2016. China's quest for a moral compass. *The Diplomat*, September. https://thediplomat.com/2016/09/chinas-quest-for-a-moral-compass/

Ledford, H. 2020. CRISPR treatment inserted directly into the body for first time. *Nature* 579 (7798): 185–185. https://doi.org/10.1038/d41586-020-00655-8.

Lee, K.-F. 2018. *AI superpowers: China, Silicon Valley, and the New World Order*. Boston: Houghton Mifflin Harcourt.

Lei, R., X. Zhai, W. Zhu, and R. Qiu. 2019. Reboot ethics governance in China. *Nature* 569 (7755): 184–186. https://doi.org/10.1038/d41586-019-01408-y.

Leng, S. 2018. China has new three-year plan to clean up environment. *South China Morning Post*, March 18. https://www.scmp.com/news/china/policies-politics/article/2137666/china-has-new-three-year-plan-clean-environment

Lewis, D. 2019. *Social credit case study: City citizen scores in Xiamen and Fuzhou*. Medium, October 8. https://medium.com/berkman-klein-center/social-credit-case-study-city-citizen-scores-in-xiamen-and-fuzhou-2a65feb2bbb3.

Li, A. 2016. Centralisation of power in the pursuit of law-based governance. Legal reform in China under the xi administration. *China Perspectives* 2016 (2016/2): 63–68.

Li, M. 2019. *Where does the winning mechanism for intelligent wars change 李明海\智能化战争的制胜机理变在哪里?_决策*. China Military Network.

Li, H., and L.-A. Zhou. 2005. Political turnover and economic performance: The incentive role of personnel control in China. *Journal of Public Economics* 89 (9): 1743–1762. https://doi.org/10.1016/j.jpubeco.2004.06.009.

Li, S., H. Sato, and T. Sicular. 2013. *Rising inequality in China: Challenges to a harmonious society*. Cambridge: Cambridge University Press.

Li, B., J. Li, Y. Jiang, and X. Lan. 2019a. Experience and reflection from China's Xiangya medical big data project. *Journal of Biomedical Informatics* 93: 103149. https://doi.org/10.1016/j.jbi.2019.103149.

Li, Q., L. Lan, N. Zeng, L. You, J. Yin, X. Zhou, and Q. Meng. 2019b. A framework for big data governance to advance RHINs: A case study of China. *IEEE Access* 7: 50330–50338. https://doi.org/10.1109/ACCESS.2019.2910838.

Liao, R. 2019. Two former Qualcomm engineers are using AI to fix China's healthcare problem. *TechCrunch*, February. https://social.techcrunch.com/2019/02/10/12-sigma-profile/

Lieutenant General Liu Guozhi, deputy to the National People's Congress and director of the Military Commission's Science and Technology Commission: Artificial intelligence will accelerate the process of military reform 人大代表、军委科技委主任刘国治中将:人工智能将加速军事变革进程-新华网. 2017. *Xinhua*, March. http://www.xinhuanet.com/mil/2017-03/08/c_129504550.htm

Liu, C. 2019. Multiple social credit Systems in China. *SocArXiv*. https://doi.org/10.31235/osf.io/v9frs.

Lucas, L. 2018. *China's artificial intelligence ambitions hit hurdles*, November 15. https://www.ft.com/content/8620933a-e0c5-11e8-a6e5-792428919cee

———. 2019. *China's WeDoctor shelves overseas listing over data concerns*, June 19. https://www.ft.com/content/4993fbd8-91a4-11e9-b7ea-60e35ef678d2

Martina, M., and B. Blanchard. 2019. Rise in China's defence budget to outpace economic growth target. *Reuters*, March 5. https://uk.reuters.com/article/uk-china-paliament-defence-idUKKCN1QM036

Martin-Laffon, J., M. Kuntz, and A.E. Ricroch. 2019. Worldwide CRISPR patent landscape shows strong geographical biases. *Nature Biotechnology* 37 (6): 613–620. https://doi.org/10.1038/s41587-019-0138-7.

McBride, J., and A. Chatzky. 2019. *Is 'made in China 2025' a threat to global trade?* Council on Foreign Relations, May. https://www.cfr.org/backgrounder/made-china-2025-threat-global-trade.

Mearsheimer, J.J. 2010. The gathering storm: China's challenge to US power in Asia. *The Chinese Journal of International Politics* 3 (4): 381–396. https://doi.org/10.1093/cjip/poq016.

Meyer, P. 2020. Norms of responsible state behaviour in cyberspace. In *The ethics of cybersecurity*, ed. M. Christen, B. Gordijn, and M. Loi, 347–360. Springer. https://doi.org/10.1007/978-3-030-29053-5_18.

Morris, H. 2019. China's march to be the world's first cashless society: China Daily contributor [Text]. *The Straits Times*, April 8. https://www.straitstimes.com/asia/east-asia/chinas-march-to-be-the-worlds-first-cashless-society-china-daily-contributor

Mozur, P. 2017. Beijing wants A.I. to be made in China by 2030 (Published 2017). *The New York Times*, July 20. https://www.nytimes.com/2017/07/20/business/china-artificial-intelligence.html

Naughton, B. 2007. *The Chinese economy: Transitions and growth*. London: MIT Press.

Naughton, B., and K.S. Tsai. 2015. *State capitalism, institutional adaptation, and the Chinese miracle*. Cambridge: Cambridge University Press.

Nittas, V., M. Mütsch, F. Ehrler, and M.A. Puhan. 2018. Electronic patient-generated health data to facilitate prevention and health promotion: A scoping review protocol. *BMJ Open* 8 (8): e021245. https://doi.org/10.1136/bmjopen-2017-021245.

Pasquale, F. 2015. *The black box society: The secret algorithms that control money and information*. Harvard University. https://doi.org/10.13140/RG.2.2.32841.75367.

Pecotic, A. 2019. Whoever predicts the future will win the AI arms race. *Foreign Policy*, March. https://foreignpolicy.com/2019/03/05/whoever-predicts-the-future-correctly-will-win-the-ai-arms-race-russia-china-united-states-artificial-intelligence-defense/

Pei, M. 2015. The twilight of Communist Party rule in China. *The American Interest*, November 12. https://www.the-american-interest.com/2015/11/12/the-twilight-of-communist-party-rule-in-china/

Peng, G., and Y. Yao. 2005. *The science of military strategy*. Beijing: Military Science Publishing House.

Persson, P., and E. Zhuravskaya. 2016. The limits of career concerns in federalism: Evidence from China. *Journal of the European Economic Association* 14 (2): 338–374. https://doi.org/10.1111/jeea.12142.

Porsdam Mann, S., J. Savulescu, and B.J. Sahakian. 2016. Facilitating the ethical use of health data for the benefit of society: Electronic health records, consent and the duty of easy rescue. *Philosophical Transactions of the Royal Society A: Mathematical, Physical and Engineering Sciences* 374 (2083): 20160130. https://doi.org/10.1098/rsta.2016.0130.

Principal contradiction facing Chinese society has evolved in new era: Xi. 2017. *Xinhua*, October. http://www.xinhuanet.com/english/2017-10/18/c_136688132.htm

Ray, A. 2002. The Chinese economic miracle: Lessons to be learnt. *Economic and Political Weekly* 37 (37): 3835–3848. JSTOR.

Ringen, S., and K. Ngok. 2017. What kind of welfare state is emerging in China? In *Towards universal health care in emerging economies: Opportunities and challenges*, ed. I. Yi, 213–237. Palgrave Macmillan UK. https://doi.org/10.1057/978-1-137-53377-7_8.

Rozelle, S., J. Huang, and L. Zhang. 1997. Poverty, population and environmental degradation in China. *Food Policy* 22 (3): 229–251. https://doi.org/10.1016/S0306-9192(97)00011-0.

Sacks, S. 2018. *New China data privacy standard looks more far-reaching than GDPR*. Center for Strategic and International Studies, January. https://www.csis.org/analysis/new-china-data-privacy-standard-looks-more-far-reaching-gdpr.

Sacks, S., and L. Laskai 2019. China is having an unexpected privacy awakening. *Slate Magazine*, February 7. https://slate.com/technology/2019/02/china-consumer-data-protection-privacy-surveillance.html

Sacks, S., and M.K. Li. 2018. *How Chinese cybersecurity standards impact doing business in China*. Center for Strategic and International Studies, August. https://www.csis.org/analysis/how-chinese-cybersecurity-standards-impact-doing-business-china.

Schneider, M. 2009. The nuclear doctrine and forces of the People's Republic of China. *Comparative Strategy* 28 (3): 244–270. https://doi.org/10.1080/01495930903025276.

Selden, M., and L. You. 1997. The reform of social welfare in China. *World Development* 25 (10): 1657–1668. https://doi.org/10.1016/S0305-750X(97)00055-7.

Sheehan, M. 2018. *How China's massive AI plan actually works*. MacroPolo, February. https://macropolo.org/analysis/how-chinas-massive-ai-plan-actually-works/.

Shoham, Y., R. Perrault, E. Brynjolfsson, J. Clark, J. Manyika, J.C. Niebles, T. Lyons, J. Etchemendy, B. Grosz, and Z. Bauer. 2018. *AI index 2018 annual report*. Stanford University. http://cdn.aiindex.org/2018/AI%20Index%202018%20Annual%20Report.pdf.

Si, J. 2019. *These rules could save humanity from the threat of rogue AI.* World Economic Forum, May. https://www.weforum.org/agenda/2019/05/these-rules-could-save-humanity-from-the-threat-of-rogue-ai/.

Simonite, T. 2019. How health care data and lax rules help China prosper in AI. *Wired.* https://www.wired.com/story/health-care-data-lax-rules-help-china-prosper-ai/

Sizing the Prize: What's the real value of AI for your business and how can you capitalise? 2017. *PwC.* https://www.pwc.com/gx/en/issues/analytics/assets/pwc-ai-analysis-sizing-the-prize-report.pdf

Song, B. 2018. Opinion | The West may be wrong about China's social credit system. *Washington Post*, November. https://www.washingtonpost.com/news/theworldpost/wp/2018/11/29/social-credit/

Sonnad, N. 2017. *In China you now have to provide your real identity if you want to comment online.* Quartz. https://qz.com/1063073/in-china-you-now-have-to-provide-your-real-identity-if-you-want-to-comment-online/.

Stevens, L.A. 2017. *Public interest approach to data protection law: The meaning, value and utility of the public interest for research uses of data.* Edinburgh Research Archive. https://era.ed.ac.uk/handle/1842/25772.

Sun, Y. 2018. *China's citizens do care about their data privacy, actually.* MIT Technology Review. https://www.technologyreview.com/2018/03/28/67113/chinas-citizens-do-care-about-their-data-privacy-actually/.

Taddeo, M. 2012. Information warfare: A philosophical perspective. *Philosophy & Technology* 25 (1): 105–120. https://doi.org/10.1007/s13347-011-0040-9.

———. 2016. On the risks of relying on analogies to understand cyber conflicts. *Minds and Machines* 26 (4): 317–321. https://doi.org/10.1007/s11023-016-9408-z.

Taddeo, M., and L. Floridi. 2018. Regulate artificial intelligence to avert cyber arms race. *Nature* 556 (7701): 296–298. https://doi.org/10.1038/d41586-018-04602-6.

Tatlow, D.K. 2015. A scientific ethical divide Between China and West (Published 2015). *The New York Times*, June 29. https://www.nytimes.com/2015/06/30/science/a-scientific-ethical-divide-between-china-and-west.html

The 13th Five Year Plan for Economic and Social Development of the People's Republic of China (2016). (Trans. Compilation and Translation Bureau). 2016. *Communist Party of China.* https://en.ndrc.gov.cn/policyrelease_8233/201612/P020191101482242850325.pdf

The Automation Readiness Index 2018. 2018. *The economist intelligence unit.* http://automationreadiness.eiu.com

The position paper submitted by the Chinese delegation to CCW 5th Review Conference. 2016. https://www.unog.ch/80256EDD006B8954/(httpAssets)/DD1551E60648CEBBC125808A005954FA/$file/China%27s+Position+Paper.pdf

UN seeks to retain human control over force. 2018. *The campaign to stop killer robots*, August. https://www.stopkillerrobots.org/2018/08/unsg/

Wang, Maya. 2019. China's Algorithms of Repression. *Human Rights Watch*, May 1. https://www.hrw.org/report/2019/05/02/chinas-algorithms-repression/reverse-engineering-xinjiang-police-mass

Wang, Z. 2019. Data integration of electronic medical record under administrative decentralization of medical insurance and healthcare in China: A case study. *Israel Journal Health Policy Research* 8 (1): 24. https://doi.org/10.1186/s13584-019-0293-9.

Wang, J., and S. Liu. 2019. Ordering power under the party: A relational approach to law and politics in China. *Asian Journal of Law and Society* 6 (1): 1–18. https://doi.org/10.1017/als.2018.40.

Webb, A. 2019. *China and the AI edge.* Nieman Reports, March. https://niemanreports.org/articles/china-and-the-ai-edge/.

Webster, G., R. Creemers, P. Triolo, and E. Kania. 2017a. *China's plan to 'lead' in AI: Purpose, prospects, and problems.* New America, August. http://newamerica.org/cybersecurity-initiative/blog/chinas-plan-lead-ai-purpose-prospects-and-problems/.

Webster, G., R. Creemers, P. Triolo, and E.B. Kania. 2017b. *Full translation: China's 'new generation artificial intelligence development plan' (2017)*. New America. http://newamerica.org/cybersecurity-initiative/digichina/blog/full-translation-chinas-new-generation-artificial-intelligence-development-plan-2017/.

What will be the net impact of AI and related technologies on jobs in China? 2018. *PwC*. https://www.pwc.com/gx/en/issues/artificial-intelligence/impact-of-ai-on-jobs-in-china.pdf

Whiting, S.H. 2017. *Authoritarian "rule of law" and regime legitimacy: Comparative political studies*. https://doi.org/10.1177/0010414016688008.

Wong, L. 2005. *Marginalization and social welfare in China*. London: Routledge.

Wong, P. 2013. Confucian social media: An oxymoron? *Dao* 12 (3): 283–296. https://doi.org/10.1007/s11712-013-9329-y.

Wu, J., and Y. Mao. 2017. Liberty in health care: A comparative study between Hong Kong and Mainland China. *The Journal of Medicine and Philosophy* 42 (6): 690–719. https://doi.org/10.1093/jmp/jhx026.

Yan, Y. 2009. The Good Samaritan's new trouble: A study of the changing moral landscape in contemporary China1. *Social Anthropology* 17 (1): 9–24. https://doi.org/10.1111/j.1469-8676.2008.00055.x.

Yang, Y. 2018. *China's data privacy outcry fuels case for tighter rules*, October 2. https://www.ft.com/content/fdeaf22a-c09a-11e8-95b1-d36dfef1b89a

Yang, H. 2019. *China—The privacy, data protection and cybersecurity law review—Edition 6*. The Law Reviews, October. https://thelawreviews.co.uk/edition/the-privacy-data-protection-and-cybersecurity-law-review-edition-6/1210009/china.

Yang, Y., and X. Liu. 2018. China's AI push raises fears over widespread job cuts. *Financial Times*, August 30. https://www.ft.com/content/1e2db400-ac2d-11e8-94bd-cba20d67390c

Yu, M., and G. Du. 2019. Why are Chinese courts turning to AI? *The Diplomat*, January. https://thediplomat.com/2019/01/why-are-chinese-courts-turning-to-ai/

Yuan, S. 2019. AI-assisted sentencing speeds up cases in judicial system—Chinadaily.com.cn. *China Daily*, April. http://www.chinadaily.com.cn/cndy/2019-04/18/content_37459601.htm

Zhang, Z. 2018. China to focus on innovation to boost economy. *China Daily*, January. http://www.chinadaily.com.cn/a/201801/09/WS5a543bd5a31008cf16da5fa9.html

Zhang, Y., and W. Han. 2019. *In depth: China's burgeoning social credit system stirs controversy—Caixin global*. Caixin, April. https://www.caixinglobal.com/2019-04-01/in-depth-chinas-burgeoning-social-credit-system-stirs-controversy-101399430.html.

Zhang, P., and Y. Liang. 2018. China's national health guiding principles: A perspective worthy of healthcare reform. *Primary Health Care Research & Development* 19 (1): 99–104. https://doi.org/10.1017/S1463423617000421.

Zhang, G., and K. Yin 2019. *More updates on the Chinese data protection regime in 2019 [IAPP]*, February. https://iapp.org/news/a/more-positive-progress-on-chinese-data-protection-regime-in-2019/

Zhang, L., H. Wang, Q. Li, M.-H. Zhao, and Q.-M. Zhan. 2018. Big data and medical research in China. *BMJ*: j5910. https://doi.org/10.1136/bmj.j5910.

Zhao, S. 2015. A new model of big power relations? China–US strategic rivalry and balance of power in the Asia–Pacific. *Journal of Contemporary China* 24 (93): 377–397. https://doi.org/10.1080/10670564.2014.953808.

Zhu, V. 2019. *China trends #2—Large but not strong: The challenges for China's domestic ports*. Institut Montaigne, June. https://www.institutmontaigne.org/en/blog/china-trends-2-large-not-strong-challenges-chinas-domestic-ports.

Chapter 6
Translating Principles into Practices of Digital Ethics: Five Risks of Being Unethical

Luciano Floridi

Abstract In this chapter, I argue that in translating ethical principles for digital technologies into ethical practices, even the best efforts may be undermined by some unethical risks. Five of them are already encountered or foreseeable in the international debate about digital ethics: (1) ethics shopping; (2) ethics bluewashing; (3) ethics lobbying; (4) ethics dumping; and (5) ethics shirking.

Keywords Artificial intelligence · AI ethical principles · Digital technologies · Translational ethics · Unethical risks

6.1 Introduction

It has taken a very long time,[1] but today, the debate on the ethical impact and implications of digital technologies has reached the front pages of newspapers. This is understandable: digital technologies—from web-based services to Artificial Intelligence (AI) solutions—increasingly affect the daily lives of billions of people, so there are many hopes but also concerns about their design, development, and deployment (Cath et al. 2018).

After more than half a century of academic research,[2] the recent public reaction has been a flourishing of initiatives to establish *what* principles, guidelines, codes, or frameworks can ethically guide digital innovation, particularly in AI, to benefit

[1] See (Floridi 2015) for references.

[2] In the ethics of AI, see for example (Wiener 1960; Samuel 1960).

L. Floridi (✉)
Oxford Internet Institute, University of Oxford, Oxford, UK
e-mail: luciano.floridi@oii.ox.ac.uk

L. Floridi (ed.), *Ethics, Governance, and Policies in Artificial Intelligence*,
Philosophical Studies Series 144, https://doi.org/10.1007/978-3-030-81907-1_6

humanity and the whole environment. This is a positive development that shows aware- ness of the importance of the topic and interest in tackling it systematically. Yet, it is time that debate evolves from the *what* to the *how*: not just *what* ethics is needed but also *how* ethics can be effectively and successfully applied and implemented in order to make a positive difference. For example, the European *Ethics Guidelines for Trustworthy AI*[3,4] establish a benchmark for what may or may not qualify as ethically good AI in the EU. Their publication is currently being followed by practical efforts of testing, application, and implementation.

The move from a first, more theoretical *what* chapter, to a second, more practical *how* chapter, so to speak, is reasonable and commendable. However, in translating principles into practices, even the best efforts may be undermined by some unethical risks.

6.2 Ethics Shopping

A very large number of ethical principles, codes, guidelines, or frameworks have been proposed over the past few years. There are currently more than 70 recommendations, published in the last 2 years, just about the ethics of AI (Algorithm Watch 9 April 2019; Winfield 18 April Winfield 2019). This mushrooming of documents is generating inconsistency and confusion among stakeholders regarding which one may be preferable. It also puts pressure on private and public actors—that design, develop, or deploy digital solutions—to produce their own declarations for fear of appearing to be left behind, thus further contributing to the redundancy of information. In this case, the main, unethical risk is that all this hyperactivity creates a "market of principles and values", where private and public actors may shop for the kind of ethics that is best retrofitted to justify their current behaviours, rather than revising their behaviours to make them consistent with a socially accepted ethical framework (Floridi and Lord Clement-Jones 20 March Floridi and Clement-Jones 2019). Here is a more compact definition:

> *Digital ethics shopping* $=_{\text{def.}}$ The malpractice of choosing, adapting, or revising ("mixing and matching") ethical principles, guidelines, codes, frameworks, or other similar standards

[3] See (European Commission 8 April 2019), published by the High-Level Expert Group (HLEG) on Artificial Intelligence (AI) appointed by the European Commission (disclosure: I am a member of the HLEG).

[4] See for example the debates about (a) the "Human Rights Impact Assessment of Facebook in Myanmar" published by the Business for Social Responsibility, https://www.bsr.org/en/our-insights/blog-view/facebook-in-myanmar-human-rights-impact-assessment; (b) the closure of Google's Advanced Technology External Advisory Council https://blog.google/technology/ai/external-advisory-council-help-advance-responsible-development-ai/; and (c) the Ethics guidelines for trustworthy AI, published by the High-Level Expert Group of the European Commission https://ec.europa.eu/digital-single-market/en/news/ethics-guidelines-trustworthy-ai (disclosure: I was a member of the former, and still am a member of the latter).

(especially but not only in the ethics of AI), from a variety of available offers, in order to retrofit some pre-existing behaviours (choices, processes, strategies, etc.), and hence justify them a posteriori, instead of implementing or improving new behaviours by benchmarking them against public, ethical standards.

Admittedly, in a recent meta-analysis, we showed that much of the diversity "in the ethics market" is apparent and more due to wording and vocabulary rather than actual content (Floridi et al. 2018; Floridi and Cowls forthcoming). However, the potential risk of "mixing and matching" the list of ethical principles one prefers remains real, because semantic looseness and redundancy enable interpretative relativism. Ethics shopping then causes incompatibility of standards (it is hard to understand whether two companies follow the same ethical principles in developing AI solutions, for example), and with that a lower chance of comparison, competition, and accountability.

The strategy to deal with digital ethics shopping is to establish clear, shared, and publicly accepted ethical standards. This is why I recently argued (Floridi 2019) that the publication of the *Ethics Guidelines for Trustworthy AI* is a significant improvement, given that it is the closest thing available in the European Union (EU) to a comprehensive, authoritative, and public standard of what may count as socially good AI.[5] Now that the *Guidelines* are available, the malpractice of digital ethics shopping should be at least more obvious if not more difficult to indulge in, because anyone in the EU may simply subscribe to them, rather than shop for (or even cook) their own "ethics".

6.3 Ethics Bluewashing

In environmental ethics, *greenwashing* (Delmas and Burbano 2011) is the malpractice of a private or public actor seeking to appear greener, more sustainable, or ecologically friendlier than it actually is. By "ethics bluewashing", I mean to refer to the digital version of greenwashing. As there is no specific colour associated with ethically good practices in digital technologies, "blue" may serve to remind one that we are not talking about ecological sustainability but mere digital ethics cosmetics[6]:

Ethics bluewashing = def. the malpractice of making unsubstantiated or misleading claims about, or implementing superficial measures in favour of, the ethical values and benefits of digital processes, products, services, or other solutions in order to appear more digitally ethical than one is.

---

[5]See also (Mazzini forthcoming).

[6]This is not to be confused with the term bluewashing "[...] used to criticize the corporate partnerships formed under the United Nations Global Compact initiative (some say this association with the UN helps to improve the corporations' reputations) and to disparage dubious sustainable water-use projects" (Schott 4 February 2010).

Ethics greenwashing and bluewashing are forms of misinformation, often achieved by spending a fraction of the resources that would be needed to tackle the ethical problems they pretend to address. They concentrate on mere marketing, advertising, or other public relations activities (e.g. sponsoring), including the setting up of advisory groups that may be powerless or insufficiently critical. Both mal-practices are tempting because, in each case, the goals are many and all compatible:

(a) Distract the receiver of the message—usually the public, but any shareholders or stakeholders may be the target—from anything that is going wrong, could go better, or is not happening but should;
(b) Mask and leave unchanged any behaviour that ought to be improved;
(c) Achieve economic savings; and
(d) Gain some advantage, e.g. competitive or social, for example in terms of "good will".

However, contrary to what happens with greenwashing, bluewashing can more easily be combined with digital ethics shopping: a private or public actor shops for the principles that best fit its current practices, publicises them as widely as possible and then proceeds to bluewash its technological innovation without any real improvement, much lower costs, and some potential social benefits. These days, ethics bluewashing is especially tempting in the context of AI, where the ethical issues are many, the costs of doing the right thing may be high, and normative uncertainity or sometimes confusion are widespread.

The best strategy against bluewashing is the same already adopted against greenwashing: *transparency* and *education*. Public, accountable, and evidence-based trans- parency about good practices and ethical claims should be a priority on the side of the actors wishing to avoid the appearance of engaging in any bluewashing malpractice. Public and factual education, on the side of any target of bluewashing—not just the general public but also members of executive boards and advisory councils, for example—about whether and what effective ethical practices are actually implemented means that actors may be less likely to (be tempted to) distract public attention away from the ethical challenges they are facing.

As we recommended in Floridi et al. (2018), the development of metrics for the trustworthiness of AI products and services (and of digital solutions in general) would enable the user-driven benchmarking of all marketed offerings and facilitate the detection of mere bluewashing, improving public understanding, and engendering competitiveness around the development of safer, more socially and environmentally beneficial products and services. In the longer term, a system of certification for digital products and services could achieve what other similar solutions have achieved in environmental ethics: make bluewashing as visible and shameful as greenwashing.

6.4 Ethics Lobbying

Sometimes, private actors (are at least suspected to) try to use self-regulation about the ethics of AI in order to lobby against the introduction of legal norms, or in favour of their watering down or weakening their enforcement, or in order to provide an excuse for limited compliance. This specific malpractice affects many sectors, but it seems more likely in the digital one (Benkler 2019), where ethics may be exploited as if it were an alternative to legislation, and in the name of technological innovation and its positive impact on economic growth, a line of reasoning that cannot be easily supported in environmental or biomedical contexts. Here is a more general definition:

> *Digital ethics lobbying* $=_{\text{def.}}$ the malpractice of exploiting digital ethics to delay, revise, replace, or avoid good and necessary legislation (or its enforcement) about the design, development, and deployment of digital processes, products, services, or other solutions.

One may argue that digital ethics lobbying is a poor strategy, likely to fail in the long run because it is at best short-sighted: sooner or later legislation tends to catch up. Whether this argument is convincing or not, digital ethics lobbying as a short-term tactic may still cause much damage, by delaying the introduction of necessary legislation, by helping manoeuvre around or by-pass more demanding interpretations of current legislation, thus making compliance easier but also misaligned with the spirit of the law, or by influencing law-makers to pass legislation that is more favourable to the lobbyist than would otherwise be expected. Furthermore, and very importantly, the malpractice, or the suspicion of it, risks undermining the value of any digital ethical self-regulation *tout court*.

This collateral damage is deeply regrettable because self-regulation is one of the main valuable tools available for policy-making. In itself, it cannot replace the law, but if properly implemented, it can be crucially complementary (Floridi 2018), when:

- Legislation is unavailable (for example, in experimentations about augmented reality products) or
- Legislation is available, but also in need of an ethical interpretation (for example, in terms of understanding a right to explanation in the GDPR) or
- Legislation is available, but also in need of some ethical counterbalancing:
- If it is better not to do something, even if it is not (yet) illegal to do it (for example, to automate entirely and fully some medical procedure without any human supervision) or
- If it is better to do something, even if it is not (yet) legally required (for example, to implement better labour market conditions in the Gig Economy).

The strategy against digital ethics lobbying is twofold. On the one hand, it must be counteracted by good legislation and effective enforcement. This is easier if the lobbying actor (private or public) is less influential on law-makers or whenever public opinion can exercise the right level of ethical pressure. On the other hand, digital ethics lobbying must be exposed whenever it occurs and be clearly distinguished from genuine forms of self-regulation. This may happen more credibly if the process is also in itself part of a self-regulatory code of conduct of a whole industrial sector, in our case the digital tech industry, which has a more general interest in maintaining a healthy context where genuine self-regulation is both socially welcome and efficacious and ethics lobbying is exposed as unacceptable.

6.5 Ethics Dumping

"Ethics dumping" is an expression coined in 2013 by the European Commission to describe the export of unethical research practices to countries where there are weaker (or laxer, or perhaps just different, in the case of digital ethics) legal and ethical frameworks and enforcing mechanisms. It applies to any kind of research—including research in computer science, data science, machine learning, robotics, and other kinds of AI—but it is most serious in health-related and biological contexts. Fortunately, biomedical and environmental ethics may be considered universal and global; there are international agreements and frameworks and international institutions monitoring their application or enforcement, so "ethics dumping" may be fought more effectively and coherently when research involves biomedical and ecological contexts. However, in digital contexts, the variety (or indeed the lack of) of legal regimes and ethical frameworks facilitates the export of (what are considered within the original context where the "dumper" operates) unethical (or even illegal) practices, and the import of the outcomes of such practices. In other words, the problem is twofold, about *research ethics* and *consumption ethics*. So, here is a definition:

> *Digital ethics dumping* = def. the malpractice of (a) exporting research activities about digital processes, products, services, or other solutions, in other contexts or places (e.g. by European organisations outside the EU) in ways that would be ethically unacceptable in the context or place of origin and (b) importing the outcomes of such unethical research activities.

Both (a) and (b) are important. To offer a concrete, if distant, example, it is not unusual for countries to ban the cultivation of genetically modified organisms, but allow their import. This asymmetry of ethical (and legal) treatment between a practice (unethical and/ or illegal research) and its output (ethically and legally acceptable consumption of the output of the unethical research) means that ethics dumping may affect digital ethics not only in terms of unethical export of research activities but also in terms of unethical import of the outcomes of such activities. For example, a company may export its research and then design, develop, and train algorithms, e.g. for face recognition, on local personal data in a non-EU country with a different or weaker ethical and legal framework for personal data protection, which would be unethical and illegal in the EU because of the GDPR. Once trained, the algorithms may then be imported to the EU and deployed without incurring any penalty or even be frowned upon. Whereas the first step (a) may be more easily blocked, at least in terms of research ethics (Nordling 2018); the second step (b), involving the consumption of unethical research results, is fuzzier, less visibly problematic, and hence more difficult to monitor and curtail.

Unfortunately, it is likely that, in the near future, the problem of digital ethics dumping will become increasingly serious, due to the profound impact of digital technologies on health and social care as well as defence, policing and security, the ease of their global portability, the complexity of the productions processes (some stages of which may involve ethical dumping), and the immense economic interests at play. For example, especially in AI, where the EU is a net importer of solutions from the USA and China, private and public actors risk not just exporting unethical practises but also (and independently) importing solutions that may have been developed in ways that would not have been ethically acceptable within the EU.

In this case too, the strategy is twofold. One must concentrate on research ethics *and* the ethics of consumption. If one wishes to be coherent, both need to receive equal attention.

In terms of research ethics, it is slightly easier to exercise control at the source, through the ethical management of public funding for research. In this, the EU is in a leading position. However, there remains the significant problem that much R&D about digital solutions is done by the private sector, where funding may be less constrained by geographical borders (a private actor can more easily relocate its R&D to an ethically less demanding place, a geographical variation of the ethics shopping seen in Sect. 6.2) and is not ethically scrutinised in the same way as publicly funded research.

In terms of consumption ethics, especially of digital products and services, much can be done both by the establishment of a system of certification for products and services that could inform procurement, as well as public and private use. As in the case of bluewashing, the reliable and ethically acceptable provenance of digital systems and solutions will have to play an increasing role in the following years if one wishes to avoid the hypocrisy of being careful about research ethics in digital contexts and yet relaxed about the unethical use of its outcomes.

6.6 Ethics Shirking

Ethicists are well acquainted with the old malpractice of applying double standards in moral evaluations. By applying a lenient and a strict approach, one can evaluate and treat agents (or their actions, or the consequences of their actions) differently than similar agents (actions or consequences), when in fact they should all be treated equally. Usually, a risk of double standards is based, even inadvertently, on bias, unfairness, or selfish interest. The risk I wish to highlight here belongs to the same family, but it has a different genesis. To highlight its specificity, I shall borrow the expression "ethics shirking" from the financial sector[7] and define it thus:

Ethics shirking = def. the malpractice of doing increasingly less "ethical work" (such as fulfilling duties, respecting rights, and honouring commitments) in a given context the lower the return of such ethical work in that context is mistakenly perceived to be.

Ethics shirking, like ethics dumping, has historical roots and often follows geopolitical outlines. Actors are more likely to engage in ethics dumping and shirking in contexts where disadvantage populations, weaker institutions, legal uncertainties, corrupted regimes, unfair power distributions, and other economic, legal, political, or social ills prevail. It is not unusual to map, correctly, both malpractices along the divide between Global North and Global South, or to see both as affecting above all Low- and Middle-Income Countries. The colonial past still exerts a disgraceful role. It is also important to recall that, in digital contexts, these malpractices can affect segments of a population within the Global North. The Gig Economy may be seen as a case of ethics shirking within developed countries. And the development of self-driving cars may be interpreted as an instance of research dumping in some states of the USA. In this case, the 1968 Vienna Convention on Road Traffic, which establishes international principles to govern traffic laws, requires that a driver is always fully in control and responsible for the behaviour of a vehicle in traffic. However, the USA is not a signatory country and the requirement does not apply, meaning state vehicle codes do not prohibit automated vehicles, and several states have enacted laws for automated vehicles. This is also why research on self-driving cars happens mostly in the USA—as well as the related incidents and human suffering.

The strategy against ethics shirking consists in tackling its origin, which is a lack of clear allocation of responsibility. Agents may be more tempted to shirk their ethical work in a given context the more they (think they) can relocate responsibilities elsewhere. This happens more likely and easily in "D contexts", where one's own responsibility may be perceived (mistakenly) to be lower because it is *distant, diminished*, *delegated*, or *distributed* (Floridi 2013). Thus, ethics shirking is an agency unethical cost of deresponsabilisation. It is this genesis that makes it a special case of the ethical problem of double standards. This is why more fairness and less bias are necessary—insofar as ethics shirking is a special case of the problem of double standards—but they are also insufficient to remove the incentive to engage in ethics shirking. To uproot such a malpractice, one also needs an ethics of distributed responsibility (Floridi 2016) that relocates responsibilities—and hence praise and blame, reward and punishment, and ultimately causal accountability and legal liability—where they rightly belong.

[7] https://www.nasdaq.com/investing/glossary/s/shirking I owe the suggestion to include "ethics shirking" as a significant risk in digital ethics and to use the expression itself to capture it to (Cowls et al. unpublished).

6.7 Conclusion

I hope this short article may work as a map for those who wish to avoid or minimise some of the most obvious and significant ethical risks, when navigating from principles to practices in digital ethics. From a Socratic perspective, a malpractice is often the result of a misjudged solution or a mistaken opportunity. Understanding as early as possible that shortcuts, postponements, or quick fixes do not lead to better ethical solutions but to more serious problems, which become increasingly difficult to solve the later one deals with them, does not guarantee that the five malpractices analysed in this article will disappear, but it does mean that they will be reduced insofar as they are genuinely based on misunderstanding and misjudgements. Not knowing better is the source of a lot of evil.[8] So, the solution is often more and better information for all.

References

Algorithm Watch. 2019. *The AI ethics guidelines global inventory*. https://algorithmwatch.org/en/project/ai-ethics-guidelines-global-inventory/.

Benkler, Y. 2019. Don't let industry write the rules for AI. *Nature* 569 (161). https://doi.org/10.1038/d41586-019-01413-1.

Cath, C., S. Wachter, B. Mittelstadt, M. Taddeo, and L. Floridi. 2018. Artificial intelligence and the 'good society': The US, EU, and UK approach. *Science and Engineering Ethics* 24 (2): 505–528.

Floridi, Luciano, and Josh Cowls. forthcoming. *A unified framework of principles for AI in society*.

Cowls, Josh, Marie-Thérèse Png, and Yung Au. unpublished. *Some tentative foundations for* "Global" *algorithmic ethics*.

Delmas, M.A., and V.C. Burbano. 2011. The drivers of greenwashing. *California Management Review* 54 (1): 64–87. https://doi.org/10.1525/cmr.2011.54.1.64.

European Commission. 2019. *Ethics guidelines for trustworthy AI*. https://ec.europa.eu/digital-single-market/en/news/ethics-guidelines-trustworthy-ai.

Floridi, L. 2013. Distributed morality in an information society. *Science and Engineering Ethics* 19 (3): 727–743.

———. 2015. *The ethics of information*. Oxford: Oxford University Press.

———. 2016. Faultless responsibility: On the nature and allocation of moral responsibility for distributed moral actions. *Philosophical Transactions of the Royal Society A: Mathematical, Physical and Engineering Sciences* 374 (2083): 20160112.

———. 2018. Soft ethics, the governance of the digital and the general data protection regulation. *Philosophical Transactions of the Royal Society A: Mathematical, Physical and Engineering Sciences* 376 (2133): 20180081. https://doi.org/10.1098/rsta.2018.0081.

Floridi, Luciano. 2019. Establishing the rules for building trustworthy AI. *Nature – Machine Intelligence*.

Floridi, Luciano, and Tim Lord Clement-Jones. 2019. The five principles key to any ethical framework for AI. *New Statesman*. https://tech.newstatesman.com/policy/ai-ethics-framework.

[8]Many thanks to Josh Cowls, Jessica Morley, David Sutcliffe, and David Watson for their very valuable feedback on an earlier version of this article.

Floridi, Luciano, Josh Cowls, Monica Beltrametti, Raja Chatila, Patrice Chazerand, Virginia Dignum, Christoph Luetge, Robert Madelin, Ugo Pagallo, Francesca Rossi, Burkhard Schafer, Peggy Valcke, Effy Vayena, and J. Minds Machines. 2018. AI4People—An ethical framework for a good. *AI society: Opportunities, Risks, Principles, and Recommendations* 28 (4): 689–707. https://doi.org/10.1007/s11023-018-9482-5.

Mazzini, Gabriele. forthcoming. A system of governance for artificial intelligence through the lens of emerging intersections between AI and EU law. In *Digital revolution – New challenges for law*, ed. A. De Franceschi, R. Schulze, M. Graziadei, O. Pollicino, F. Riente, S. Sica, and P. Sirena. SSRN: https://ssrn.com/abstract=3369266.

Nordling, L. 2018. Europe's biggest research fund cracks down on 'ethics dumping'. *Nature* 559 (7712): 17.

Samuel, A.L. 1960. Some moral and technical consequences of automation—A refutation. *Science* 132 (3429): 741–742.

Schott, Ben. 2010. Bluewashing. *The New York Times*.

Wiener, N. 1960. Some moral and technical consequences of automation. *Science* 131 (3410): 1355–1358.

Winfield, Alan. 2019. *An updated round up of ethical principles of robotics and AI*. http://alanwinfield.blogspot.com/2019/04/an-updated-round-up-of-ethical.html.

Chapter 7
How AI Can Be a Force for Good – An Ethical Framework to Harness the Potential of AI While Keeping Humans in Control

Mariarosaria Taddeo and Luciano Floridi

Abstract The article has the goal of indicating how to harness the potential for good of artificial intelligence (AI) – defined as a distinct form of autonomous and self-learning agency and thus raises unique ethical challenges – while mitigating its ethical challenges. The analyses focuses first on uses of AI that may lead to undue discrimination, lack of explainability, the responsibility gap, and the nudging potential of AI and its negative impact on human self-determination. It then turns on the role that ethical analyses in harnessing the potential for good of AI and argues that existing guidelines for the ethics design, development and use of AI will be effective insofar as they are translated into viable guidelines to shape AI-based innovation and that this is the task of digital ethics as a translational ethics.

Keywords Artificial intelligence · Digital ethics · Ethics of AI · Explainability · Responsibility · Self-determination · Translational ethics

Artificial intelligence (AI) is not just a new technology that requires regulation. It is a powerful force that is reshaping daily practices, personal and professional interactions, and environments. For the well-being of humanity it is crucial that this power is used as a force of good. Ethics plays a key role in this process by ensuring that regulations of AI harness its potential while mitigating its risks.

AI may be defined in many ways. Get its definition wrong, and any assessment of the ethical challenges of AI becomes science fiction at best or an irresponsible distraction at worst, as in the case of the singularity debate. A scientifically sound approach is to draw on its classic definition (McCarthy et al. 2006) as a growing

M. Taddeo (✉)
Oxford Internet Institute, University of Oxford, Oxford, UK

Alan Turing Institute, London, UK
e-mail: mariarosaria.taddeo@oii.ox.ac.uk

L. Floridi
Oxford Internet Institute, University of Oxford, Oxford, UK
e-mail: luciano.floridi@oii.ox.ac.uk

L. Floridi (ed.), *Ethics, Governance, and Policies in Artificial Intelligence*, Philosophical Studies Series 144, https://doi.org/10.1007/978-3-030-81907-1_7

resource of interactive, autonomous, self-learning agency, which enables computational artifacts to perform tasks that otherwise would require human intelligence to be executed successfully (Samuel 1960). AI can then be further defined in terms of features such as the computational models on which it relies or the architecture of the technology.

But when it comes to ethical and policy-related issues, the latter distinctions are unnecessary (Yang et al. 2018). On the one hand, AI is fueled by data and therefore faces ethical challenges related to data governance, including consent, ownership, and privacy. These data-related challenges may be exacerbated by AI, but would occur even without AI. On the other hand, AI is a distinct form of autonomous and self- learning agency and thus raises unique ethical challenges. The latter are the focus of this article.

The ethical debate on AI as a new form of agency dates to the 1960s (Samuel 1960; Wiener 1960). Since then, many of the relevant problems have concerned delegation and responsibility. As AI is used in ever more contexts, from recruitment to health care, understanding which tasks and decisions to entrust (delegate) to AI and how to ascribe responsibility for its performance are pressing ethical problems. At the same time, as AI becomes invisibly ubiquitous, new ethical challenges emerge. The protection of human self-determination is one of the most relevant and must be addressed urgently. The application of AI to profile users for targeted advertising, as in the case of online service providers, and in political campaigns, as unveiled by the Cambridge Analytica case, offer clear examples of the potential of AI to capture users' preferences and characteristics and hence shape their goals and nudge their behavior to an extent that may undermine their self-determination.

7.1 Delegation and Responsibility

AI applications are becoming pervasive. Users rely on them to deal with a variety of tasks, from delivering goods to ensuring national defense (Taddeo and Floridi 2018). Assigning these tasks to AI brings huge benefits to societies. It lowers costs, reduces risks, increases consistency and reliability, and enables new solutions to complex problems. For example, AI applications can lower diagnostic errors by 85% in breast cancer patients (Wang et al. 2016), and AI cybersecurity systems can reduce the aver- age time to identify and neutralize cyberattacks from 101 days to a few hours (Taddeo and Floridi 2018). However, delegation may also lead to harmful, unintended consequences, especially when it involves sensitive decisions or tasks (Asaro 2012; Russell 2015) and excludes or even precludes human supervision (Yang et al. 2018). The case of COMPAS, an AI legal system that discriminated against African-American and Hispanic men when making decisions about granting parole (Jeff Larson 2016), has become infamous. Robust procedures for human oversight are needed to minimize such unintended consequences and redress any unfair impacts of AI. Still, human oversight is insufficient if it deals with problems only after they occur.

Techniques to explain AI and predict its outcomes are also needed. The Explainable Artificial Intelligence program of DARPA (Defense Advanced Research Project

Agency) is an excellent example. The goal of this pro- gram is to define new techniques to explain the decision-making processes of AI systems. This will enable users to understand how AI systems work, and designers and developers to improve the systems to avoid mistakes and mitigate the risks of misuse. To be successful, similar projects must include an ethical impact analysis from the beginning, to assess AI′s benefits and risks and define guiding principles for an ethically sound design and use of AI.

The effects of decisions or actions based on AI are often the result of countless interactions among many actors, including designers, developers, users, software, and hardware. This is known as distributed agency (JO). With distributed agency comes distributed responsibility. Existing ethical frameworks address individual, human responsibility, with the goal of allocating punishment or reward based on the actions and intentions of an individual. They were not developed to deal with distributed responsibility.

Only recently have new ethical theories been defined to take distributed agency into account. The proposed theories rely on contractual and tort liability (Pagallo 2013) or on strict liability (Floridi 2016) and adopt a faultless responsibility model. This model separates responsibility of an agent from their intentions to perform a given action or their ability to control its outcomes, and holds all agents of a distributed system, such as a company, responsible. This is key when considering the case of AI, because it distributes moral responsibility among designers, regulators, and users. In doing so, the model plays a central role in preventing evil and fostering good, because it nudges all involved agents to adopt responsible behaviors.

Establishing good practices for delegation and defining new models to ascribe moral responsibility are essential to seize the opportunities created by AI and address the related challenges, but they are still not enough. Ethical analyses must be extended to account for the invisible influence exercised by AI on human behavior.

7.2 Invisibility and Influence

AI supports services, platforms, and devices that are ubiquitous and used on a daily basis. In 2017, the International Federation of Robotics suggested that by 2020, more than 1.7 million new AI-powered robots will be installed in factories worldwide. In the same year, the company Juniper Networks issued a report estimating that, by 2022, 55% of households worldwide will have a voice assistant, like Amazon Alexa.

As it matures and disseminates, AI blends into our lives, experiences, and environments and becomes an invisible facilitator that mediates our interactions in a convenient, barely noticeable way. While creating new opportunities, this invisible integration of AI into our environments poses further ethical issues. Some are domain-dependent. For example, trust and transparency are crucial when embedding AI solutions in homes, schools, or hospitals, whereas equality, fairness, and the protection of creativity and rights of employees are essential in the integration of AI in the workplace (Primiero and Taddeo 2012). But the integration of AI also poses another fundamental risk: the erosion of human self-determination due to the invisibility and influencing power of AI.

This invisibility enhances the influencing power of AI. With their predictive capabilities and relentless nudging, ubiquitous but imperceptible, AI systems can shape our choices and actions easily and quietly. This is not necessarily detrimental. For example, it may foster social interaction and cooperation (Shirado and Christakis 2017). However, AI may also exert its influencing power beyond our wishes or understanding, undermining our control on the environment, societies, and ultimately on our choices, projects, identities, and lives. The improper design and use of invisible AI may threaten our fragile, and yet constitutive, ability to determine our own lives and identities and keep our choices open.

7.3 Translational Ethics

To deal with the risks posed by AI, it is imperative to identify the right set of fundamental ethical principles to inform the design, regulation, and use of AI and leverage it to benefit as well as respect individuals and societies. It is not an easy task, as ethical principles may vary depending on cultural contexts and the domain of analysis. This is a problem that the IEEE Global Initiative on Ethics of Autonomous and Intelligent Systems (IEEE Standards Association n.d.) tackles with the aim of advancing public debate on the values and principles that should under- pin ethical uses of AI.

More important, some agreement on the fundamental principles is emerging. A re- cent comparative analysis (Cowls and Floridi 2018) of the main international initiatives focusing on AI ethics highlights substantive overlap of the principles endorsed by these initiatives and some of the key principles of bioethics, namely beneficence, nonmaleficence, autonomy, and justice. There is reason to be optimistic about further convergence, as other principles may be extracted from the Universal Declaration of Human Rights. This convergence will foster coherence, and hence compatibility, of different ethical frameworks for AI and provide overarching ethical guidance for the design, regulations, and uses of this technology.

Once identified, ethical principles must be translated into viable guidelines to shape AI-based innovation. Such translation has precedents, especially in medicine, where translational research goes “from bench to bedside;’ building on research advances in biology to develop new therapies and treatments. Likewise, translational ethics builds on academic advances to shape regulatory and governance approaches. This approach underpins the forthcoming recommendations for the ethical design and regulation of AI to be issued by the AI4People project.

Launched in the European Parliament in February 2018, AI4People was set up to help orient AI toward the good of society and everyone in it. The initiative combines efforts of a scientific committee of international experts and a forum of stakeholders, in consultation with the High-Level Expert Group on Artificial Intelligence of the European Com- mission, to propose a series of concrete and actionable recommendations for the ethical and socially preferable development of AI.

A translational ethics of AI needs to formulate foresight methodologies to indicate ethical risks and opportunities and prevent unwanted consequences. Impact assessment analyses are an example of this methodology. They provide a step-by-step evaluation of the impact of practices or technologies deployed in a given organization on aspects such as privacy, transparency, or liability.

Foresight methodologies can never map the entire spectrum of opportunities, risks, and unintended consequences of AI systems, but may identify preferable alternatives, valuable courses of action, likely risks, and mitigating strategies. This has a dual advantage. As an opportunity strategy, foresight methodologies can help leverage ethical solutions. As a form of risk management, they can help prevent or mitigate costly mistakes, by avoiding decisions or actions that are ethically unacceptable. This will lower the opportunity costs of choices not made or options not seized for lack of clarity or fear of backlash.

Ethical regulation of the design and use of AI is a complex but necessary task. The alter- native may lead to devaluation of individual rights and social values, rejection of AI-based innovation, and ultimately a missed opportunity to use AI to improve individual well- being and social welfare. Humanity learned this lesson the hard way when it did not regulate the impact of the industrial revolution on labor forces, and also when it recognized too late the environmental impact of massive industrialization and global consumerism. It has taken a very long time, social unrest, and even revolutions to protect workers' rights and establish sustainability frameworks.

The AI revolution is equally significant, and humanity must not make the same mistake again. It is imperative to address new questions about the nature of post-AI societies and the values that should underpin the design, regulation, and use of AI in these societies. This is why initiatives like the above- mentioned AI4People and IEEE projects, the European Union (EU) strategy for AI, the EU Declaration of Cooperation on Artificial Intelligence, and the Partnership on Artificial Intelligence to Benefit People and Society are so important (see the supplementary materials for suggested further reading). A coordinated effort by civil society, politics, business, and academia will help to identify and pursue the best strategies to make AI a force for good and unlock its potential to foster human flourishing while respecting human dignity.

Acknowledgments M.T. and L.F.are members of the Partnership on Artificial Intelligence to Benefit People and Society; L.F. is also chair of the scientific committee of Al4People.

References and Notes

Asaro, Peter. 2012. On banning autonomous weapon systems: Human rights, automation, and the dehumanization of lethal decision-making. *International Review of the Red Cross* 94 (886): 687–709. https://doi.org/10.1017/S1816383112000768.

Cowls, Josh, and Luciano Floridi. 2018. *Prolegomena to a white paper on an ethical framework for a good Ai society*, SSRN scholarly paper ID 3198732. Rochester, NY: Social Science Research Network. https://papers.ssrn.com/abstract=3198732.

Floridi, Luciano. 2016. Faultless responsibility: On the nature and allocation of moral responsibility for distributed moral actions. *Philosophical Transactions of the Royal Society A: Mathematical, Physical and Engineering Sciences* 374 (2083): 20160112. https://doi.org/10.1098/rsta.2016.0112.

IEEE Standards Association. n.d. *Ethically aligned design, version 2*. https://ethicsinaction.ieee.org/.

Jeff Larson, Julia Angwin. 2016. *How we analyzed the COMPAS recidivism algorithm*. Text/html.

McCarthy, John, Marvin L. Minsky, Nathaniel Rochester, and Claude E. Shannon. 2006. A proposal for the Dartmouth summer research project on artificial intelligence. *AI Magazine* 27 (4): 12. https://doi.org/10.1609/aimag.v27i4.1904.

Pagallo, Ugo. 2013. What robots want: Autonomous machines, codes and new Frontiers of legal responsibility. In *Human law and computer law: Comparative perspectives*, ed. Mireille Hildebrandt and Jeanne Gaakeer, 47–65. Dordrecht: Springer Netherlands. https://doi.org/10.1007/978-94-007-6314-2_3.

Primiero, Giuseppe, and Mariarosaria Taddeo. 2012. A modal type theory for formalizing trusted communications. *Journal of Applied Logic* 10 (1): 92–114. https://doi.org/10.1016/j.jal.2011.12.002.

Russell, Stuart. 2015. Robotics: Ethics of artificial intelligence: Take a stand on AI weapons. *Nature* 521 (7553): 415–418. https://doi.org/10.1038/521415a.

Samuel, Arthur L. 1960. Some moral and technical consequences of automation–a refutation. *Science* 132 (3429): 741–742. https://doi.org/10.1126/science.132.3429.741.

Shirado, Hirokazu, and Nicholas A. Christakis. 2017. Locally Noisy autonomous agents improve global human coordination in network experiments. *Nature* 545 (7654): 370–374. https://doi.org/10.1038/nature22332.

Taddeo, Mariarosaria, and Luciano Floridi. 2018. Regulate artificial intelligence to avert cyber arms race. *Nature* 556 (7701): 296–298. https://doi.org/10.1038/d41586-018-04602-6.

Wang, Dayong, Aditya Khosla, Rishab Gargeya, Humayun Irshad, and Andrew H. Beck. 2016. Deep learning for identifying metastatic breast cancer. *ArXiv:1606.05718 [Cs, q-Bio]*, June. http://arxiv.org/abs/1606.05718.

Wiener, N. 1960. Some moral and technical consequences of automation. *Science* 131 (3410): 1355–1358. https://doi.org/10.1126/science.131.3410.1355.

Yang, Guang-Zhong, Jim Bellingham, Pierre E. Dupont, Peer Fischer, Luciano Floridi, Robert Full, Neil Jacobstein, et al. 2018. The grand challenges of *science robotics*. *Science robotics* 3 (14): eaar7650. https://doi.org/10.1126/scirobotics.aar7650.

Chapter 8
The Ethics of Algorithms: Key Problems and Solutions

Andreas Tsamados, Nikita Aggarwal, Josh Cowls , Jessica Morley , Huw Roberts, Mariarosaria Taddeo , and Luciano Floridi

Abstract Research on the ethics of algorithms has grown substantially over the past decade. Alongside the exponential development and application of machine learning algorithms, new ethical problems and solutions relating to their ubiquitous use in society have been proposed. This article builds on a review of the ethics of algorithms published in 2016 (Mittelstadt et al. Big Data Soc 3(2). https://doi.org/10.1177/2053951716679679, 2016). The golas are to contribute to the debate on the identification and analysis of the ethical implications of algorithms, to provide an updated analysis of epistemic and normative concerns, and to offer actionable guidance for the governance of the design, development and deployment of algorithms.

Keywords Algorithm · Artificial intelligence · Autonomy · Digital ethics · Explainability · Fairness · Machine learning · Privacy · Responsibility · Transparency · Trust

A. Tsamados · J. Morley · H. Roberts · L. Floridi (✉)
Oxford Internet Institute, University of Oxford, Oxford, UK
e-mail: andreas.tsamados@oii.ox.ac.uk; Jessica.morley@kellogg.ox.ac.uk; jessica.morley@phc.ox.ac.uk; huw.roberts@oii.ox.ac.uk; luciano.floridi@oii.ox.ac.uk

N. Aggarwal
Faculty of Law, Oxford Internet Institute, University of Oxford, Oxford, UK
e-mail: nikita.aggarwal@law.ox.ac.uk

J. Cowls · M. Taddeo
Oxford Internet Institute, University of Oxford, Oxford, UK

Alan Turing Institute, London, UK
e-mail: josh.cowls@oii.ox.ac.uk; mariarosaria.taddeo@oii.ox.ac.uk

L. Floridi (ed.), *Ethics, Governance, and Policies in Artificial Intelligence*,
Philosophical Studies Series 144, https://doi.org/10.1007/978-3-030-81907-1_8

8.1 Introduction

Algorithms have become a key element underpinning crucial services and infrastructures of information societies. Individuals interact with recommender systems—algorithmic systems that make suggestions about what a user may like—on a daily basis, be it to choose a song, a movie, a product or even a friend (Paraschakis 2017; Perra and Rocha 2019; Milano et al. 2020). At the same time, schools and hospitals (Obermeyer et al. 2019; Zhou et al. 2019; Morley et al. 2019b), financial institutions (Lee and Floridi 2020; Aggarwal 2020) courts (Green and Chen 2019; Yu and Du 2019), local governmental bodies (Eubanks 2017; Lewis 2019), and national governments (Labati et al. 2016; Hauer 2019; Taddeo and Floridi 2018a; Taddeo et al. 2019; Roberts et al. 2019), all increasingly rely on algorithms to make significant decisions.

The potential for algorithms to improve individual and social welfare comes with significant ethical risks (Floridi and Taddeo 2016). Algorithms are not ethically neutral. Consider, for example, how the outputs of translation and search engine algorithms are largely perceived as objective, yet frequently encode language in gendered ways (Larson 2017; Prates et al. 2019). Bias has also been reported in algorithmic advertisement, with opportunities for higher paying jobs and jobs within the field of science and technology advertised to men more often than to women (Datta et al. 2015; Lambrecht and Tucker 2019). Likewise, prediction algorithms used to manage the health data of millions of patients in the United States exacerbate existing problems, with white patients given measurably better care than comparably similar, black patients (Obermeyer et al. 2019). While solutions to these issues are being discussed and designed, the number of algorithmic systems exhibiting ethical problems continues to grow.

Since 2012, artificial intelligence (AI) has been experiencing a new 'summer', both in terms of the technical advances being made and the attention that the field has received from academics, policy makers, technologists, and investors (Perrault et al. 2019). Within this, there has been a growing body of research on the ethical implications of algorithms, particularly in relation to *fairness*, *accountability*, and *transparency* (Lee 2018; Hoffmann et al. 2018; Shin and Park 2019). In 2016, our research group at the Digital Ethics Lab published a comprehensive study that sought to map these ethical concerns (Mittelstadt et al. 2016). However, this is a fast-changing field and both novel ethical problems and ways to address them have emerged, making it necessary to improve and update that study. In particular, work on the ethics of algorithms has increased significantly since 2016, when national governments, non-governmental organisations, and private companies started to take a prominent role in the conversation on "fair" and "ethical" AI and algorithms (Sandvig et al. 2016; Binns 2018a; Selbst et al. 2019; Wong 2019; Ochigame 2019). Both the quantity and the quality of the research available on the topic have expanded enormously. Given these changes, this article updates our previous work in light of new insights into the ethics of algorithms, updates the initial analysis, includes references to the literature that were missed by the original review, and

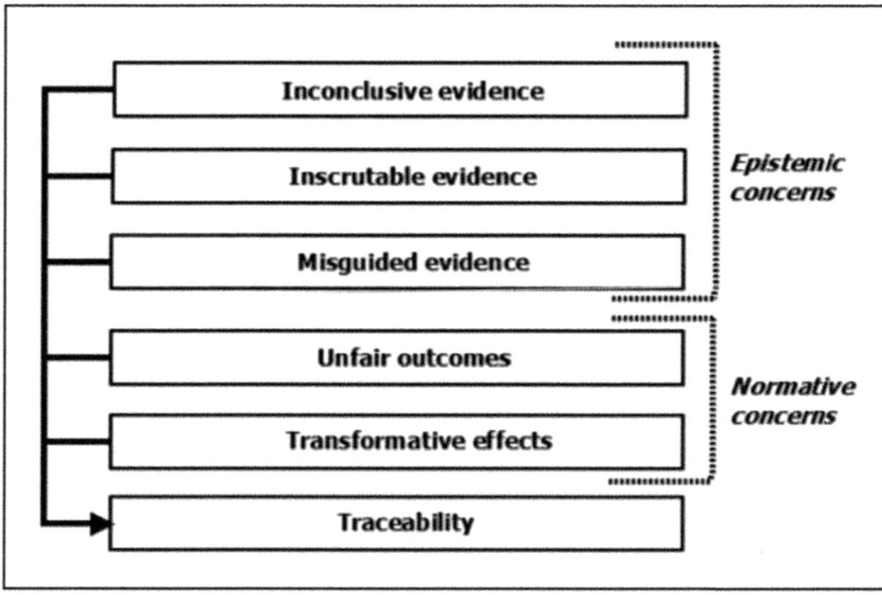

Fig. 8.1 Six types of ethical concerns raised by algorithms. (Mittelstadt et al. 2016, 4)

extends the analysed topics, including for example work on AI for social good (see the Conclusion). At the same time, the conceptual map proposed in 2016 (see Fig. 8.1) remains a fruitful framework for reviewing the current debate on the ethics of algorithms, identifying the ethical problems that algorithms give rise to, and the solutions that have been proposed in recent relevant literature. Specifically, in Sect. 2, we summarise the conceptual map. In Sects. 3, 4, 5, 6, 7 and 8 we offer a meta-analysis of the current debate on the ethics of algorithms and draw links with the types of ethical concerns previously identified. Section 9 concludes the article with an overview.

8.2 Map of the Ethics of Algorithms

There is little agreement in the relevant literature on the definition of an algorithm. The term is often used to indicate both the formal definition of an algorithm as a mathematical construct, with "a finite, abstract, effective, compound control structure, imperatively given, accomplishing a given purpose under given provisions" (Hill 2016, 47), as well as domain-specific understandings which focus on the implementation of these mathematical constructs into a technology configured for a specific task. In this article, we decided to maintain the same approach adopted in the 2016 article and to focus on the ethical issues posed by algorithms as mathematical constructs, their implementations as programs and configurations (applications), and the ways in which these can be addressed. We consider algorithms that are used to (1) turn data into evidence for a given outcome, which is used to (2) trigger and motivate an action that may have ethical consequences. Actions (1) and (2) may be performed by (semi-)autonomous algorithms—such as Machine Learning (ML) algorithms—and this complicates (3) the attribution of responsibility for the effects of actions that an algorithm may trigger. Here, ML is of particular interest, as a field which includes deep learning architectures. Computer systems deploying ML algorithms may be described as "autonomous" or "semi-autonomous", to the extent that their outputs are induced from data and thus non-deterministic.

Based on this approach, we used the conceptual map shown in Fig. 8.1 to identify the ethical issues that algorithms pose. The map identifies six ethical concerns, which define the conceptual space of the ethics of algorithms as a field of research. Three of the ethical concerns refer to epistemic factors, specifically: inconclusive, inscrutable, and misguided evidence. Two are explicitly normative: unfair outcomes and transformative effects; while one—traceability—is relevant both for epistemic and normative purposes.

The epistemic factors in the map highlight the relevance of the quality and accuracy of the data for the justifiability of the conclusions that algorithms reach and which, in turn, may shape morally-loaded decisions affecting individuals, societies, and the environment. The normative concerns identified in the map refer explicitly to the ethical impact of algorithmically-driven actions and decisions, including lack of transparency (opacity) of algorithmic processes, unfair outcomes, and unintended consequences. Epistemic and normative concerns, together with the distribution of the design, development, and deployment of algorithms make it hard to trace the chain of events and factors leading to a given outcome, thus hindering the possibility of identifying its cause, and of attributing moral responsibility for it. This is what the sixth ethical concern, traceability, refers to.

It is important to stress that this conceptual map can be interpreted at both a micro- and macro-ethical level. At the micro-ethical level, it sheds light on the ethical problems that particular algorithms may pose. By highlighting how these issues are inseparable from those related to data and responsibilities, it shows the need to take a macro-ethical approach to addressing the ethics of algorithms as part of a wider conceptual space, namely, digital ethics (Floridi and Taddeo 2016). As Floridi and Taddeo argue:

> While they are distinct lines of research, the ethics of data, algorithms and practices are obviously intertwined ... [Digital] ethics must address the whole conceptual space and hence all three axes of research together, even if with different priorities and focus (Floridi and Taddeo 2016, 4).

In the remainder of this article we address each of these six ethical concerns in turn, offering an updated analysis of the ethics of algorithms literature (at a micro level), with the goal of contributing to the debate on digital ethics (at a macro level).

8.3 Inconclusive Evidence Leading to Unjustified Actions

Research focusing on *inconclusive evidence* refers to the way in which non-deterministic, ML algorithms produce outputs that are expressed in probabilistic terms (James et al. 2013; Valiant 1984). These types of algorithms generally identify association and correlation between variables in the underlying data, but not causal connections. As such, they encourage the practice of *apophenia*: "seeing patterns where none actually exist, simply because massive quantities of data can offer connections that radiate in all directions" (boyd and Crawford 2012, 668). This is highly problematic, as patterns identified by algorithms may be the result of inherent properties of the system modelled by the data, of the datasets (that is, of the model itself, rather than the underlying system), or of skillful manipulation of datasets (properties neither of the model nor of the system). This is the case, for example, of Simpson's paradox, when trends that are observed in different groups of data reverse when the data is aggregated (Blyth 1972). In the last two cases, poor quality of the data leads to inconclusive evidence to support human decisions.

Recent research has underlined the concern that inconclusive evidence can give rise to serious ethical risks. For example, focusing on non-causal indicators may distract attention from the underlying causes of a given problem (Floridi et al. 2020). Even with the use of causal methods, the available data may not always contain enough information to justify an action or make a decision fair (Olhede and Wolfe 2018, 7). Data quality—the timeliness, completeness and correctness of a dataset—constrains the questions that can be answered using a given dataset (Olteanu et al. 2016). Additionally, the insights that can be extracted from datasets are fundamentally dependent on the assumptions that guided the data collection process itself (Diakopoulos and Koliska 2017). For example, algorithms designed to predict patient outcomes in clinical settings rely entirely on data inputs that can be quantified (e.g. vital signs and previous success rates of comparative treatments), whilst ignoring other emotional facts (e.g. the willingness to live) which can have a significant impact on patient outcomes, and thus undermine the accuracy of the algorithmic prediction (Buhmann et al. 2019). This example highlights how insights stemming from algorithmic data processing can be uncertain, incomplete, and time-sensitive (Diakopoulos and Koliska 2017).

One may embrace a naïve, inductivist approach and assume that inconclusive evidence can be avoided if algorithms are fed enough data, even if a causal explanation for these results cannot be established. Yet, recent research rejects this view. In particular, literature focusing on the ethical risks of racial profiling using algorithmic systems has demonstrated the limits of this approach highlighting, among other things, that long-standing structural inequalities are often deeply embedded in the algorithms' datasets and are rarely, if ever, corrected for (Hu 2017; Turner Lee 2018; Noble 2018; Benjamin 2019; Richardson et al. 2019; Abebe et al. 2020). More data by themselves do not lead to greater accuracy or greater representation. On the contrary, they may exacerbate issues of inconclusive data by enabling correlations to be found where there really are none. As Ruha

Benjamin (2020) put it "computational depth without historical or sociological depth is just superficial learning [not deep learning]". These limitations pose serious constraints on the justifiability of algorithmic outputs, which could have a negative impact on individuals or an entire population due to suboptimal inferences or, in the case of the physical sciences, even tip the evidence for or against "a specific scientific theory" (Ras et al. 2018, 10). This is why it is crucial to ensure that data fed to algorithms are validated independently, and data retention and reproducibility measures are in place to mitigate inconclusive evidence leading to unjustified actions, along to auditing processes to identify unfair outcomes and unintended consequences (Henderson et al. 2018; Rahwan 2018; Davis and Marcus 2019; Brundage et al. 2020).

The danger arising from inconclusive evidence and erroneous actionable insights also stems from the perceived mechanistic objectivity associated with computer-generated analytics (Karppi 2018; Lee 2018; Buhmann et al. 2019). This can lead to human decision-makers ignoring their own experienced assessments—so-called 'automation bias' (Cummings 2012)— or even shirking part of their responsibility for decisions (see Traceability below) (Grote and Berens 2020). As we shall see in Sects. 4 and 8, a lack of understanding of how algorithms generate outputs exacerbates this problem.

8.4 Inscrutable Evidence Leading to Opacity

Inscrutable evidence focuses on problems related to the lack of transparency that often characterise algorithms (particularly ML algorithms and models); the socio-technical infrastructure in which they exist; and the decisions they support. Lack of transparency—whether inherent due to the limits of technology or acquired by design decisions and obfuscation of the underlying data (Lepri et al. 2018; Dahl 2018; Ananny and Crawford 2018; Weller 2019)—often translates into a lack of scrutiny and/or accountability (Oswald 2018; Webb et al. 2019), and leads to a lack of "trustworthiness" (see AI HLEG 2019).

According to the recent literature, factors contributing to the overall lack of algorithmic transparency include: the cognitive impossibility for humans to interpret massive algorithmic models and datasets; a lack of appropriate tools to visualise and track large volumes of code and data; code and data that are so poorly structured that they are impossible to read; and ongoing updates and human influence over a model (Diakopoulos and Koliska 2017; Stilgoe 2018; Zerilli et al. 2019; Buhmann et al. 2019). Lack of transparency is also an inherent characteristic of self-learning algorithms, which alter their decision logic (produce new sets of rules) during the learning process, making it difficult for developers to maintain a detailed understanding of why certain changes were made (Burrell 2016; Buhmann et al. 2019). However, this does not necessarily translate into opaque outcomes, as even without understanding each logical step, developers can adjust hyperparameters, the parameters that govern the training process, to test for various outputs. In this respect,

Martin (2019) stresses that, while the difficulty of explaining ML algorithms' outputs is certainly real, it is important not to let this difficulty incentivise organisations to develop complex systems in order to shirk responsibility.

Lack of transparency can also result from the malleability of algorithms, whereby algorithms can be reprogrammed in a continuous, distributed, and dynamic way (Sandvig et al. 2016). Algorithmic malleability allows developers to monitor and improve an already-deployed algorithm, but it may also be abused to blur the history of its evolution and leave end-users in a state of confusion about the affordances of a given algorithm (Ananny and Crawford 2018). Consider for example Google's main search algorithm. Its malleability enables the company to make continuous revisions, suggesting a permanent state of destabilisation (Sandvig et al. 2016). This requires those affected by the algorithm to monitor it constantly and update their understanding accordingly –an impossible task for most (Ananny and Crawford 2018).

As Floridi and Turilli (2009, 105) note, transparency is not an "ethical principle in itself but a pro-ethical condition for enabling or impairing other ethical practices or principles". And indeed, complete transparency can itself cause distinct ethical problems (Ananny and Crawford 2018): transparency can provide users with some critical information about the features and limitations of an algorithm, but it can also overwhelm users with information and thus render the algorithm more opaque (Kizilcec 2016; Ananny and Crawford 2018). Other research stress that excessive focus on transparency can be detrimental to innovation and unnecessarily divert resources that could instead be used to improving safety, performance and accuracy (Danks and London 2017; Oswald 2018; Ananny and Crawford 2018; Weller 2019). For example, the debate over prioritising transparency (and explainability) is especially contentious in the context of medical algorithms (Robbins 2019).

Transparency can enable individuals to game the system (Martin 2019; Magalhães 2018; Cowls et al. 2019). Knowledge about the source of a dataset, the assumptions under which sampling was done, or the metrics that an algorithm uses to sort new inputs, may be used to figure out ways to take advantage of an algorithm (Szegedy et al. 2014; Yampolskiy 2018). Yet, the ability to game algorithms is only within reach for some groups of the population—those with higher digital literacy for example—thus creating another form of social inequality (Martin 2019; Bambauer and Zarsky 2018). Therefore, confusing transparency for an end in itself, instead of a pro-ethical factor (Floridi 2017) enabling crucial ethical practices, may not solve existing ethical problems related to the use of algorithms and, indeed, pose new ones. This is why it is important to distinguish between the different factors that may hinder transparency of algorithms, identify their cause, and nuance the call for transparency by specifying which factors are required and at which layers of algorithmic systems they should be addressed (Diakopoulos and Koliska 2017).

There are different ways of addressing the problems related to lack of transparency. For example, Gebru et al. propose that the constraints on transparency posed by the malleability of algorithms can be addressed, in part, by using standard documentary procedures similar to those deployed in the electronics industry, where

> every component, no matter how simple or complex, is accompanied with a datasheet describing its operating characteristics, test results, recommended usage, and other information (Gebru et al. 2020, 2).

Unfortunately, publicly available documentation is currently uncommon in the development of algorithmic systems and there is no agreed-upon format for what should be included when documenting the origin of a dataset (Arnold et al. 2019; Gebru et al. 2020).

Although relatively nascent, another potentially promising approach to enforcing algorithmic transparency is the use of technical tools to test and audit algorithmic systems and decision-making. Testing whether algorithms exhibit negative tendencies, like unfair discrimination, and auditing a prediction or decision trail in detail, can help maintain a high level of transparency (Weller 2019; Malhotra et al. 2018; Brundage et al. 2020). To this end, discursive frameworks have been developed to help businesses and public sector organisations understand the potential impacts of opaque algorithms, thus encouraging good practices (ICO 2020). For instance, the AI Now Institute at New York University has produced algorithmic impact assessment guidance, which seeks to raise awareness and improve dialogue over potential harms of ML algorithms (Reisman et al. 2018). This includes the two aims of enabling developers to design more transparent, and therefore more trustworthy ML algorithms, and of improving the public understanding and control of algorithms. In the same vein, Diakopoulos and Koliska have provided a comprehensive list of "transparency factors" across four layers of algorithmic systems: data, model, inference, and interface. Factors include, *inter alia*

> uncertainty (e.g. error margins), timeliness (e.g. when was the data collected), completeness or missing elements, sampling method, provenance (e.g. sources), and volume (e.g. of training data used in machine learning) (Diakopoulos and Koliska 2017, 818).

Effective transparency procedures are likely, and indeed ought to, involve an *interpretable explanation* of the internal processes of these systems. Buhmann et al. (2019) argue that while a lack of transparency is an inherent feature of many ML algorithms, this does not mean that improvements cannot be made. For example, companies like Google and IBM have increased their efforts to make ML algorithms more interpretable and inclusive by making tools such as Explainable AI, AI Explainability 360, and the What-If Tool publicly available. These tools provide developers and also the general public with interactive visual interfaces that improve human readability, explore various model results, provide case-based reasoning, directly interpretable rules, and even identify and mitigate unwanted biases in datasets and algorithmic models (Mojsilovic 2018; Wexler 2018).

However, explanations for ML algorithms are constrained by the type of explanation sought, the fact that decisions are often multi-dimensional in their nature, and that different users may require different explanations (Edwards and Veale 2017). Identifying appropriate methods for providing explanations has been a problem since the late 1990s (Tickle et al. 1998), but contemporary efforts can be categorised into two main approaches: subject-centric explanations and model-centric explanations (Doshi-Velez and Kim 2017; Lee et al. 2017; Baumer 2017; Buhmann et al.

2019). In the former, the accuracy and length of the explanation is tailored to users and their specific interactions with a given algorithm (see for example [(Green and Viljoen 2020) and the game-like model proposed by [Watson and Floridi 2020]); in the latter, explanations concern the model as a whole and do not depend on their audience.

Explainability is particularly important when considering the rapidly growing number of open source and easy-to-use models and datasets. Increasingly, non-experts are experimenting with state-of-the-art algorithmic models widely available via online libraries or platforms, like GitHub, without always fully grasping their limits and properties (Hutson 2019). This has prompted scholars to suggest that, to tackle the issue of technical complexity, it is necessary to invest more heavily in public education to enhance computational and data literacy (Lepri et al. 2018). Doing so would seem to be an appropriate long-term solution to the multi-layered issues introduced by ubiquitous algorithms, and open source software is often cited as critical to the solution (Lepri et al. 2018).

8.5 Misguided Evidence Leading to Unwanted Bias

Developers are predominantly focused on ensuring that their algorithms perform the tasks for which they were designed. Thus, the type of thinking that guides developers is essential to understanding the emergence of bias in algorithms and algorithmic decision-making. Some scholars refer to the dominant thinking in the field of algorithm development as being defined by "algorithmic formalism"—an adherence to prescribed rules and form (Green and Viljoen 2020, 21). While this approach is useful for abstracting and defining analytical processes, it tends to ignore the social complexity of the real world (Katell et al. 2020). Indeed, this approach leads to algorithmic interventions that strive to be 'neutral' but in doing so, it risks entrenching existing social conditions (Green and Viljoen 2020, 20), while creating the illusion of precision (Karppi 2018; Selbst et al. 2019). For these reasons, the use of algorithms in some settings is questioned altogether (Selbst et al. 2019; Mayson 2019; Katell et al. 2020; Abebe et al. 2020). For example, a growing number of scholars criticise the use of algorithm-based risk assessment tools in court settings (Berk et al. 2018; Abebe et al. 2020).

Some scholars affirm the limits of abstractions with regard to unwanted bias in algorithms and argue for the need to develop a sociotechnical frame to address and improve the fairness of algorithms (Edwards and Veale 2017; Selbst et al. 2019; Wong 2019; Katell et al. 2020; Abebe et al. 2020). In this respect, Selbst et al. (2019, 60–63) point to five abstraction "traps", or failures to account for the social context in which algorithms operate, which persist in algorithmic design due to the absence of a sociotechnical frame, namely:

(i) a failure to model the entire system over which a social criterion, such as fairness, will be enforced;

(ii) a failure to understand how repurposing algorithmic solutions designed for one social context may be misleading, inaccurate, or otherwise do harm when applied to a different context;
(iii) a failure to account for the full meaning of social concepts such as fairness, which can be procedural, contextual, and contestable, and cannot be resolved through mathematical formalisms;
(iv) a failure to understand how the insertion of technology into an existing social system changes the behaviours and embedded values of the pre-existing system; and
(v) a failure to recognize the possibility that the best solution to a problem may not involve technology.

The term ‘bias’ often comes with a negative connotation, but it is used here to denote a “deviation from a standard” (Danks and London 2017, 4692), which can occur at any stage of the design, development, and deployment process. The data used to train an algorithm is one of the main sources from which bias emerges (Shah 2018), through preferentially sampled data or from data reflecting existing societal bias (Diakopoulos and Koliska 2017; Danks and London 2017; Binns 2018b; Malhotra et al. 2018). For example, morally problematic structural inequalities that disadvantage certain ethnicities may not be apparent in data and thus not corrected for (Nobles 2018; Benjamin 2019). Additionally, data used to train algorithms are seldom obtained “according to any specific experimental design” (Olhede and Wolfe 2018, 3) and are used even though they may be inaccurate, skewed, or systemically biased, offering a poor representation of a population under study (Richardson et al. 2019).

One possible approach to mitigating this problem is to exclude intentionally some specific data variables from informing algorithmic decision-making. Indeed, the processing of statistically relevant sensitive or “protected variables”—such as gender or race—is typically limited or prohibited under anti-discrimination and data protection law, in order to limit the risks of unfair discrimination. Unfortunately, even if protections for specific classes can be encoded in an algorithm, there could always be biases that were not considered *ex ante*, as in the case, for example, of language models reproducing heavily male-focused texts (Fuster et al. 2017; Doshi-Velez and Kim 2017). Even while bias may be anticipated and protected variables excluded from the data, unanticipated proxies for these variables could still be used to reconstruct biases, leading to “bias by proxy” that is difficult to detect and avoid (Fuster et al. 2017; Gillis and Spiess 2019).

At the same time, there may be good reasons to rely on statistically biased estimators in algorithmic processing, as they can be used to mitigate training data bias. In this way, one type of problematic algorithmic bias is counterbalanced by another type of algorithmic bias or by introducing compensatory bias when interpreting algorithmic outputs (Danks and London 2017). Simpler approaches to mitigating bias in data involve piloting algorithms in different contexts and with various datasets (Shah 2018). Having a model, its datasets, and metadata (on provenance) published to enable external scrutiny can also help correct unseen or unwanted bias (Shah 2018). It is also worth noting that so-called 'synthetic data', or AI-generated data, produced via reinforcement learning or generative adversarial networks (GANs) offer an opportunity to address issues of data bias (Floridi 2019a; Xu et al. 2018). Fair data generation with GANs may help diversify datasets used in computer vision algorithms (Xu et al. 2018). For example, StyleGAN2 (Karras et al. 2019) is able to produce high-quality images of non-existing human faces, and has proven to be especially useful in creating diverse datasets of human faces, something that many algorithmic systems for facial recognition currently lack (Obermeyer et al. 2019; Kortylewski et al. 2019; Harwell 2020).

Unwanted bias also occurs due to improper deployment of an algorithm. Consider transfer context bias: the problematic bias that emerges when a functioning algorithm is used in a new environment. For example, if a research hospital's healthcare algorithm is used in a rural clinic and assumes that the same level of resources are available to the rural clinic as the research hospital, the healthcare resource allocation decisions generated by the algorithm will be inaccurate and flawed (Danks and London 2017).

In the same vein, Grgić-Hlača et al. (2018) warn of vicious cycles when algorithms make misguided chain assessments. For example, in the context of the COMPAS risk-assessment algorithm, one of the assessment criteria for predicting recidivism is the criminal history of a defendant's friends. It follows that having friends with a criminal history would create a vicious cycle in which a defendant with convicted friends will be deemed more likely to offend, and therefore sentenced to prison, hence increasing the number of people with criminal records in a given group on the basis of mere correlation (Grgić-Hlača et al. 2018; Richardson et al. 2019).

High-profile examples of algorithmic bias in recent years—not least investigative reporting around the COMPAS system (Angwin et al. 2016)— have led to a growing focus on issues of algorithmic fairness. The definition and operationalisation of algorithmic fairness have become "urgent tasks in academia and industry" (Shin and Park 2019), as the significant uptick in the number of papers, workshops and conferences dedicated to 'fairness, accountability and transparency' (FAT) highlights (Hoffmann et al. 2018; Ekstrand and Levy 2018; Shin and Park 2019). We analyse key topics and contributions in this area in the next section.

8.6 Unfair Outcomes Leading to Discrimination

There is widespread agreement on the need for algorithmic fairness, particularly to mitigate the risks of direct and indirect discrimination (under US law, 'disparate treatment' and 'disparate impact', respectively) due to algorithmic decisions (Barocas and Selbst 2016; Grgić-Hlača et al. 2018; Green and Chen 2019). Yet there remains a lack of agreement among researchers on the definition, measurements and standards of algorithmic fairness (Gajane and Pechenizkiy 2018; Saxena et al. 2019; Lee 2018; Milano et al. 2020). Wong (2019) identifies up to

21 definitions of fairness across the literature and such definitions are often mutually inconsistent (Doshi-Velez and Kim 2017).

There are many nuances in the definition, measurement, and application of different standards of algorithmic fairness. For instance, algorithmic fairness can be defined both in relation to groups as well as individuals (Doshi-Velez and Kim 2017). Four main definitions of algorithmic fairness have gained prominence in the recent literature (see for example [Kleinberg et al. 2016; Corbett-Davies and Goel 2018]):

(i) *anti-classification*, which refers to protected categories, such as race and gender, and their proxies not being explicitly used in decision making;
(ii) *classification parity*, which regards a model as being fair if common measures of predictive performance, including false positive and negative rates, are equal across protected groups;
(iii) *calibration*, which considers fairness as a measure of how well-calibrated an algorithm is between protected groups;
(iv) *statistical parity*, which defines fairness as an equal average probability estimate over all members of protected groups.

However, each of these commonly used definitions of fairness has drawbacks and are generally mutually incompatible (Kleinberg et al. 2016). Taking anti-classification as an example, protected characteristics, such as race, gender and religion, cannot simply be removed from training data in order to prevent discrimination, as noted above (Gillis and Spiess 2019). Structural inequalities mean that formally non-discriminatory data points such as postcodes can act as proxies for, and be used, either intentionally or unintentionally, to infer protected characteristics, like race (Edwards and Veale 2017).

There are important cases where it is appropriate to consider protected characteristics to make equitable decisions. For example, lower female reoffending rates mean that excluding gender as an input in recidivism algorithms would leave women with disproportionately high risk ratings (Corbett-Davies and Goel 2018). Because of this, Binns (2018a) stresses the importance of considering the historical and sociological context that cannot be captured in the data presented to algorithms but that can inform contextually appropriate approaches to fairness in algorithms. It is also critical to note that algorithmic models can often produce unexpected outcomes, contrary to human intuitions and perturb their understanding. For example, as Grgić-Hlača et al. (2018) highlight, using features that people believe to be fair can in some cases increase the racism exhibited by algorithms and decrease accuracy.

Regarding methods for improving algorithmic fairness, Veale and Binns (2017) and Katell et al. (2020) offer two approaches. The first envisages a third-party intervention, whereby an entity external to the provider of algorithms would hold data on sensitive or protected characteristics and attempt to identify and reduce discrimination caused by the data and models. The second approach proposes a collaborative knowledge-based method which would focus on community-driven data resources containing practical experiences of ML and modelling (Veale and

Binns 2017; Katell et al. 2020). The two approaches are not mutually exclusive, they may bring different benefits depending on contexts of application, and their combination may also be beneficial.

Given the significant impact that algorithmic decisions have on people's lives and the importance of context for choosing appropriate measures of fairness, it is surprising that there has been little effort to capture public views on algorithmic fairness (Lee et al. 2017; Saxena et al. 2019; Binns 2018a). Examining public perceptions of different definitions of algorithmic fairness, Saxena et al. (2019, 3) note that in the context of loan decisions people exhibit a preference for a "calibrated fairness definition", or merit-based selection, as compared to "treating similar people similarly" and argue in favour of the principle of affirmative action. In a similar study, Lee (2018) offers evidence suggesting that, when considering tasks that require uniquely human skills, people consider algorithmic decisions to be less fair and algorithms to be less trustworthy.

Reporting on empirical work conducted on algorithmic interpretability and transparency, Webb et al. (2019) reveal that moral references, particularly on fairness, are consistent across participants discussing their preferences on algorithms. The study notes that people tend to go beyond personal preferences to focus instead on "right and wrong behaviour", as a way to indicate the need to understand the context of deployment of the algorithm and the difficulty of understanding the algorithm and its consequences (Webb et al. 2019). In the context of recommender systems, Burke (2017) proposes a multi-stakeholder and multi-sided approach to defining fairness, moving beyond user-centric definitions to include the interests of other system stakeholders.

It has become clear that understanding the public view on algorithmic fairness would help technologists in developing algorithms with fairness principles that align with the sentiments of the general public on prevailing notions of fairness (Saxena et al. 2019, 1). Grounding the design decisions of the providers of an algorithm "with reasons that are acceptable by the most adversely affected" as well as being "open to adjustments in light of new reasons" (Wong 2019, 15) is crucial to improving the social impact of algorithms. It is important to appreciate, however, that measures of fairness are often completely inadequate when they seek to validate models that are deployed on groups of people that are already disadvantaged in society because of their origin, income level, or sexual orientation. We simply cannot "optimise around" existing economic, social, and political power dynamics (Winner 1980; Benjamin 2019).

8.7 Transformative Effects Leading to Challenges for Autonomy and Informational Privacy

The collective impact of algorithms has spurred discussions on the autonomy afforded to end users. (Ananny and Crawford 2018; Beer 2017; Taddeo and Floridi 2018b; Möller et al. 2018; Malhotra et al. 2018; Shin and Park 2019; Hauer 2019). Algorithm-based services are increasingly featured "within an ecosystem of complex, socio-technical issues" (Shin and Park 2019), which can hinder the autonomy of users. Limits to users' autonomy stem from three sources:

(i) pervasive distribution and proactivity of (learning) algorithms to inform users' choice (Yang et al. 2018; Taddeo and Floridi 2018b);
(ii) users' limited understanding of algorithms;
(iii) lack of second-order power (or appeals) over algorithmic outcomes (Rubel et al. 2019).

In considering the ethical challenges of AI, Yang et al. (2018, 11) focus on the impact of autonomous, self-learning algorithms on human self-determination and stress that "AI's predictive power and relentless nudging, even if unintentional, should foster and not undermine human dignity and self-determination".

The risks that algorithmic systems may hinder human autonomy by shaping users' choices has been widely reported in the literature and has taken centre stage in most of the high-level ethical principles for AI, including, *inter alia*, those of the European Commission's European Group on Ethics in Science and Technologies, and the UK's House of Lords Artificial Intelligence Committee (Floridi and Cowls 2019). In their analysis of these high-level principles, Floridi and Cowls (2019) note that it does not suffice that algorithms promote people's autonomy: rather, the autonomy of algorithms should be constrained and reversible. Looking beyond the West, the Beijing AI Principles—developed by a consortium of China's leading companies and universities for guiding AI research and development—also emphasise that human autonomy should be respected (Roberts et al. 2020).

Human autonomy can also be limited by the inability of an individual to understand some information or make the appropriate decisions. As Shin and Park suggest, algorithms "do not have the affordance that would allow users to understand them or how best to utilize them to achieve their goals" (Shin and Park 2019, 279). As such, a key issue identified in debates over users' autonomy is the difficulty of striking an appropriate balance between people's own decision-making and that which they delegate to algorithms (Floridi et al. 2018). This is further complicated by a lack of transparency over the decision-making process by which particular decisions are delegated to algorithms. Ananny and Crawford (2018) note that often this process does not account for all stakeholders, and is not void of structural inequalities.

As a method of Responsible Research and Innovation (RRI), 'participatory design' is often mentioned for its focus on the design of algorithms to promote the values of end users and protect their autonomy (Whitman et al. 2018; Katell et al.

2020). Participatory design aims at "bringing participants' tacit knowledge and embodied experience into the design process" (Whitman et al. 2018, 2). For example, Rahwan's 'Society-in-the-Loop' (2018) conceptual framework seeks to enable different stakeholders in society to design algorithmic systems before deployment, and to amend and reverse the decisions of algorithmic systems that already underlie social activities. This framework aims to maintain a well-functioning "algorithmic social contract", defined as "a pact between various human stakeholders, mediated by machines" (Rahwan 2018, 1). It accomplishes this by identifying and negotiating the values of different stakeholders affected by algorithmic systems as the basis for monitoring adherence to the social contract.

Informational privacy is intimately linked with user autonomy (Cohen 2000; Rössler 2015). Informational privacy guarantees peoples' freedom to think, communicate, and form relationships, among other essential human activities (Rachels 1975; Allen 2011). However, people's increasing interaction with algorithmic systems has effectively reduced their ability to control who has access to information that concerns them and what is being done with it. The vast amounts of sensitive data required in algorithmic profiling and predictions, central to recommender systems, pose multiple issues regarding individuals' informational privacy.

Algorithmic profiling takes place over an indefinite period of time, in which individuals are categorised according to a system's internal logic, and their profiles are updated as new information is obtained about them. This information is typically obtained directly, from when a person interacts with a given system, or indirectly, inferred from algorithmically assembled groups of individuals (Paraschakis 2018). Indeed, algorithmic profiling will also rely on information gathered about other individuals and groups of people that have been categorised in a similar manner to a targeted person. This includes information ranging from characteristics like geographical location and age, to information on specific behaviour and preferences, including what type of content a person is likely to seek the most on a given platform (Chakraborty et al. 2019). While this poses a problem of *inconclusive evidence*, it also indicates that if group privacy (Taylor et al. 2017) is not ensured, it may be impossible for individuals to ever remove themselves from the process of algorithmic profiling and predictions (Milano et al. 2020). In other words, individuals' informational privacy cannot be secured without securing group privacy.

Users may not always be aware of, or may not have the ability to gain awareness about, the type of information that is being held about them and what that information is used for. Considering that recommender systems contribute to the dynamic construction of individuals' identities by intervening in their choices, a lack of control over one's information translates in a loss of autonomy.

Giving individuals the ability to contribute to the design of a recommender system can help create more accurate profiles that account for attributes and social categories that would have otherwise not been included in the labelling used by the system to categorise users (Milano et al. 2020). While the desirability of improving algorithmic profiling will vary with the context, improving algorithmic design by including feedback from the various stakeholders of the algorithm falls in line with the aforementioned scholarship on RRI and improves users' ability for self-determination (Whitman et al. 2018).

Knowledge about who owns one's data and what is done with them can also help inform trade-offs between informational privacy and information-processing benefits (Sloan and Warner 2018, 21). For example, in medical contexts, individuals are more likely to be willing to share information that can help inform their, or others' diagnostics, less so in the context of job recruitment. Information coordination norms, as Sloan and Warner (2018) argue, can serve to ensure that these trade-offs adapt correctly to different contexts and do not place an excessive amount of responsibility and effort on single individuals. For example, personal information ought to flow differently in the context of law enforcement procedures as compared to a job recruitment process. The European Union's General Data Protection Regulation has played an important role in instituting the basis of such norms (Sloan and Warner 2018).

Finally, a growing scholarship on differential privacy is providing new privacy protection methods for organisations looking to protect their users' privacy while also keeping good model quality, as well as manageable software costs and complexity, striking a balance between utility and privacy (Abadi et al. 2016; Wang et al. 2017; Xian et al. 2017). Technical advancements of this kind, which allow organisations to share publicly a dataset while keeping information about individuals secret (preventing re-identification), and can ensure provable privacy protection on sensitive data, such as genomic data (Wang et al. 2017). Indeed, differential privacy was recently used by Social Science One and Facebook to release safely one of the largest datasets (38 million URLs shared publicly on Facebook) for academic research on the societal impacts of social media (King and Persily 2020).

8.8 Traceability Leading to Moral Responsibility

The technical limitations of various ML algorithms, such as lack of transparency and lack of explainability, undermine their scrutability and highlight the need for novel approaches to tracing moral responsibility and accountability for the actions performed by ML algorithms. Regarding moral responsibility, Reddy et al. (2019) note a common blurring between technical limitations of algorithms and the broader legal, ethical, and institutional boundaries in which they operate. Even for non-learning algorithms, traditional, linear conceptions of responsibility prove to offer limited guidance in contemporary sociotechnical contexts. Wider sociotechnical structures make it difficult to trace back responsibility for actions performed by distributed, hybrid systems of human and artificial agents (Floridi 2012; Crain 2018).

Additionally, due to the structure and operation of the data brokerage market, it is in many cases impossible to "trace any given datum to its original source" once it has been introduced to the marketplace (Crain 2018, 93). Reasons for this include trade secret protection; complex markets that "divorce" the data collection process from the selling and buying process; and the mix of large volumes of computationally generated information with "no 'real' empirical source" combined with genuine data (Crain 2018, 94).

The technical complexity and dynamism of ML algorithms make them prone to concerns of "agency laundering": a moral wrong which consists in distancing oneself from morally suspect actions, regardless of whether those actions were intended or not, by blaming the algorithm (Rubel et al. 2019). This is practiced by organisations as well as by individuals. Rubel et al. provide a straightforward and chilling example of agency laundering by Facebook:

> Using Facebook's automated system, the ProPublica team found a user-generated category called "Jew hater" with over 2200 members. [. . .] To help ProPublica find a larger audience (and hence have a better ad purchase), Facebook suggested a number of additional categories. [. . .] ProPublica used the platform to select other profiles displaying anti-Semitic categories, and Facebook approved ProPublica's ad with minor changes. When ProPublica revealed the anti-Semitic categories and other news outlets reported similarly odious categories, Facebook responded by explaining that algorithms had created the categories based on user responses to target fields [and that] "[w]e never intended or anticipated this functionality being used this way" (Rubel et al. 2019, 1024–25).

Today, the failure to grasp the unintended effects of mass personal data processing and commercialisation, a familiar problem in the history of technology (Wiener 1950; Klee 1996; Benjamin 2019), is coupled with the limited explanations that most ML algorithms provide. This approach risks to favour avoidance of responsibility through "the computer said so" type of denial (Karppi 2018). This can lead field experts, such as clinicians, to avoid questioning the suggestion of an algorithm even when it may seem odd to them. The interplay between field experts and ML algorithms can prompt "epistemic vices" (Grote and Berens 2020), like dogmatism or gullibility (Hauer 2019), and hinder the attribution of responsibility in distributed systems (Floridi 2016). To address this issue, Shah's analysis (2018) stresses that the risk that some stakeholders may breach their responsibilities can be addressed, for example, by establishing separate bodies for the ethical oversight of algorithms (e.g. DeepMind Health established an Independent Review Panel with unfettered access to the company until Google halted it in 2019) (Murgia 2018). However, expecting a single oversight body, like a research ethics committee or institutional review board, to "be solely responsible for ensuring the rigour, utility, and probity of big data" is unrealistic (Lipworth et al. 2017, 8). Indeed, some have argued that these initiatives lack any sort of consistency and can rather lead to "ethics bluewashing", understood as

> implementing superficial measures in favour of, the ethical values and benefits of digital processes, products, services, or other solutions in order to appear more digitally ethical than one is. (Floridi 2019b, 187).

Faced with strict legal regimes, resourceful actors may also resort to so-called "ethics dumping" whereby unethical "processes, products or services" are exported to countries with weaker frameworks and enforcement mechanisms, after which the outcomes of such unethical activities are "imported back" (Floridi 2019b, 190).

There are a number of detailed approaches to establishing algorithmic accountability in the reviewed literature. While ML algorithms do require a level of technical intervention to improve their explainability, most approaches focus on normative interventions. For example, Ananny and Crawford argue that, at least, providers of algorithms ought to facilitate public discourse about their technology (Ananny and Crawford 2018). Similarly, to address the issue of *ad hoc* ethical actions, some have claimed that accountability should first and foremost be addressed as a matter of convention (Dignum et al. 2018; Reddy et al. 2019).

Looking to fill the convention "gap", Buhmann et al. (2019) borrow from the seven principles for algorithms set out by the Association for Computing Machinery, claiming that through, *inter alia*, awareness of their algorithms, validation, and testing, an organisation should take responsibility for their algorithms regardless of how opaque they are (Malhotra et al. 2018). Decisions regarding the deployment of algorithms should incorporate factors such as desirability and the wider context in which they will operate, which should then lead to a more accountable "algorithmic culture" (Vedder and Naudts 2017, 219). In order to capture such considerations, "interactive and discursive fora and processes" with relevant stakeholders, as suggested by Buhmann et al., may prove a useful means (Buhmann et al. 2019, 13).

In the same vein, Binns (2018b) focuses on the political-philosophical concept of "public reason". Considering that the processes for ascribing responsibility for the actions of an algorithm differ, both in nature and scope, in the public versus private sector, Binns calls for the establishment of a publicly shared framework (Binns 2018b; see also Dignum et al. 2018), according to which algorithmic decisions should be able to withstand the same level of public scrutiny that human decision-making would receive. This approach has been echoed by many others in the reviewed literature (Ananny and Crawford 2018; Blacklaws 2018; Buhmann et al. 2019).

Problems relating to 'agency laundering' and 'ethics shirking' arise from the inadequacy of existing conceptual frameworks to trace and ascribe moral responsibility. As Floridi points out, when considering algorithmic systems and the impact of their actions

> we are dealing with DMAs [distributed moral actions] arising from morally neutral interactions of (potentially hybrid) networks of agents? In other words, who is responsible (*distributed moral responsibility*, DMR) for DMAs?, (Floridi 2016, 2).

Floridi's analysis suggests ascribing full moral responsibility "by default and overridably" to *all* the agents in the network which are causally relevant to the given action of the network. The proposed approach builds on the concepts of back-propagation from network theory, strict liability from jurisprudence, and common knowledge from epistemic logic. Notably, this approach decouples moral responsibility from intentionality of the actors and from the very idea of punishment and reward for performing a given action, to focus instead on the need to rectify mistakes (back-propagation) and improve the ethical working of all the agents in the network.

8.9 Conclusion

This article builds on, and updates, previous research conducted by our group (Mittelstadt et al. 2016) to review relevant literature published since 2016 on the ethics of algorithms. Although that article is now inevitably outdated in terms of specific references and detailed information about the literature reviewed, the map, and the six categories that it provides, have withstood the test of time and remain a valuable tool to scope ethics of algorithms as an area of research, with a growing body of literature focusing on each of the six categories contributing either to refine our understanding of existing problems or to provide solutions to address them.

Since 2016, the ethics of algorithms has become a central topic of discussion among scholars, technology providers, and policy makers. The debate has gained traction also because of the so-called "summer of AI", and with it the pervasive use of ML algorithms. Many of the ethical questions analysed in this article and the literature it reviews have been addressed in national and international ethical guidelines and principles, like the aforementioned European Commission's European Group on Ethics in Science and Technologies, the UK's House of Lords Artificial Intelligence Committee (Floridi and Cowls 2019), and the OECD principles on AI (OECD 2019).

One aspect that was not explicitly captured by the original map, and which is becoming a central point of discussion in the relevant literature, is the increasing focus on the use of algorithms, AI and digital technologies more broadly, to deliver socially good outcomes (Hager et al. 2019) (Cowls et al. 2019). While it is true, at least in principle, that any initiative aimed at using algorithms for social good should address satisfactorily the risks that each of the six categories in the map identifies, there is also a growing debate on the principles and criteria that should inform the design and governance of algorithms, and digital technologies more broadly, for the explicit purpose of social good.

Ethical analyses are necessary to mitigate the risks while harnessing the potential for good of these technologies, insofar as they serve the twin goals of clarifying the nature of the ethical risks and of the potential for good of algorithms and digital technologies and translating (Taddeo and Floridi 2018b; Morley et al. 2019a; b) this understanding into sound, actionable guidance for the governance of the design and use of digital artefacts.

References

Abadi, Martin, Andy Chu, Goodfellow Ian, H. Brendan McMahan, Ilya Mironov, Kunal Talwar, and Li Zhang. 2016. Deep learning with differential privacy. In *Proceedings of the 2016 ACM SIGSAC conference on computer and communications security*, 308–318. Vienna: ACM. https://doi.org/10.1145/2976749.2978318.

Abebe, Rediet, Solon Barocas, Jon Kleinberg, Karen Levy, Manish Raghavan, and David G. Robinson. 2020. Roles for computing in social change. *ArXiv:1912.04883 [Cs]*, January. https://doi.org/10.1145/3351095.3372871.

Aggarwal, Nikita. 2020. The norms of algorithmic credit scoring. *SSRN Electronic Journal*. https://doi.org/10.2139/ssrn.3569083.

AI HLEG. 2019. *Ethics guidelines for trustworthy AI*, available online.

Allen, Anita. 2011. Unpopular privacy what must we Hide? *Oxford University Press*. https://doi.org/10.1093/acprof:oso/9780195141375.001.0001.

Ananny, Mike, and Kate Crawford. 2018. Seeing without knowing: Limitations of the transparency ideal and its application to algorithmic accountability. *New Media & Society* 20 (3): 973–989. https://doi.org/10.1177/1461444816676645.

Angwin, Julia, Jeff Larson, Surya Mattu, and Kirchner Lauren. 2016. *'Machine Bias', 2016*. https://www.propublica.org/article/machine-bias-risk-assessments-in-criminal-sentencing.

Arnold, Matthew, Rachel K. E. Bellamy, Michael Hind, Stephanie Houde, Sameep Mehta, Aleksandra Mojsilovic, Ravi Nair, et al. 2019. FactSheets: Increasing trust in AI services through supplier's declarations of conformity. *ArXiv:1808.07261 [Cs]*, February. http://arxiv.org/abs/1808.07261.

Bambauer, Jame, and Tal Zarsky. 2018. The algorithmic game. *Notre Dame Law Review* 94 (1): 1–47.

Barocas, Solon, and Andrew D. Selbst. 2016. Big data's disparate impact. *SSRN Electronic Journal*. https://doi.org/10.2139/ssrn.2477899.

Baumer, Eric P.S. 2017. Toward human-centered algorithm design. *Big Data & Society* 4 (2): 205395171771885. https://doi.org/10.1177/2053951717718854.

Beer, David. 2017. The social power of algorithms. *Information, Communication & Society* 20 (1): 1–13. https://doi.org/10.1080/1369118X.2016.1216147.

Benjamin, Ruha. 2019. *Race after technology: Abolitionist tools for the new Jim code*. Medford: Polity.

———. 2020. *2020 vision: Reimagining the default settings of technology & society*. https://iclr.cc/virtual_2020/speaker_3.html.

Berk, Richard, Hoda Heidari, Shahin Jabbari, Michael Kearns, and Aaron Roth. 2018. 'Fairness in criminal justice risk assessments: The state of the art'. *Sociological Methods & Research*, July, 004912411878253. https://doi.org/10.1177/0049124118782533.

Binns, Reuben. 2018a. 'Fairness in machine learning: Lessons from political philosophy'. *ArXiv:1712.03586 [Cs]*, January. http://arxiv.org/abs/1712.03586.

———. 2018b. Algorithmic accountability and public reason. *Philosophy & Technology* 31 (4): 543–556. https://doi.org/10.1007/s13347-017-0263-5.

Blacklaws, Christina. 2018. Algorithms: Transparency and accountability. *Philosophical Transactions of the Royal Society A: Mathematical, Physical and Engineering Sciences* 376 (2128): 20170351. https://doi.org/10.1098/rsta.2017.0351.

Blyth, Colin R. 1972. On Simpson's paradox and the sure-thing principle. *Journal of the American Statistical Association* 67 (338): 364–366. https://doi.org/10.1080/01621459.1972.10482387.

Boyd, Danah, and Kate Crawford. 2012. Critical questions for big data. *Information, Communication & Society* 15 (5): 662–679. https://doi.org/10.1080/1369118X.2012.678878.

Brundage, Miles, Shahar Avin, Jasmine Wang, Haydn Belfield, Gretchen Krueger, Gillian Hadfield, Heidy Khlaaf, et al. 2020. Toward trustworthy AI development: Mechanisms for supporting verifiable claims. *ArXiv:2004.07213 [Cs]*, April. http://arxiv.org/abs/2004.07213.

Buhmann, Alexander, Johannes Paßmann, and Christian Fieseler. 2019. Managing algorithmic accountability: Balancing reputational concerns, engagement strategies, and the potential of rational discourse. *Journal of Business Ethics*, June. https://doi.org/10.1007/s10551-019-04226-4.

Burke, Robin. 2017. 'Multisided Fairness for Recommendation'. *ArXiv:1707.00093 [Cs]*, July. http://arxiv.org/abs/1707.00093.

Burrell, Jenna. 2016. How the machine "thinks": Understanding opacity in machine learning algorithms. *Big Data & Society* 3 (1): 205395171562251. https://doi.org/10.1177/2053951715622512.

Chakraborty, Abhijnan, Gourab K. Patro, Niloy Ganguly, Krishna P. Gummadi, and Patrick Loiseau. 2019. Equality of voice: Towards fair representation in crowdsourced top-K recommendations. In *Proceedings of the conference on fairness, accountability, and transparency – FAT* '19*, 129–138. Atlanta: ACM Press. https://doi.org/10.1145/3287560.3287570.

Cohen, Julie. 2000. Examined lives: Informational privacy and the subject as object. *Georgetown Law Faculty Publications and Other Works*, January. https://scholarship.law.georgetown.edu/facpub/810.

Corbett-Davies, Sam, and Sharad Goel. 2018. The measure and mismeasure of fairness: A critical review of fair machine learning. *ArXiv:1808.00023 [Cs]*, August. http://arxiv.org/abs/1808.00023.

Cowls, Josh, Thomas King, Mariarosaria Taddeo, and Luciano Floridi. 2019. Designing AI for social good: Seven essential factors. *SSRN Electronic Journal.* https://doi.org/10.2139/ssrn.3388669.

Crain, Matthew. 2018. The limits of transparency: Data brokers and commodification. *New Media & Society* 20 (1): 88–104. https://doi.org/10.1177/1461444816657096.

Cummings, Mary. 2012. Automation Bias in intelligent time critical decision support systems. In *In AIAA 1st intelligent systems technical conference*. Chicago: American Institute of Aeronautics and Astronautics. https://doi.org/10.2514/6.2004-6313.

Dahl, E.S. 2018. Appraising black-boxed technology: The positive prospects. *Philosophy & Technology* 31 (4): 571–591. https://doi.org/10.1007/s13347-017-0275-1.

Danks, David, and Alex John London. 2017. Algorithmic Bias in autonomous systems. In *Proceedings of the twenty-sixth international joint conference on artificial intelligence*, 4691–4697. Melbourne: International Joint Conferences on Artificial Intelligence Organization. https://doi.org/10.24963/ijcai.2017/654.

Datta, Amit, Michael Carl Tschantz, and Anupam Datta. 2015. Automated experiments on Ad privacy settings. *Proceedings on Privacy Enhancing Technologies* 2015 (1): 92–112. https://doi.org/10.1515/popets-2015-0007.

Davis, Ernest, and Gary Marcus. 2019. *Rebooting AI: Building artificial intelligence we can trust.* Pantheon Books.

Diakopoulos, Nicholas, and Michael Koliska. 2017. Algorithmic transparency in the news media. *Digital Journalism* 5 (7): 809–828. https://doi.org/10.1080/21670811.2016.1208053.

Dignum, Virginia, Maite Lopez-Sanchez, Roberto Micalizio, Juan Pavón, Marija Slavkovik, Matthijs Smakman, Marlies van Steenbergen, et al. 2018. Ethics by design: Necessity or curse? In *Proceedings of the 2018 AAAI/ACM conference on AI, ethics, and society – AIES '18*, 60–66. New Orleans: ACM Press. https://doi.org/10.1145/3278721.3278745.

Doshi-Velez, Finale, and Been Kim. 2017. Towards a rigorous science of interpretable machine learning. *ArXiv:1702.08608 [Cs, Stat]*, March. http://arxiv.org/abs/1702.08608.

Edwards, Lilian, and Michael Veale. 2017. Slave to the algorithm? Why a right to explanation is probably not the remedy you are looking for. *SSRN Electronic Journal.* https://doi.org/10.2139/ssrn.2972855.

Ekstrand, Michael, and Karen Levy. 2018. *'FAT* Network'. 2018.* https://fatconference.org/network.

Eubanks, Virginia. 2017. *Automating inequality: How high-tech tools profile, police, and punish the poor.* 1st ed. New York: St. Martin's Press.

Floridi, Luciano. 2012. Distributed morality in an information society. *Science and Engineering Ethics* 19 (3): 727–743. https://doi.org/10.1007/s11948-012-9413-4.

———. 2016. Faultless responsibility: On the nature and allocation of moral responsibility for distributed moral actions. *Philosophical Transactions of the Royal Society A: Mathematical, Physical and Engineering Sciences* 374 (2083): 20160112. https://doi.org/10.1098/rsta.2016.0112.

———. 2017. Infraethics–on the conditions of possibility of morality. *Philosophy & Technology* 30 (4): 391–394. https://doi.org/10.1007/s13347-017-0291-1.

———. 2019a. What the near future of artificial intelligence could be. *Philosophy & Technology* 32 (1): 1–15. https://doi.org/10.1007/s13347-019-00345-y.

———. 2019b. Translating principles into practices of digital ethics: Five risks of being unethical. *Philosophy & Technology* 32 (2): 185–193. https://doi.org/10.1007/s13347-019-00354-x.

Floridi, Luciano, and Josh Cowls. 2019. A unified framework of five principles for AI in society. *Harvard Data Science Review*, June. https://doi.org/10.1162/99608f92.8cd550d1.

Floridi, Luciano, Josh Cowls, Monica Beltrametti, Raja Chatila, Patrice Chazerand, Virginia Dignum, Christoph Luetge, et al. 2018. AI4People—An ethical framework for a good AI society: Opportunities, risks, principles, and recommendations. *Minds and Machines* 28 (4): 689–707. https://doi.org/10.1007/s11023-018-9482-5.

Floridi, Luciano, Josh Cowls, Thomas C. King, and Mariarosaria Taddeo. 2020. How to design AI for social good: Seven essential factors. *Science and Engineering Ethics* 26 (3): 1771–1796. https://doi.org/10.1007/s11948-020-00213-5.

Floridi, Luciano, and Mariarosaria Taddeo. 2016. 'What is data ethics?' *Philosophical transactions of the Royal Society a: Mathematical. Physical and Engineering Sciences* 374 (2083): 20160360. https://doi.org/10.1098/rsta.2016.0360.

Fuster, Andreas, Paul Goldsmith-Pinkham, Tarun Ramadorai, and Ansgar Walther. 2017. Predictably unequal? The effects of machine learning on credit markets. *SSRN Electronic Journal*. https://doi.org/10.2139/ssrn.3072038.

Gajane, Pratik, and Mykola Pechenizkiy. 2018. On formalizing fairness in prediction with machine learning. *ArXiv:1710.03184 [Cs, Stat]*, May. http://arxiv.org/abs/1710.03184.

Gebru, Timnit, Jamie Morgenstern, Briana Vecchione, Jennifer Wortman Vaughan, Hanna Wallach, Hal Daumé III, and Kate Crawford. 2020. Datasheets for datasets. *ArXiv:1803.09010 [Cs]*, March. http://arxiv.org/abs/1803.09010.

Gillis, Talia B., and Jann Spiess. 2019. Big data and discrimination. *University of Chicago Law Review* 459.

Green, Ben, and Yiling Chen. 2019. Disparate interactions: An algorithm-in-the-loop analysis of fairness in risk assessments. In *Proceedings of the conference on fairness, accountability, and transparency – FAT* '19*, 90–99. Atlanta: ACM Press. https://doi.org/10.1145/3287560.3287563.

Green, Ben, and Salomé Viljoen. 2020. Algorithmic realism: Expanding the boundaries of algorithmic thought. In *Proceedings of the 2020 conference on fairness, accountability, and transparency*, 19–31. Barcelona: ACM. https://doi.org/10.1145/3351095.3372840.

Grgić-Hlača, Nina, Elissa M. Redmiles, Krishna P. Gummadi, and Adrian Weller. 2018. Human perceptions of fairness in algorithmic decision making: A case study of criminal risk prediction. *ArXiv:1802.09548 [Cs, Stat]*, February. http://arxiv.org/abs/1802.09548.

Grote, Thomas, and Philipp Berens. 2020. On the ethics of algorithmic decision-making in healthcare. *Journal of Medical Ethics* 46 (3): 205–211. https://doi.org/10.1136/medethics-2019-105586.

Hager, Gregory D., Ann Drobnis, Fei Fang, Rayid Ghani, Amy Greenwald, Terah Lyons, David C. Parkes, et al. 2019. Artificial intelligence for social good. *ArXiv:1901.05406 [Cs]*, January. http://arxiv.org/abs/1901.05406.

Harwell, Drew. 2020. Dating apps need women. Advertisers need diversity. AI companies offer a solution: Fake people. *Washington Post*, 2020.

Hauer, Tomas. 2019. Society caught in a labyrinth of algorithms: Disputes, promises, and limitations of the new order of things. *Society* 56 (3): 222–230. https://doi.org/10.1007/s12115-019-00358-5.

Henderson, Peter, Koustuv Sinha, Nicolas Angelard-Gontier, Nan Rosemary Ke, Genevieve Fried, Ryan Lowe, and Joelle Pineau. 2018. Ethical challenges in data-driven dialogue systems. In *Proceedings of the 2018 AAAI/ACM conference on AI, ethics, and society*, 123–129. New Orleans: ACM. https://doi.org/10.1145/3278721.3278777.

Hill, Robin K. 2016. What an algorithm is. *Philosophy & Technology* 29 (1): 35–59. https://doi.org/10.1007/s13347-014-0184-5.

Hoffmann, Anna Lauren, Sarah T. Roberts, Christine T. Wolf, and Stacy Wood. 2018. Beyond fairness, accountability, and transparency in the ethics of algorithms: Contributions and perspectives from LIS. *Proceedings of the Association for Information Science and Technology* 55 (1): 694–696. https://doi.org/10.1002/pra2.2018.14505501084.

Hu, Margaret. 2017. Algorithmic Jim Crow. *Fordham Law Review*. https://ir.lawnet.fordham.edu/flr/vol86/iss2/13/.

Hutson, Matthew. 2019. Bringing machine learning to the masses. *Science* 365 (6452): 416–417. https://doi.org/10.1126/science.365.6452.416.

ICO. 2020. *ICO and the Turing consultation on explaining AI decisions guidance*. ICO. 30 March 2020. https://ico.org.uk/about-the-ico/ico-and-stakeholder-consultations/ico-and-the-turing-consultation-on-explaining-ai-decisions-guidance/.

James, Gareth, Daniella Witten, Trevor Hastie, and Robert Tibshirani. 2013. *An introduction to statistical learning*.

Karppi, Tero. 2018. "The computer said so": On the ethics, effectiveness, and cultural techniques of predictive policing. *Social Media + Society* 4 (2). https://doi.org/10.1177/2056305118768296.

Karras, Tero, Samuli Laine, and Timo Aila. 2019. A style-based generator architecture for generative adversarial networks. *ArXiv:1812.04948 [Cs, Stat]*, March. http://arxiv.org/abs/1812.04948.

Katell, Michael, Meg Young, Dharma Dailey, Bernease Herman, Vivian Guetler, Aaron Tam, Corinne Binz, Daniella Raz, and P.M. Krafft. 2020. Toward situated interventions for algorithmic equity: Lessons from the field. In *Proceedings of the 2020 conference on fairness, accountability, and transparency*, 45–55. Barcelona: ACM. https://doi.org/10.1145/3351095.3372874.

King, Gary, and Nathaniel Persily. 2020. *Unprecedented Facebook URLs Dataset Now Available for Academic Research through Social Science One*. 2020. Unprecedented Facebook URLs Dataset now Available for Academic Research through Social Science One.

Kizilcec, René. 2016. How much information? | proceedings of the 2016 CHI conference on human factors in computing systems. In *Proceedings of the 2016 CHI conference on human factors in computing systems*, 2390–2395. https://dl.acm.org/doi/abs/10.1145/2858036.2858402.

Klee, Robert. 1996. *Introduction to the philosophy of science: Cutting nature at its seams*. Oxford University Press.

Kleinberg, Jon, Sendhil Mullainathan, and Manish Raghavan. 2016. Inherent trade-offs in the fair determination of risk scores. *ArXiv:1609.05807 [Cs, Stat]*, November. http://arxiv.org/abs/1609.05807.

Kortylewski, Adam, Bernhard Egger, Andreas Schneider, Thomas Gerig, Andreas Morel-Forster, and Thomas Vetter. 2019. *Analyzing and reducing the damage of dataset Bias to face recognition with synthetic data*. http://openaccess.thecvf.com/content_CVPRW_2019/html/BEFA/Kortylewski_Analyzing_and_Reducing_the_Damage_of_Dataset_Bias_to_Face_CVPRW_2019_paper.html.

Labati, Ruggero Donida, Angelo Genovese, Enrique Muñoz, Vincenzo Piuri, Fabio Scotti, and Gianluca Sforza. 2016. Biometric recognition in automated border control: A survey. *ACM Computing Surveys* 49 (2): 1–39. https://doi.org/10.1145/2933241.

Lambrecht, Anja, and Catherine Tucker. 2019. Algorithmic Bias? An empirical study of apparent gender-based discrimination in the display of STEM career ads. *Management Science* 65 (7): 2966–2981. https://doi.org/10.1287/mnsc.2018.3093.

Larson, Brian. 2017. Gender as a variable in natural-language processing: Ethical considerations. In *Proceedings of the first ACL workshop on ethics in natural language processing*, 1–11. Valencia: Association for Computational Linguistics. https://doi.org/10.18653/v1/W17-1601.

Lee, Michelle Seng Ah., and Luciano Floridi. 2020. Algorithmic fairness in mortgage lending: From absolute conditions to relational trade-offs. *SSRN Electronic Journal*. https://doi.org/10.2139/ssrn.3559407.

Lee, Min Kyung. 2018. Understanding perception of algorithmic decisions: Fairness, trust, and emotion in response to algorithmic management. *Big Data & Society* 5 (1): 205395171875668. https://doi.org/10.1177/2053951718756684.

Lee, Min Kyung, Ji Tae Kim, and Leah Lizarondo. 2017. A human-Centered approach to algorithmic services: Considerations for fair and motivating smart community service management that allocates donations to non-profit organizations. In *Proceedings of the 2017 CHI conference on human factors in computing systems – CHI '17*, 3365–3376. Denver: ACM Press. https://doi.org/10.1145/3025453.3025884.

Lepri, Bruno, Nuria Oliver, Emmanuel Letouzé, Alex Pentland, and Patrick Vinck. 2018. Fair, transparent, and accountable algorithmic decision-making processes: The premise, the proposed solutions, and the open challenges. *Philosophy & Technology* 31 (4): 611–627. https://doi.org/10.1007/s13347-017-0279-x.

Lewis, Dev. 2019. *Social credit case study: City citizen scores in Xiamen and Fuzhou*. Medium: Berkman Klein Center Collection. 8 October 2019. https://medium.com/berkman-klein-center/social-credit-case-study-city-citizen-scores-in-xiamen-and-fuzhou-2a65feb2bbb3.

Lipworth, Wendy, Paul H. Mason, Ian Kerridge, and John P.A. Ioannidis. 2017. Ethics and epistemology in big data research. *Journal of Bioethical Inquiry* 14 (4): 489–500. https://doi.org/10.1007/s11673-017-9771-3.

Magalhães, João Carlos. 2018. Do algorithms shape character? Considering algorithmic ethical subjectivation. *Social Media + Society* 4 (2): 205630511876830. https://doi.org/10.1177/2056305118768301.

Malhotra, Charru, Vinod Kotwal, and Surabhi Dalal. 2018. Ethical framework for machine learning. In *2018 ITU kaleidoscope: Machine learning for a 5G future (ITU K)*, 1–8. Santa Fe: IEEE. https://doi.org/10.23919/ITU-WT.2018.8597767.

Martin, Kirsten. 2019. Ethical implications and accountability of algorithms. *Journal of Business Ethics* 160 (4): 835–850. https://doi.org/10.1007/s10551-018-3921-3.

Mayson, Sandra G. 2019. Bias In, Bias Out. *Yale Law Journal*, 128. https://papers.ssrn.com/sol3/papers.cfm?abstract_id=3257004.

Milano, Silvia, Mariarosaria Taddeo, and Luciano Floridi. 2020. Recommender systems and their ethical challenges. *AI & SOCIETY*, February. https://doi.org/10.1007/s00146-020-00950-y.

Mittelstadt, Brent Daniel, Patrick Allo, Mariarosaria Taddeo, Sandra Wachter, and Luciano Floridi. 2016. The ethics of algorithms: Mapping the debate. *Big Data & Society* 3 (2). https://doi.org/10.1177/2053951716679679.

Mojsilovic, Aleksandra. 2018. *Introducing AI explainability 360*. 2018. https://www.ibm.com/blogs/research/2019/08/ai-explainability-360/.

Möller, Judith, Damian Trilling, Natali Helberger, and Bram van Es. 2018. Do not blame it on the algorithm: An empirical assessment of multiple recommender systems and their impact on content diversity. *Information, Communication & Society* 21 (7): 959–977. https://doi.org/10.1080/1369118X.2018.1444076.

Morley, Jessica, Luciano Floridi, Libby Kinsey, and Anat Elhalal. 2019a. 'From what to how: An initial review of publicly available AI ethics tools, methods and research to translate principles into practices'. *Science and Engineering Ethics*, December. https://doi.org/10.1007/s11948-019-00165-5.

Morley, Jessica, Caio Machado, Christopher Burr, Josh Cowls, Mariarosaria Taddeo, and Luciano Floridi. 2019b. The debate on the ethics of AI in health care: A reconstruction and critical review. *SSRN Electronic Journal*. https://doi.org/10.2139/ssrn.3486518.

Murgia, Madhumita. 2018. 'DeepMind's move to transfer health unit to Google stirs data fears'. *Financial Times*, 2018.

Noble, Safiya Umoja. 2018. *Algorithms of oppression: How search engines reinforce racism*. New York: New York University Press.

Obermeyer, Ziad, Brian Powers, Christine Vogeli, and Sendhil Mullainathan. 2019. Dissecting racial Bias in an algorithm used to manage the health of populations. *Science* 366 (6464): 447–453. https://doi.org/10.1126/science.aax2342.

Ochigame, Rodrigo. 2019. *'The invention of "ethical AI"', 2019.* https://theintercept.com/2019/12/20/mit-ethical-ai-artificial-intelligence/.

OECD. 2019. *Recommendation of the council on artificial intelligence.* https://legalinstruments.oecd.org/en/instruments/OECD-LEGAL-0449.

Olhede, S.C., and P.J. Wolfe. 2018. The growing ubiquity of algorithms in society: Implications, impacts and innovations. *Philosophical Transactions of the Royal Society A: Mathematical, Physical and Engineering Sciences* 376 (2128): 20170364. https://doi.org/10.1098/rsta.2017.0364.

Olteanu, Alexandra, Carlos Castillo, Fernando Diaz, and Emre Kiciman. 2016. Social data: Biases, methodological pitfalls, and ethical boundaries. *SSRN Electronic Journal.* https://doi.org/10.2139/ssrn.2886526.

Oswald, Marion. 2018. Algorithm-assisted decision-making in the public sector: Framing the issues using administrative law rules governing discretionary power. *Philosophical Transactions of the Royal Society A: Mathematical, Physical and Engineering Sciences* 376 (2128): 20170359. https://doi.org/10.1098/rsta.2017.0359.

Paraschakis, Dimitris. 2017. Towards an ethical recommendation framework. In *2017 11th international conference on research challenges in information science (RCIS)*, 211–220. Brighton: IEEE. https://doi.org/10.1109/RCIS.2017.7956539.

———. 2018. *Algorithmic and ethical aspects of recommender systems in E-commerce.* Malmö: Malmö universitet.

Perra, Nicola, and Luis E.C. Rocha. 2019. Modelling opinion dynamics in the age of algorithmic personalisation. *Scientific Reports* 9 (1): 7261. https://doi.org/10.1038/s41598-019-43830-2.

Perrault, Raymond, Shoham Yoav, Erik Brynjolfsson, Clark Jack, John Etchmendy, Barbara Grosz, Lyons Terah, Manyika James, Mishra Saurabh, and Niebles Juan Carlos. 2019. *Artificial Intelligence Index Report* 2019.

Prates, Marcelo O. R., Pedro H. Avelar, and Luís C. Lamb. 2019. Assessing gender bias in machine translation: A case study with Google translate. *Neural Computing and Applications*, March. https://doi.org/10.1007/s00521-019-04144-6.

Rachels, James. 1975. Why privacy is important. *Philosophy & Public Affairs* 4 (4): 323–333.

Rahwan, Iyad. 2018. Society-in-the-loop: Programming the algorithmic social contract. *Ethics and Information Technology* 20 (1): 5–14. https://doi.org/10.1007/s10676-017-9430-8.

Ras, Gabrielle, Marcel van Gerven, and Pim Haselager. 2018. Explanation methods in deep learning: Users, values, concerns and challenges. *ArXiv:1803.07517 [Cs, Stat]*, March. http://arxiv.org/abs/1803.07517.

Reddy, Elizabeth, Baki Cakici, and Andrea Ballestero. 2019. Beyond mystery: Putting algorithmic accountability in context. *Big Data & Society* 6 (1): 205395171982685. https://doi.org/10.1177/2053951719826856.

Reisman, Dillon, Jason Schultz, Kate Crawford, and Meredith Whittaker. 2018. Algorithmic impact assessments: A practical framework for public agency accountability. *AI Now Institute*, April. https://ainowinstitute.org/aiareport2018.pdf.

Richardson, Rashida, Jason Schultz, and Kate Crawford. 2019. *Dirty data, bad predictions: How civil rights violations impact police data, predictive policing systems, and justice.* https://papers.ssrn.com/sol3/papers.cfm?abstract_id=3333423.

Robbins, Scott. 2019. A misdirected principle with a catch: Explicability for AI. *Minds and Machines* 29 (4): 495–514. https://doi.org/10.1007/s11023-019-09509-3.

Roberts, Huw, Josh Cowls, Jessica Morley, Mariarosaria Taddeo, Vincent Wang, and Luciano Floridi. 2019. The Chinese approach to artificial intelligence: An analysis of policy and regulation. *SSRN Electronic Journal.* https://doi.org/10.2139/ssrn.3469784.

———. 2020. The Chinese approach to artificial intelligence: An analysis of policy, ethics, and regulation. *AI & SOCIETY*, June. https://doi.org/10.1007/s00146-020-00992-2.

Rössler, Beate. 2015. *The value of privacy.*

Rubel, Alan, Clinton Castro, and Adam Pham. 2019. Agency laundering and information technologies. *Ethical Theory and Moral Practice* 22 (4): 1017–1041. https://doi.org/10.1007/s10677-019-10030-w.

Sandvig, Christian, Kevin Hamilton, Kerry Karahalios, and Cedric Langbort. 2016. When the algorithm itself is a racist: Diagnosing ethical harm in the basic components of software. *International Journal of Communication* 10: 4972–4990.

Saxena, Nripsuta, Karen Huang, Evan DeFilippis, Goran Radanovic, David Parkes, and Yang Liu. 2019. How do fairness definitions fare? Examining Public Attitudes Towards Algorithmic Definitions of Fairness. *ArXiv:1811.03654 [Cs]*, January. http://arxiv.org/abs/1811.03654.

Selbst, Andrew D., Danah Boyd, Sorelle A. Friedler, Suresh Venkatasubramanian, and Janet Vertesi. 2019. Fairness and abstraction in sociotechnical systems. In *Proceedings of the conference on fairness, accountability, and transparency – FAT* '19*, 59–68. Atlanta: ACM Press. https://doi.org/10.1145/3287560.3287598.

Shah, Hetan. 2018. Algorithmic accountability. *Philosophical Transactions of the Royal Society A: Mathematical, Physical and Engineering Sciences* 376 (2128): 20170362. https://doi.org/10.1098/rsta.2017.0362.

Shin, Donghee, and Yong Jin Park. 2019. Role of fairness, accountability, and transparency in algorithmic affordance. *Computers in Human Behavior* 98 (September): 277–284. https://doi.org/10.1016/j.chb.2019.04.019.

Sloan, Robert H., and Richard Warner. 2018. When is an algorithm transparent? Predictive analytics, privacy, and public policy. *IEEE Security & Privacy* 16 (3): 18–25. https://doi.org/10.1109/MSP.2018.2701166.

Stilgoe, Jack. 2018. Machine learning, social learning and the governance of self-driving cars. *Social Studies of Science* 48 (1): 25–56. https://doi.org/10.1177/0306312717741687.

Szegedy, Christian, Wojciech Zaremba, Ilya Sutskever, Joan Bruna, Dumitru Erhan, Ian Goodfellow, and Rob Fergus. 2014. Intriguing properties of neural networks. *ArXiv:1312.6199 [Cs]*, February. http://arxiv.org/abs/1312.6199.

Taddeo, Mariarosaria, and Luciano Floridi. 2018a. Regulate artificial intelligence to avert cyber arms race. *Nature* 556 (7701): 296–298. https://doi.org/10.1038/d41586-018-04602-6.

———. 2018b. How AI can be a force for good. *Science* 361 (6404): 751–752. https://doi.org/10.1126/science.aat5991.

Taddeo, Mariarosaria, Tom McCutcheon, and Luciano Floridi. 2019. Trusting artificial intelligence in cybersecurity is a double-edged sword. *Nature Machine Intelligence* 1 (12): 557–560. https://doi.org/10.1038/s42256-019-0109-1.

Taylor, Linnet, Luciano Floridi, and Bart van der Sloot, eds. 2017. *Group privacy: New challenges of data technologies*. New York: Springer Berlin Heidelberg.

Tickle, A.B., R. Andrews, M. Golea, and J. Diederich. 1998. The truth will come to light: Directions and challenges in extracting the knowledge embedded within trained artificial neural networks. *IEEE Transactions on Neural Networks* 9 (6): 1057–1068. https://doi.org/10.1109/72.728352.

Turilli, Matteo, and Luciano Floridi. 2009. The ethics of information transparency. *Ethics and Information Technology* 11 (2): 105–112. https://doi.org/10.1007/s10676-009-9187-9.

Turner Lee, Nicol. 2018. Detecting racial Bias in algorithms and machine learning. *Journal of Information, Communication and Ethics in Society* 16 (3): 252–260. https://doi.org/10.1108/JICES-06-2018-0056.

Valiant, L.G. 1984. A theory of the learnable. *Communications of the ACM* 27 (11): 1134–1142. https://doi.org/10.1145/1968.1972.

Veale, Michael, and Reuben Binns. 2017. Fairer machine learning in the real world: Mitigating discrimination without collecting sensitive data. *Big Data & Society* 4 (2): 205395171774353. https://doi.org/10.1177/2053951717743530.

Vedder, Anton, and Laurens Naudts. 2017. Accountability for the use of algorithms in a big data environment. *International Review of Law, Computers & Technology* 31 (2): 206–224. https://doi.org/10.1080/13600869.2017.1298547.

Wang, Shuang, Xiaoqian Jiang, Siddharth Singh, Rebecca Marmor, Luca Bonomi, Dov Fox, Michelle Dow, and Lucila Ohno-Machado. 2017. Genome privacy: Challenges, technical approaches to mitigate risk, and ethical considerations in the United States: Genome privacy in biomedical research. *Annals of the New York Academy of Sciences* 1387 (1): 73–83. https://doi.org/10.1111/nyas.13259.

Watson, David, and Luciano Floridi. 2020. The explanation game: A formal framework for interpretable machine learning. *SSRN Electronic Journal*. https://doi.org/10.2139/ssrn.3509737.

Webb, Helena, Menisha Patel, Michael Rovatsos, Alan Davoust, Sofia Ceppi, Ansgar Koene, Liz Dowthwaite, Virginia Portillo, Marina Jirotka, and Monica Cano. 2019. "It would be pretty immoral to choose a random algorithm": Opening up algorithmic interpretability and transparency. *Journal of Information, Communication and Ethics in Society* 17 (2): 210–228. https://doi.org/10.1108/JICES-11-2018-0092.

Weller, Adrian. 2019. Transparency: Motivations and challenges. *ArXiv:1708.01870 [Cs]*, August. http://arxiv.org/abs/1708.01870.

Wexler, James. 2018. *The what-if tool: Code-free probing of machine learning models*. 2018. https://ai.googleblog.com/2018/09/the-what-if-tool-code-free-probing-of.html.

Whitman, Madisson, Chien-yi Hsiang, and Kendall Roark. 2018. Potential for participatory big data ethics and algorithm design: A scoping mapping review. In *Proceedings of the 15th participatory design conference on short papers, situated actions, workshops and tutorial – PDC '18*, 1–6. Hasselt and Genk: ACM Press. https://doi.org/10.1145/3210604.3210644.

Wiener, Norbert. 1950. The human use of human beings.

Winner, Langdon. 1980. Do artifacts have politics? *Modern Technology: Problem or Opportunity?* 109 (1): 121–136.

Wong, Pak-Hang. 2019. Democratizing algorithmic fairness. *Philosophy & Technology*, June. https://doi.org/10.1007/s13347-019-00355-w.

Xian, Zhengzheng, Qiliang Li, Xiaoyu Huang, and Lei Li. 2017. New SVD-based collaborative filtering algorithms with differential privacy. *Journal of Intelligent & Fuzzy Systems* 33 (4): 2133–2144. https://doi.org/10.3233/JIFS-162053.

Xu, Depeng, Shuhan Yuan, Zhang Lu, and Xintao Wu. 2018. FairGAN: Fairness-aware generative adversarial networks. In *2018 IEEE international conference on big data (big data)*, 570–575. Seattle: IEEE. https://doi.org/10.1109/BigData.2018.8622525.

Yampolskiy, Roman V. 2018. *Artificial intelligence safety and security*. Chapman and Hall/CRC.

Yang, Guang-Zhong, Jim Bellingham, Pierre E. Dupont, Peer Fischer, Luciano Floridi, Robert Full, Neil Jacobstein, et al. 2018. The grand challenges of *science robotics*. *Science robotics* 3 (14): eaar7650. https://doi.org/10.1126/scirobotics.aar7650.

Yu, Meng, and Guodong Du. 2019. Why are Chinese courts turning to AI? *The Diplomat*, 19 January 2019. https://thediplomat.com/2019/01/why-are-chinese-courts-turning-to-ai/.

Zerilli, John, Alistair Knott, James Maclaurin, and Colin Gavaghan. 2019. Transparency in algorithmic and human decision-making: Is there a double standard? *Philosophy & Technology* 32 (4): 661–683. https://doi.org/10.1007/s13347-018-0330-6.

Zhou, Na, Chuan-Tao Zhang, Hong-Ying Lv, Chen-Xing Hao, Tian-Jun Li, Jing-Juan Zhu, Hua Zhu, et al. 2019. Concordance study between IBM Watson for oncology and clinical practice for patients with cancer in China. *The Oncologist* 24 (6): 812–819. https://doi.org/10.1634/theoncologist.2018-0255.

Chapter 9
How to Design AI for Social Good: Seven Essential Factors

Luciano Floridi, Josh Cowls, Thomas C. King, and Mariarosaria Taddeo

Abstract The idea of Artificial Intelligence for Social Good (henceforth AI4SG) is gaining traction within information societies in general and the AI community in particular. It has the potential to tackle social problems through the development of AI-based solutions. Yet, to date, there is only limited understanding of what makes AI socially good in theory, what counts as AI4SG in practice, and how to reproduce its initial successes in terms of policies. This article addresses this gap by identifying seven ethical factors that are essential for future AI4SG initiatives. The analysis is supported by 27 case examples of AI4SG projects. Some of these factors are almost entirely novel to AI, while the significance of other factors is heightened by the use of AI. From each of these factors, corresponding best practices are formulated which, subject to context and balance, may serve as preliminary guidelines to ensure that well-designed AI is more likely to serve the social good.

Keywords AI4SG · Artificial Intelligence · Ethics · Social Good · Transparency · Privacy · Safety

Authors "Luciano Floridi and Josh Cowls" have equally contributed to this chapter.

L. Floridi
Oxford Internet Institute, University of Oxford, Oxford, UK
e-mail: luciano.floridi@oii.ox.ac.uk

J. Cowls (✉) · M. Taddeo
Oxford Internet Institute, University of Oxford, Oxford, UK

Alan Turing Institute, London, UK
e-mail: josh.cowls@oii.ox.ac.uk; mariarosaria.taddeo@oii.ox.ac.uk

T. C. King
Oxford Internet Institute, University of Oxford, Oxford, UK

Amherst, Cheltenham, UK

L. Floridi (ed.), *Ethics, Governance, and Policies in Artificial Intelligence*, Philosophical Studies Series 144, https://doi.org/10.1007/978-3-030-81907-1_9

9.1 Introduction

The idea of "Artificial Intelligence (AI) for Social Good" (henceforth AI4SG) is becoming popular in many information societies and gaining traction within the AI community (Hager et al. 2017). Projects seeking to use AI for social good vary significantly. They range from models to predict septic shock (Henry et al. 2015) to game-theoretic models to prevent poaching (Fang et al. 2016); from online reinforcement learning to target HIV-education at homeless youths (Yadav et al. 2016a, b) to probabilistic models to prevent harmful policing (Carton et al. 2016) and support student retention (Lakkaraju et al. 2015). Indeed, new applications of AI4SG appear almost daily, making possible socially good outcomes that were once less easily achievable, unfeasible, or unaffordable.

Several frameworks for the design, development, and deployment of ethical AI in general have recently emerged (see Floridi et al. 2018 for a comparative analysis and synthesis). However, there is still only limited understanding about what constitutes AI "for the social good" (Taddeo and Floridi 2018a). Approaching AI4SG ad hoc, by analysing specific areas of application—like famine-relief or disaster management—as an annual summit for AI industry and government has done ("AI for Good Global Summit" 2017, 2018, 2019) indicates the presence of a phenomenon, but neither explains it, nor does it suggest how other AI4SG solutions could and should be designed to harness AI's full potential. Furthermore, many projects that generate socially good outcomes using AI are not (self-)described as such (Moore 2019).

Lacking a clear understanding of what makes AI socially good in theory, what may be described as AI4SG in practice, and how to reproduce its initial successes in terms of policies is a problem because designers of AI4SG face at least two main challenges: unnecessary failures and missed opportunities. AI software is shaped by human values which, if not carefully selected, may lead to "good-AI-gone-bad" scenarios. For example, consider the failure of IBM's oncology-support software, which attempts to use machine learning to identify cancerous tumours, but which was rejected by medical practitioners "on the ground" (Ross and Swetlitz 2017). The system was trained using synthetic data and was not sufficiently refined to interpret ambiguous, nuanced, or otherwise "messy" patient health records (Strickland 2019). It also relied on US medical protocols, which are not applicable worldwide. The heedless deployment and the poor design of the software led to misdiagnoses and erroneous treatment suggestions, breaching the trust of doctors. Context-specific design and deployment could help prevent such value misalignment and deliver successful AI4SG projects on a more consistent basis.

At the same time, the genuinely socially good outcomes of AI may arise merely by chance, for example through an accidental application of an AI solution in a different context. This was the case with the use of a different version of IBM's cognitive system. In this case, the Watson system was originally designed to identify biological mechanisms, but when used in a classroom setting, it inspired engineering students to solve design problems (Goel et al. 2015). In this instance, AI provided a unique mode of education. But lacking a clear understanding of AI4SG means that this success is accidental and cannot be repeated systematically, whilst for each "accidental success" there may be countless examples of missed opportunities to exploit the benefits of AI for advancing socially good outcomes in different settings.

In order to avoid unnecessary failures and missed opportunities, AI4SG would benefit from an analysis of the essential factors that support and underwrite the design and deployment of successful AI4SG. In this article, we provide the first, fine-grained analysis of these factors. Our aim here is not to document every single ethical consideration for an AI4SG project. For example, it is essential, and hopefully self-evident, that an AI4SG project ought not to advance the proliferation of weapons of mass destruction, an imperative which we do not discuss here (Taddeo and Floridi 2018b). Likewise, it is important to acknowledge at the outset that there are myriad circumstances in which AI will not be the most effective way to address a particular social problem. This could be due to the existence of alternative approaches that are more efficacious (i.e., "Not AI For Social Good") or because of the unacceptable risks that the deployment of AI would introduce (i.e., "AI For Insufficient Social Good" as weighed against its risks). Nor do we foresee many (or perhaps any) cases in which AI is a "silver bullet"—the single-handed solution to an entrenched social problem (i.e., "Only AI for Social Good"). What is therefore essential about the factors and the corresponding best practices is not their incorporation in every circumstance; we note several examples where it would be morally defensible not to incorporate a particular factor. Instead, what is essential is that each best practice is (i) considered proactively, and (ii) not incorporated if and only if there is a clear, demonstrable, and morally defensible reason why it should not be.

In this article, we focus on identifying factors that are particularly relevant to AI as a technological infrastructure, to the extent that it is designed and used for the advancement of social good. To anticipate, these seven factors are: (1) falsifiability and incremental deployment; (2) safeguards against the manipulation of predictors; (3) receiver-contextualised intervention; (4) receiver-contextualised explanation and transparent purposes; (5) privacy protection and data subject consent; (6) situational fairness; and (7) human-friendly semanticisation. With these factors identified, the questions that are likely to arise in turn are: how these factors ought to be evaluated and resolved, by whom, and with what supporting mechanism (e.g. regulation or codes of conduct). These questions, which are not within the scope of this article and will be addressed in the next stage of this research, are intertwined with wider ethical and political issues regarding the legitimacy of decision-making with, and about, AI.

The rest of the article is structured as follows. In Sect. 9.2, we explain how we identified the seven factors. In Sect. 9.3, we analyse the seven factors individually. We elucidate each of them by reference to one or more case studies, and we derive from each factor a corresponding best practice for AI4SG creators to follow. In the concluding section, we discuss the factors and suggest how tensions between them may be resolved.

9.2 Methodology

AI4SG initiatives are successful insofar as they help to reduce, mitigate or eradicate a given problem of moral significance. Thus, our analysis of the essential factors for successful AI4SG is based on the following working definition:

> AI4SG =def. the design, development, and deployment of AI systems in ways that (i) prevent, mitigate or resolve problems adversely affecting human life and/or the wellbeing of the natural world, and/or (ii) enable socially preferable and/or environmentally sustainable developments.[1]

Following this definition, we analysed a set of 27 projects, obtained via a systematic review of relevant literature undertaken by the authors, to identify clear and significant cases of successful and unsuccessful examples of AI4SG. The literature analysis that underpins this article involved searching five databases (Google Scholar, PhilPapers, Scopus, SSRN, and Web of Science), between October 2018 and May 2019. We initially conducted a broad search for AI for Social Good on each of these search engines. This general search returned many results on AI's application for good. Hence, we searched for uses of AI in areas related to human life and the wellbeing of the natural world, like 'healthcare', 'education', 'equality', 'climate change', and 'environmental protection'. This provided disjoined keywords from which we derived chosen synonyms to perform area-specific searches. Each research-area search used the query: < area and synonyms> AND ("Artificial Intelligence" OR "Machine Learning" OR "AI") AND "Social Good". From the set of 27 cases, we identified 7 cases (see Appendix for a list) as being most representative in terms of scope, variety, impact, and for their potentiality to corroborate the essential factors that we argue should characterise the design of AI4SG projects.

The factors described in this article have been identified in coherence with more general work in the field of AI ethics. Each factor relates to at least one of five ethical principles of AI—beneficence, nonmaleficence, justice, autonomy, and explicability—identified in the comparative analysis mentioned above (Floridi et al. 2018). This coherence is crucial: AI4SG cannot be inconsistent with the ethical framework guiding the design and evaluation of AI in general. The principle of beneficence is of particular relevance. It states that the use of AI should provide benefit to people and the natural world, and indeed AI4SG projects should not just comply with but reify this principle, such that the benefits of AI4SG should be preferable and sustainable, in line with the definition above. Beneficence is thus a necessary condition of AI4SG, yet it is insufficient, not least because the beneficent impact of an AI4SG project may be "offset" by the creation or amplification of other

[1] While it is beyond present scope to adjudicate this for any particular case, it is important to acknowledge at the outset that in practice there is likely to be considerable disagreement and contention regarding what would constitute a socially good outcome.

risks or harms.[2] Moreover, while others of these ethical principles, such as autonomy and explicability, indeed recur throughout our discussion, the factors we evince below are more closely associated with design considerations that are specific to AI4SG, and may be operationalised in the form of the corresponding best practices provided for each. In this way, ethical analysis informing the design and the deployment of AI4SG initiatives has a central role in mitigating foreseeable risks of unintended consequences and possible misuses of the technology.

Before discussing the factors, it is important to clarify three general features of the whole set: dependency, order, and coherence. The seven factors are often intertwined and co-dependent, but for the sake of simplicity we discuss them separately. Nothing should be inferred from this choice. In the same way, the factors are all essential, none of them is "more important" than any other, so we shall introduce them not in terms of priority, but somewhat historically, starting with factors that pre-date AI, and yet take on greater importance when AI technologies are used, owing to the particular capabilities and risks of AI (Yang et al. 2018).[3] These include falsifiability and incremental deployment and safeguards against the manipulation of data. There are also factors that relate more intrinsically to the sociotechnical characteristics of AI as it exists today, like situational fairness and human-friendly semanticisation.

The factors are ethically robust and pragmatically applicable, in the sense that they give rise to design considerations in the form of best practices that should be ethically endorsed. It is crucial to stress here that the seven factors we identify are not by themselves sufficient for socially good AI, but careful consideration of each of them is, we argue, necessary. Hence, the set of factors we identify should not be taken as a "checklist" which, if merely complied with, guarantees socially good outcomes from the use of AI in a particular domain. In the same vein, we highlight the need to strike a balance between the different factors, and indeed between tensions that may arise even within a single factor. It follows that seeking to frame a project as "for social good" or "not for social good" in a binary way seems needlessly reductive, not to mention subjective. The aim of the article is not to identify, or offer the means to identify AI4SG projects; our goal is to identify ethically important characteristics of projects that could feasibly be described as AI4SG.

[2] This should not be taken as necessitating a utilitarian calculation: the beneficial impact of a given project may be "offset" by the violation of some categorical imperative. Therefore even if an AI4SG project would do "more good than harm", the harm may be ethically intolerable. In such a hypothetical case, one would not be morally obliged to develop and deploy the project in question.

[3] As noted in the introduction, we cannot hope to document every single ethical consideration for a social good project, so even the least novel factors here are those that take on new relevance in the context of AI.

9.3 Seven Essential Factors for Successful AI4SG

As we anticipated, the factors are (1) falsifiability and incremental deployment; (2) safeguards against the manipulation of predictors; (3) receiver-contextualised intervention; (4) receiver-contextualised explanation and transparent purposes; (5) privacy protection and data subject consent; (6) situational fairness; and (7) human-friendly semanticisation. We shall elucidate each factor with one or more examples of projects in the sample, and offer a corresponding best practice.

9.3.1 *Falsifiability and Incremental Deployment*

Trustworthiness is essential for technology in general (Taddeo and Floridi 2011; Taddeo 2017), and for AI4SG applications in particular, to be adopted and have a meaningful positive impact on human life and environmental wellbeing. Trustworthiness of an AI application entails a high probability that the application will respect the principle of beneficence, or at the very least the principle of nonmaleficience. While there is no universal rule or guideline that can ensure or guarantee trustworthiness, falsifiability is an essential factor to improve the trustworthiness of technological applications in general, and AI4SG applications in particular.

Falsifiability entails the specification, and the possibility of empirical testing, of one or more critical requirements, that is, an essential condition, resource, or means for a capability to be fully operational, such that something could or should not work without it. Safety is an obvious critical requirement. Hence, for an AI4SG system to be trustworthy, its safety should be falsifiable.[4] If falsifiability is not possible, then the critical requirements cannot be checked, and then the system should not be deemed trustworthy. This is why falsifiability is an essential factor for all conceivable AI4SG projects.

Unfortunately, we cannot know for sure that a given AI4SG application is safe unless we can test the application in all possible contexts. In this case, the map of testing would simply equate to the territory of deployment. As this reductio ad absurdum makes clear, complete certainty is out of reach. What is within reach, in an uncertain and fuzzy world with many unforeseen situations, is the possibility to know when a given critical requirement is not implemented or may be failing to work properly. Hence, if the critical requirements are falsifiable, we can know when the AI4SG application is not trustworthy, but not whether it is trustworthy.

Critical requirements should be tested with an incremental deployment cycle. Unintended hazardous effects may only reveal themselves after testing. At the same

[4] It is of course likely that in practice, an assessment of the safety of an AI system must also take into account wider societal values and cultural beliefs, for example, which may necessitate different trade-offs between the requirements of critical requirements like safety and other, potentially competing norms and expectations.

time, software should only be tested in the real world if it is safe to do so. This requires adoption of a deployment cycle whereby developers: (a) ensure that the application's most critical requirements or assumptions are falsifiable, (b) undertake hypothesis testing of those most critical requirements and assumptions in safe, protected contexts, and, if these hypotheses are not disproven over a small set of suitable contexts, then (c) conduct testing across increasingly wide contexts, and/or test a larger set of less critical requirements, and all this while (d) being ready to halt or modify the deployment as soon as hazardous or other unwanted effects may appear.

AI4SG applications may use formal approaches to try to test critical requirements. For example, they may include the use of formal verification to ensure that autonomous vehicles, and AI systems in other safety-critical contexts, would make the ethically preferable choice (Dennis et al. 2016). Such methods offer safety checks that, in terms of falsifiability, can be proved correct. Simulations may offer roughly similar guarantees. A simulation enables one to test whether critical requirements (again, consider safety) are met under a set of formal assumptions. Unlike a formal proof, a simulation cannot always indicate that the required properties are necessarily always satisfied. But a simulation often enables one to test a much wider set of cases that cannot be dealt with formally, e.g., due to the complexity of the proof.

It would be misguided to rely purely on formal properties or simulations to falsify an AI4SG application. The assumptions of these models cage the real-world applicability of any conclusions that one might make. And assumptions may be incorrect in reality. What one may prove to be correct via a formal proof, or likely correct via testing in simulation, may be disproved later with the real-world deployment of the system. For example, developers of a game-theoretic model for wildlife security assumed a relatively flat topography without serious obstructions. Hence, the software that they developed originally had an incorrect definition of an optimal patrol route. Incremental testing of the application enabled the refinement of the optimal patrol route by proving wrong the assumption of a flat topography (Fang et al. 2016).

If novel dilemmas in real-world contexts require the alteration of prior assumptions made in the lab, one solution is to rectify a priori assumptions after deployment. Alternatively, one may adopt an "on-the-fly" or runtime system for a constant update of a program's processing ("understanding") of its inputs. Yet, problems also abound with this approach. For example, Microsoft's infamous Twitter bot, Tay, acquired meanings, in a very loose sense, at runtime, as it learned from Twitter users how it should respond to tweets. After deployment in the real—and frequently vicious—world of social media, however, the bot's ability to adapt constantly its "conceptual understanding" became an unfortunate bug, as Tay "learned" and regurgitated offensive language and unethical associations between concepts from other users (Neff and Nagy 2016).

The use of a retrodictive approach—that is, an attempt to understand some aspect of reality through a priori information—to deal with the falsifiability of requirements presents similar problems. This is noteworthy, since retrodiction is the primary method of supervised machine learning approaches that learn from data (e.g., the learning of a continuous transformation function in the case of neural networks).

From the previous analysis it follows that the essential factor of falsifiability and incremental development comprises a cycle: engineering requirements that are falsifiable (so that it is at least possible to know whether the requirements are not met); falsification testing for incrementally improving levels of trustworthiness; adjustment of a priori assumptions; and then and only then deployment in an incrementally wider and critical context. Germany's approach to regulating autonomous vehicles offer a good example of this incremental approach. Deregulated zones allow experimentation of constrained autonomy and, after increasing the levels of trustworthiness, manufacturers may test vehicles with higher levels of autonomy (Pagallo 2017). Indeed, the creation of such deregulated zones, or teststrecken, was one recommendation to support more ethical AI policy at the European level (Floridi et al. 2018). The identification of this essential factor yields the following best practice:

1) AI4SG designers should identify falsifiable requirements and test them in incremental steps from the lab to the "outside world".

9.3.2 Safeguards Against the Manipulation of Predictors

The use of AI to predict future trends or patterns is very popular in AI4SG contexts, from applying automated prediction to redress academic failure (Lakkaraju et al. 2015), to preventing illegal policing (Carton et al. 2016), and detecting corporate fraud (Zhou and Kapoor 2011). The predictive power of AI4SG faces two risks: the manipulation of input data, and excessive reliance on non-causal indicators.

The manipulation of data is not a new problem, nor is it limited to AI systems alone. Well-established findings such as Goodhart's Law (Goodhart 1975), which is often summarised as "when a measure becomes a target, it ceases to be a good measure" (Strathern 1997, 308), long pre-date widespread adoption of AI systems. But in the case of AI, the problem of data manipulation may be exacerbated (Manheim and Garrabrant 2019) and lead to unfair outcomes that breach the principle of justice. As such, it is a noteworthy risk for any AI4SG initiative, because it can impair the predictive power of AI and lead to the avoidance of socially good interventions at the individual level. Consider the concern raised by Ghani over teachers who face being evaluated in respect to:

> the percentage of students in their class who are above a certain risk threshold. If the model was transparent — for example, heavily reliant on math GPA — the teacher could inflate math grades and reduce the intermediate risk scores of their students (Ghani 2016).

As Ghani goes on to argue, the same concern applies to predictors of adverse police officer interactions:

these systems [are] very easy to understand and interpret, but that also makes them easy to game. An officer who has had two uses of force in the past 80 days may choose to be a bit more careful over the next 10 days, until the count rolls over to zero again.

These hypothetical examples make clear that, when the model used is an easy one to understand "on the ground", it is already open to abuse or "gaming", independently of whether AI is used. The introduction of AI complicates matters, owing to the scale at which AI is typically applied. [5] As we have seen, if the information used to predict a given outcome is known, an agent with such information (that is predicted to take a particular action) can change each predictive variable's value in order to avoid an intervention. In this way, the predictive power of the overall model is reduced, as it has been shown by empirical research in the domain of corporate fraud (Zhou and Kapoor 2011). Such a phenomenon could carry across from fraud detection to the domains that AI4SG initiatives seek to address.[6]

At the same time, there is a risk that excessive reliance on non-causal indicators—that is, data which is correlated with, but not causal of, a phenomenon—may distract attention from the context in which the AI4SG designer is seeking to intervene. To be effective, any such intervention should alter the underlying causes of a given problem, such as a student's domestic problems or inadequate corporate governance, rather than non-causal predictors. To do otherwise is to risk addressing only a symptom, rather than the root cause of a problem.

These risks suggest the need to consider the use of safeguards as a design factor for AI4SG projects. Such safeguards may constrain the selection of indicators to be used in the design of AI4SG projects; the extent to which these indicators should shape interventions; and/or the level of transparency that should apply to how indicators affect decision. This yields the following best practice:

2) AI4SG designers should adopt safeguards which (i) ensure that non-causal indicators do not inappropriately skew interventions, and (ii) limit, when appropriate, knowledge of how inputs affect outputs from AI4SG systems, to prevent manipulation.

9.3.3 Receiver-Contextualised Intervention

It is essential that software intervenes in users' life only in ways that respect their autonomy. Again, this is not a problem that arises only with AI-driven interventions, but the use of AI introduces new considerations. In particular, a core challenge for AI4SG projects is to devise interventions that balance current and future benefits.

[5] While, for the sake of simplicity, our focus is on *minimising* the spread of information used to predict an outcome, we do not intend to foreclose on the suggestion, offered in Prasad (2018), that in some cases a fairer approach may be to *maximise* the available information and hence "democratise" the ability to manipulate predictors.

[6] For a discussion of the use of artificial intelligence in criminal acts more generally, see King et al. 2019.

The balancing problem, which is familiar to preference-elicitation research (Boutilier 2002; Faltings et al. 2004; Chajewska et al. 2000), boils down to a temporal choice interdependency. An intervention in the present can elicit user preferences that then enable the software to contextualise future interventions to the given user. Consequently, an intervention strategy that has no impact on user autonomy (e.g., one that lacks any interventions) may be ineffective in extracting the necessary information for correctly contextualised future interventions. Conversely, an intervention that overly infringes upon a user's autonomy may cause the user to reject the technology, making future interventions impossible.

This balancing consideration is a common one for AI4SG initiatives. Take, for example, interactive activity recognition software for people with cognitive disabilities (Chu et al. 2012). The software is designed to prompt patients to maintain a daily schedule of activities (e.g., taking medication), whilst minimising interruptions to their wider goals. Each intervention is contextualised in such a way that the software learns the timing of future interventions from responses to past interventions. Moreover, only important interventions are made, and yet all interventions are partially optional because declining one prompt leads to the same prompt later on. Here, the concern was that patients would reject an overly intrusive technology; hence a balance was sought. This balance is lacking in our second example. A game-theoretic application intervenes in wildlife security officers' patrols by offering suggested routes (Fang et al. 2016). If a route poses physical obstacles, however, then the software lacks the possibility to provide alternative suggestions. Officers may ignore the advice by taking a different route, but not without disengaging from the application. It is essential to relax such constraints, so that users can ignore an intervention, but accept subsequent, more appropriate interventions (in the form of advice) later on.

These examples point to the importance of seeing users as equal partners in both the design and deployment of autonomous decision-making systems. The adoption of this mindset might have helped prevent the tragic loss of two Boeing 737 Max airliners. It appears that the pilots of these flights struggled to reverse a software malfunction caused by faulty sensors, due in part to the absence of "optional safety features" which Boeing sold separately (Tabuchi and Gelles 2019).

The risk of false positives (unnecessary intervention, creating disillusionment) is often just as problematic as false negatives (no intervention where it is necessary, limiting effectiveness). Hence, a suitable receiver-contextualised intervention is one that achieves the right level of disruption while respecting autonomy through optionality. This contextualisation rests on information about users' capacities, preferences and goals, and the circumstances in which the intervention will take effect.

One can consider five dimensions relevant to a receiver-contextualised intervention. Four of these dimensions emerge from McFarlane's taxonomy of interdisciplinary research on disruptive computer-human interruptions (McFarlane 1999; McFarlane and Latorella 2002, 17–19). These are: the individual characteristics of the person receiving the intervention; the methods of coordination between the receiver and the system; the meaning or purpose of the intervention; and the overall effects of the intervention.[7] A fifth dimension of relevance is optionality: a user can choose either to ignore all offered advice or to drive the process and request a different intervention better suited to their needs.

We can summarise these five dimensions in the form of the following best practice for receiver-contextualised intervention:

3) AI4SG designers should build decision-making systems in consultation with users interacting with, and impacted, by these systems; with understanding of users' characteristics, of the methods of coordination, and the purposes and effects of an intervention; and with respect for users' right to ignore or modify interventions.

9.3.4 Receiver-Contextualised Explanation and Transparent Purposes

AI4SG applications should be designed to make explainable the operations and outcomes of these systems and to make transparent their purposes. These two requirements are of course intrinsically linked, as the operations and outcomes of AI systems reflect the wider purposes of human designers; in this section, we address both in turn.

Making AI systems explainable is an important ethical principle (Floridi et al. 2018). It has been a focus of research since at least 1975 (Shortliffe and Buchanan 1975). And it has gained more attention recently (Thelisson et al. 2017; Wachter et al. 2016) given the increasingly pervasive distribution of AI systems. As we saw above, AI4SG projects should offer interventions that are contextualised to the receiver. In addition, the explanation for an intervention should be contextualised in order to be adequate, and protect the autonomy of the receiver.

Designers of AI4SG projects have tried to increase the explainability of decision-making systems in various ways. For example, researchers have used machine learning to predict academic adversity (Lakkaraju et al. 2015). These predictors used concepts that the school officials interpreting the system found familiar and salient, such as GPA scores and socio-economic categorisations. Researchers have also used reinforcement-learning to help officials at homeless shelters educate homeless youths about HIV (Yadav et al. 2016a, b). The system learns how to maximise the influence of HIV education, by choosing which homeless youths to educate, on the basis that homeless youths may pass on their knowledge. One version of the system explained which youth was chosen by revealing their social network graph. However, the homeless shelter officials found that these explanations were counter-intuitive, potentially affecting the understanding of how the system

[7]The four remaining dimensions proposed by MacFarlane—the source of the interruption, the method of expression, the channel of conveyance and the human activity changed by the interruption—are not relevant for purpose of this article.

worked and, hence, users' trust in the system. These two cases exemplify the importance of the right conceptualisation when explaining an AI-based decision.

The right conceptualisation is likely to vary between AI4SG projects, because they differ greatly in their objectives, subject matter, context and stakeholders. The conceptual framework, that is, the Level of Abstraction (Floridi 2017) depends on what is being explained and to whom. A LoA is a key component of a theory, and hence of any explanation. A theory comprises five components:

1. a System, which is the referent or object analysed by a theory;
2. a Purpose, which is the "what for" that motivates the analysis of a system (note that this answers the question "what is the analysis for?" and should not be confused with a system's purpose, which answers the question "what is the system for?". Below, we use the term "goal" for system's purpose whenever there may be a risk of confusion);
3. a Level of Abstraction, which provides a lens through which a system is analysed, and generates;
4. a Model, that is, some relevant and reliable information about the analysed system, which identifies;
5. a Structure of the system, which comprises the features that belong to the system being analysed.

There is an interdependency between the choice of the specific purpose, the relevant LoA that can fulfil the purpose, the system analysed, and the model obtained by analysing the system at a specified LoA for a particular purpose. The LoA provides the conceptualisation of the system (e.g., GPA scores, and socio-economic backgrounds). But the purpose constrains the construction of LoAs. For example, if we choose to explain the decision making system itself (e.g., the use of particular machine learning techniques), then the LoA can only conceptualise those AI techniques. In turn, the LoA generates the model, which explains the system. The model identifies system structures, such as a specific student's GPA score, poor attendance rate, and their socioeconomic background being predictors of their academic failure. Consequently, designers must choose carefully the purpose and the corresponding LoA, so that the explanation model can provide the right explanation of the system in question for a given receiver.

A LoA is chosen for a specific purpose: for example, a LoA chosen to explain a decision taken on the basis of outcomes obtained through an algorithmic procedure varies depending on whether the explanation is meant for the receiver of that decision or for an engineer responsible for the design of the algorithmic procedure. This is because, depending on the purpose and its granularity (e.g. a customer-friendly vs. engineer-friendly explanation), not every LoA is appropriate for a given receiver. Sometimes, a receiver's conceptual view of the world may differ from the one on which the explanation is based. In other cases, a receiver and an explanation may be conceptually aligned, but the receiver may not agree on the level of granularity (LoA) of the information (what we called more precisely the model) provided. Conceptual disalignment means that the receiver may find the explanation irrelevant, unintelligible or, as we shall see below, questionable. In respect of (un)intelligibility, a LoA may use unknown labels (so-called observables), or labels that have different meanings for different users.

Empirical studies (Gregor and Benbasat 1999) suggest that the suitability of an explanation differs among receivers according to their expertise. Receivers may require explanations about how the AI software came to a decision, especially when they must take action based on that decision (Gregor and Benbasat 1999; Watson et al. 2019). How the AI system came to a conclusion can be just as important as the justification for that conclusion. Consequently, designers must also contextualise the method of explanation to the receiver.

The case of the software that uses influence-maximisation algorithms to target homeless youths for HIV education provides a good example of the relevance of the receiver-contextualisation of concepts (Yadav et al. 2016a, b). The researchers involved in this project considered three possible LoAs when designing the explanation model: the first LoA included utility calculations; the second LoA focused on social graph connectivity; and a third LoA focusing on pedagogic purpose. The first LoA highlighted the utility of targeting one homeless youth over another. According to the researchers, in this case, homeless shelter workers (the receivers) might have misunderstood the utility calculations or found them irrelevant. Utility calculations offer little explanatory power beyond the decision itself, because they often simply show that the "best" choice was made, and how good it was. Explanations based on the second LoA faced a different problem: the receivers assumed that the most central nodes in the network were the best for maximising the influence of education, while the optimal choice is often a set of less well-connected nodes. This disjuncture may have arisen from the nature of the connectivity between members of the network of homeless youths, which reflects real-life uncertainty about friendships. Since who counts as a "friend" is often vague and changeable over time, the researchers classified edges in the network as either "certain" or "uncertain" based on domain knowledge. For "uncertain" relationships, the probability of a friendship existing between two youths was determined by domain experts.[8] The third LoA was eventually chosen, after subsequent user testing of different explanation frameworks (Yadav et al. 2016a, b). In light of their stated goal to justify decisions in a way that would be intuitive to homeless shelter officials, the researchers considered omitting references to the Maximum Expected Utility (MEU) calculations, even though this is what actually underlies the decisions made by the system. Instead, the researchers considered justifying decisions using concepts with which officials would be more comfortable and familiar, such as the centrality of the nodes (i.e., the youths) that the system recommends officials prioritise for intervention. In this way, the researchers sought to provide the most relevant information contextualised to the receiver.

[8] Note that the significance of involving domain experts in the process was not merely to improve their experience as decision recipients, but also for their unparalleled knowledge of the domain that the researchers drew upon in the system design, helping to provide the researchers with what Pagallo (2015) calls "preventive understanding" of the field.

As this example shows, given a particular system, the purpose one chooses to pursue when seeking an explanation of it, at what LoA, and the issuing model that is obtained are crucial variables that impact the effectiveness of an explanation. Explainability breeds trust in, and fosters adoption of, AI4SG solutions (Herlocker et al. 2000; Swearingen and Sinha 2002; Bilgic and Mooney 2005). This is why it is essential that software uses persuasive argumentation for the target audience. This is likely to include information about both the general functionality and logic employed by a system and the reasons for the specific decision being made (Wachter et al. 2017).

Transparency in the goal (i.e., system's purpose) of the system is also crucial, for it follows directly from the principle of autonomy. Consider, for example, the development of AI solutions to prompt people with cognitive disabilities to take their medication (Chu et al. 2012). On its face, this application may seem invasive, involving vulnerable users, limiting the effectiveness of receiver-conceptualised explanation. However, the system is not designed to coerce the patients into a given behaviour, nor is it designed to resemble a human being. The patients have autonomy not to interact with the AI system in question. This case highlights the importance of transparency in goals, particularly in contexts in which explainable operations and outcomes are unworkable or undesirable. Transparency in goals, thus, undergirds other safeguards around the protection of target populations and may help ensure compliance with relevant legislation and precedent (Reed 2018).

Conversely, opaque goals may prompt misunderstanding and the potential for harms. For instance, when users of an AI system are unclear about what type of agent they are dealing with—human, artificial, or a hybrid combination of both—they may wrongly assume that the tacit norms of human-to-human social interaction are upheld (e.g., not recording every detail of a conversation) (R. Kerr 2003). As ever, the social context in which an AI4SG application takes place impacts the extent to which AI systems should be transparent in their operations. Because transparency is the default but not absolute position, there may be valid reasons for designers to obviate informing users of the software's goals. For example, the scientific value of a project or the health and safety conditions of a public space may justify temporarily opaque goals. Consider a study that deceived students into believing that they were interacting with a human course-assistant that was in fact, over time, realised to be a bot (Eicher et al. 2017). The bot's deception, as the authors argue, was for playing the "imitation game" without causing the students to choose simpler and less human-like natural-language queries based on preconceptions of AI capabilities. In such cases, the choice between opacity and transparency may be informed by preexisting notions of informed consent for human-subject experiments embedded in the Nuremberg Code, the Declaration of Helsinki, and the Belmont Report (Nijhawan et al. 2013).

More broadly, the ability to avoid the use of an AI system becomes more likely when AI software reveals its endogenous goals, like classifying data about a person. For example, AI software could inform staff in a hospital ward that it has the goal of classifying their hygiene levels (Haque et al. 2017). In this case, the staff may decide to avoid such classifications if there are reasonable alternative actions that they can take. In other cases, revealing a goal makes it less likely to be fulfilled.

Making transparent the goals and motivations of AI4SG developers themselves is an essential factor to the success of any project, but one that may contrast the very purpose of the system. This is why it is crucial to assess, at a design stage, what level of transparency (i.e. how much transparency, of what kind, for whom, and about what?) the project will embrace given its overall goal and the context of implementation. Taken together with the need for receiver-conceptualised explanation, this consideration yields the following set of best practices:

4) AI4SG designers should choose a Level of Abstraction for AI explanation that fulfils the desired explanatory purpose and is appropriate to the system and the receivers; then deploy arguments that are rationally and suitably persuasive for the receivers to deliver the explanation; and ensure that the goal (the system's purpose) for which an AI4SG system is developed and deployed is knowable to receivers of its outputs by default.

9.3.5 *Privacy Protection and Data Subject Consent*

Of our seven factors, privacy has perhaps the most voluminous literature. This should not be a surprise, since privacy is considered to be an essential condition for safety, human dignity, and social cohesion, among other things (Solove 2008), and because earlier waves of digital technology have already had a major impact on privacy (Nissenbaum 2009). People's safety may be compromised when a malicious actor or state gain control over individuals via privacy infringements (Taddeo 2015; Lynskey 2015). Respect for privacy is also a necessary condition of human dignity, since we can view personal information as constituting an individual, and deprivatising records without consent is likely to constitute a violation of human dignity (Floridi 2016). The conception of individual privacy as a fundamental right underlies recent legislative action in, for example, Europe (through its General Data Protection Regulation) and Japan (through its Act on Protection of Personal Information), as well as judicial decisions in jurisdictions such as India (Mohanty and Bhatia 2017). Privacy supports people in deviating from social norms without causing offense, and communities in maintaining their social structures, so privacy also undergirds social cohesion.

In the case of AI4SG, it is particularly important to emphasise the relevance of users' consent to the use of personal data. Tensions may arise between different thresholds of consent (Price and Cohen 2019). The tension is often at its most fraught in "life-or-death" situations such national emergencies and pandemics. Consider the outbreak of Ebola in West Africa in 2014, which posed a complex ethical dilemma (The Economist 2014). In this case, the rapid release and analysis of call-data records from cell phone users in the region may have allowed epidemiologists to track the spread of the deadly disease. However, the release of the data was held up over valid concerns around users' privacy, as well as the value of the data to industrial competitors.

In circumstances where haste is not so crucial, it is possible to obtain a subject's consent for—and before—the data being used. The level or type of consent sought can vary with the context. In healthcare, one may adopt an assumed consent threshold, whereby reporting a medical issue to a doctor constitutes assumed consent on the part of a patient. In other circumstances, an informed consent threshold will be more appropriate. Yet, since informed consent requires researchers to obtain a patient's specific consent before using their data for a non-consented purpose, practitioners may choose an explicit consent threshold to general data processing, i.e., for any medical usage. This threshold does not require informing the patient about all of the possible ways that researchers may use their data (Etzioni 1999). Another alternative is the evolving notion of "dynamic consent", whereby individuals can monitor and adjust their privacy preferences on a granular level (Kaye et al. 2015).

In other cases, informed consent may be waived altogether. This was the case with the recent creation of machine learning software to predict the prognosis of ovarian cancer sufferers by drawing upon retrospective analysis of anonymised images (Lu et al. 2019). The use of patient health data in the development of AI solutions without patients' consent has also attracted the attention of data protection regulators. In 2017, the UK's Information Commissioner ruled that the Royal Free NHS Foundation Trust violated the Data Protection Act when it provided patient details to Google DeepMind, for the purposes of training an AI system to diagnose acute kidney injury (Burgess 2017). The Commissioner noted as a "shortcoming" that "patients were not adequately informed that their data would be used as part of the test" ("Royal Free—Google DeepMind Trial Failed to Comply with Data Protection Law" 2017).

Striking a balance between respecting patient privacy and creating effective AI4SG is still possible, however. This was the challenge faced by the researchers in Haque et al. (2017), who wanted to create a system for tracking compliance with rules around hand hygiene in hospitals, to prevent the spread of infections. Despite the clear technical advantages of taking a computer vision-based approach to the problem, the use of video recording runs up against privacy regulations constraining it. Even in cases where video recording is allowed, access to the recordings (in order to train an algorithm) is often strict. Instead, the researchers resorted to "depth images", which de-identify subjects, preserving their privacy. While this design choice meant "losing important visual appearance cues in the process", it satisfied privacy rules, and the researchers' non-intrusive system still managed to outperform existing solutions.

Finally, consent in the online space is also problematic; users often lack the choice and are presented with a 'take it or leave it' option when accessing online services (Nissenbaum 2011; Taddeo and Floridi 2015). The relative lack of protection or consent for the second-hand use of personal data that is publicly shared online enables the development of ethically problematic AI software. For example, a recent paper used publicly available images of faces uploaded to a dating website as a way to train AI software to detect someone's sexuality based on a small number of photos (Wang and Kosinski 2018). While the study received ethics committee approval, it raises further questions around consent, since it is implausible that the users of the dating website could or necessarily would have consented to the use of their data for this particular purpose.

Privacy is not a novel problem, but the centrality of personal data to many AI (and AI4SG) applications heightens its ethical significance and creates issues around consent (Taddeo and Floridi 2018a). From this we can derive the following best practice:

5) AI4SG designers should respect the threshold of consent established for the processing of datasets of personal data.

9.3.6 *Situational Fairness*

AI developers typically rely on data, which may be biased in ways that are socially significant. This bias may carry across to the algorithmic decision-making that underpins many AI systems, in ways that are unfair to the subjects of the decision-making process (Caliskan et al. 2017) and, thus, may breach the principle of justice. These decisions may be based on factors of ethical importance (e.g., ethnic, gender, or religious grounds) and irrelevant to the decision-making at hand, or they may be relevant but legally protected as a nondiscriminatory characteristic (Friedman and Nissenbaum 1996). Moreover, AI-driven decisions may be amalgamated from factors that are not of obvious ethical importance, and yet collectively constitute unfairly biased decision-making (Pedreshi et al. 2008; Floridi 2012).

AI4SG initiatives relying on biased data may propagate this bias through a vicious cycle (Yang et al. 2018). Such a cycle would begin with a biased dataset informing a first phase of AI decision-making, resulting in discriminatory actions, leading to the collection and use of biased data in turn. Consider the use of AI to predict preterm birth in the United States, where the health outcomes of pregnant women have long been affected by their ethnicity. Longstanding bias against African-American women seeking treatment, owing to harmful historical stereotypes, contributes to a maternal morbidity rate that is over three times higher than that of white women (CDC 2019). Here, AI may offer great potential to reduce this stark racial divide, but only if the same historical discrimination is not replicated in AI systems (Banjo 2018). Or consider the use of predictive policing software. Developers may train predictive policing software on policing data that contains deeply ingrained prejudices. When discrimination affects arrest rates, it becomes embedded in prosecution data (Lum and Isaac 2016). Such biases may cause discriminatory decisions (e.g., warnings or arrests) that feed back into the increasingly biased datasets (Crawford 2016), thereby completing a vicious cycle.

These examples involve the use of AI to improve outcomes in domains where data were already collected. Yet, in many other contexts, AI4SG projects (or indeed similar initiatives) are, in effect, making citizens "visible" in ways that they previously were not, including in global South contexts (Taylor and Broeders 2015). This increased visibility stresses the importance of protecting against the potential amplification of harmful bias by AI technologies.

Clearly, designers must sanitise the datasets used to train AI. However, there is equally a risk of applying too strong a disinfectant, so to speak, by removing important contextual nuances which could improve ethical decision-making. So, designers must also ensure that AI decision-making maintains sensitivity to factors that are important for inclusiveness. For instance, we should ensure that a word processor interacts identically with a human user regardless of that user's gender and ethnicity, but also expect that it may operate in a non-equal and yet equitable way by aiding people with visual impairments.

Such expectations are not always met in the context of AI-driven reasoning. Compared to the word processor, AI makes possible a far wider range of decision-making and interaction modalities, many of which are driven by potentially biased data. Training datasets may contain natural language that carries unfair associations between genders and words which, in turn, carry normative power (Caliskan et al. 2017). In other contexts and use cases, an equitable approach may require differences in communication, based on factors such as gender. Consider the case of the virtual teaching assistant which failed to discriminate sufficiently well between men and women in its responses to being told that a user was expecting a baby, congratulating the men and ignoring the women (Eicher et al. 2017). A BBC News investigation highlighted an even more egregious example: a mental health chatbot deemed suitable for use by children was unable to understand a child explicitly reporting underage sexual abuse (White 2018). As these cases make clear, the use of AI in human-computer interactions, such as chatbots, requires the correct understanding of both the salient groups to which a user belongs and the characteristics they embody when they interact with the software.

Respecting situational fairness is essential for the successful implementation of AI4SG. To achieve it, AI4SG projects need to remove factors (and their proxies) that are of ethical importance but irrelevant to an outcome, and include the same factors when these are required, whether for the sake of inclusiveness, safety, or other ethical considerations. The problem of historical biases affecting future decision-making is an old one. What is new is the potential that these biases will be embedded in, strengthened, and perpetuated anew by erroneous reinforcement learning mechanisms. This risk is especially pronounced when considered alongside the risk of opacity in AI decision-making systems and their outcomes. We will return to this topic in the next section.

From our identification of situational fairness as an essential factor, we can yield the following best practice:

6) AI4SG designers should remove from relevant datasets variables and proxies that are irrelevant to an outcome, except when their inclusion supports inclusivity, safety, or other ethical imperatives.

9.3.7 Human-Friendly Semanticisation

AI4SG must allow humans to curate and foster their "semantic capital", that is,

> any content that can enhance someone's power to give meaning to and make sense of (semanticise) something (Floridi 2018).

This is crucial to maintain and foster human autonomy. With AI, we may often have the technical capacity to automate meaning- and sense-creation (semanticisation), but mistrust or unfairness may also arise if we do so carelessly. Two problems emerge. The first problem is that AI software may define semanticisation in a way that diverges from our own choices. This is the case if a procedure arbitrarily defines meanings (e.g., based on a coin toss). The same problem may arise if AI software support some kind of semanticisation based on preexisting uses. For example, researchers have developed an application that predicts the legal meaning of 'violation' based on past cases (Al-Abdulkarim et al. 2015). If one used the software to define the meaning of 'violation',[9] then one would end up limiting the role of judges and justices. They would no longer be able to semanticise (refine and re-define the meaning, and the possibility of making sense of) "violation", when they interpret the law. This is a problem, because past usage does not always predict how we would semanticise the same concepts or phenomena in the future.

The second problem is that, in a social setting, it would be impractical for AI software to define all meanings and senses. Some semanticisation is subjective, because who or what is involved in the semanticisation is also partly constitutive of the process and its outcome. For example, only legally empowered agents can define the legal meaning of 'violation'. Likewise, the meaning and sense of affective symbols, such as facial expressions, also depends on the type of agent showing a given expression. Affective AI can detect an emotion (Martínez-Miranda and Aldea 2005), an artificial agent may state accurately that a human appears sad, but cannot change the meaning of sadness.

The solution to these two problems rest on distinguishing between tasks that should and should not be delegated to an artificial system. AI should be deployed to facilitate human-friendly semantisation, but not to provide it itself. This is true, for example, when considering patients with Alzheimer's disease. Research into carer-patient relations highlights three points (Burns and Rabins 2000). First, carers play a critical, but burdensome, role in reminding patients of the activities in which they participate, e.g., taking medication. Second, carers also play a critical role in providing patients with meaningful interaction. And third, when carers remind patients to take their medication, the patient-carer relation may become weaker by annoying the patient, with the carer losing some capacity to provide empathy and meaningful support. Consequently, researchers have developed AI software that balances reminding the patient against annoying the patient (Chu et al. 2012). The

[9]There is no suggestion that this is the intended use.

balance is learned and optimised using reinforcement learning. The researchers designed the system so that caregivers can spend most of their time providing empathic support and preserving a meaningful relationship with the patient. As this example shows, it is possible to use AI to sweep away formulaic tasks whilst sustaining human-friendly semanticisation.

Human-centric semanticisation, as an essential factor for AI4SG, underpins our final best practice:

7) AI4SG designers should not hinder the ability for people to semanticise (that is, to give meaning to, and make sense of) something.

9.4 Conclusion: Balancing Factors for AI for Social Good

The seven factors analysed in the previous pages are summarised in Table 9.1, together with the corresponding best practices, and the five principle(s) of AI ethics identified in (Floridi and Cowls 2019) to which each factor is most closely identified. To reiterate, the principle of beneficence is assumed as a precondition for an AI4SG, so the factors relate to one or more of the other four principles: nonmaleficence, autonomy, justice and explicability.

The seven factors suggest that creating successful AI4SG requires two kinds of balances to be struck: intra and inter.

On the one hand, each single factor in and of itself may require an intrinsic balance, for example, between the risk of over-intervening and the risk of under-intervening when devising contextual interventions; or between protection-by-obfuscation and protection-by-enumeration of salient differences between people, depending on the purposes and context of a system. On the other hand, balances are not just specific to a single factor; they are also systemic, because they must also be struck between multiple factors. Consider the tension between preventing malicious actors from understanding how to "game" the input data of AI prediction systems versus enabling humans to override genuinely flawed outcomes; or the tension between ensuring the effective disclosure of the reasons behind a decision without compromising the consensual anonymity of data subjects.

The overarching question facing the AI4SG community is, for each given case, whether one is morally obliged to, or obliged not to, design, develop, and deploy a specific AI4SG project. This article does not seek to answer such a question in the abstract. Resolving the tensions that are likely to arise among and between factors is highly context-dependent, and the previous analysis is not meant to cover all potential contexts, not least because this would be inconsistent with the argument for falsifiable hypothesis testing and incremental deployment supported in this article; nor would a checklist of purely technical "dos and don'ts" suffice. Rather, our analysis has yielded a set of essential factors that need to be considered, interpreted and evaluated contextually when one is designing, developing, and deploying a specific AI4SG project. The future of AI4SG will likely provide more

Table 9.1 Summary of seven factors supporting AI4SG and the corresponding best practices

Factors	Corresponding best practices	Corresponding ethical principle
Falsifiability and incremental deployment	Identify falsifiable requirements and test them in incremental steps from the lab to the "outside world".	Nonmaleficence
Safeguards against the manipulation of predictors	Adopt safeguards which (i) ensure that non-causal indicators do not inappropriately skew interventions, and (ii) limit, when appropriate, knowledge of how inputs affect outputs from AI4SG systems, to prevent manipulation.	Nonmaleficence
Receiver-contextualised intervention	Build decision-making systems in consultation with users interacting with and impacted by these systems; with understanding of users' characteristics, the methods of coordination, the purposes and effects of an intervention; and with respect for users' right to ignore or modify interventions.	Autonomy
Receiver-contextualised explanation and transparent purposes	Choose a Level of Abstraction for AI explanation that fulfils the desired explanatory purpose and is appropriate to the system and the receivers; then deploy arguments that are rationally and suitably persuasive for the receiver to deliver the explanation; and ensure that the goal (the system's purpose) for which an AI4SG system is developed and deployed is knowable to receivers of its outputs by default.	Explicability
Privacy protection and data subject consent	Respect the threshold of consent established for the processing of datasets of personal data.	Nonmaleficence; autonomy
Situational fairness	Remove from relevant datasets variables and proxies that are irrelevant to an outcome, except when their inclusion supports inclusivity, safety, or other ethical imperatives.	Justice
Human-friendly semanticisation	Do not hinder the ability for people to semanticise (that is, to give meaning to, and make sense of) something.	Autonomy

opportunities to enrich such a set of essential factors. AI itself may help to manage its own life cycle by providing, in a meta-reflective way, tools to evaluate how best to strike the individual and systemic balances indicated above.

The most pertinent questions to arise from the factors described in this article are likely to concern this challenge of balancing the competing needs and claims that the factors and corresponding best practices introduce. This concerns what it is that legitimates decision-making with and about AI. While we leave this concern primarily to future research, we offer some remarks on it in closing. Questions such as this are inevitably intertwined with wider ethical and political challenges regarding who has the power or "standing" to participate in this process of evaluation, as well as how multiple preferences are measured and aggregated, as Baum's trichotomic framework outlines (Baum 2017). If we assume that the challenge of balancing factors ought to be at least somewhat participatory in nature, Prasad's (2018) overview of relevant social choice theorems identifies several background conditions to support efficacious group decision-making. As these analyses suggest, the incorporation of multiple perspectives into the design of AI decision-making systems is likely to be an ethically important step both for AI in general, and AI4SG in particular.

There is much work still to be done to ensure that AI4SG projects are designed in ways that not merely advance beneficial goals and address societal challenges, but that do so in socially preferable and sustainable ways. This article seeks to contribute to lay the ground for good practices and policies in this respect, as well as for further research on the ethical considerations that should undergird AI4SG projects, and hence the "AI4SG project" at large.

Funding Floridi's and Taddeo's work was supported by Privacy and Trust Stream—Social lead of the PETRAS Internet of Things research hub—PETRAS is funded by the Engineering and Physical Sciences Research Council (EPSRC), grant agreement no. EP/N023013/1—and by the Oxford Initiative on AI for SDG, which is also supported by grants from Facebook, Google, and Microsoft. Cowls is the recipient of a Doctoral Studentship from the Alan Turing Institute. King's work was supported by a grant by Google UK Limited.

Appendix: Representative AI4SG Examples

In the table below, we list the seven initiatives from our wider sample that are especially representative in terms of scope, variety, impact, and for their potentiality to evince the factors that should characterise the design of AI4SG projects. This includes the factor(s) that were identified as a result of our analysis of each project.

	Name	References	Areas	Relevant factor(s)
A	Field Optimization of the Protection Assistant for Wildlife Security.	Fang et al. (2016)	Environmental sustainability	1), 3)
B	Identifying Students at Risk of Adverse Academic Outcomes	Lakkaraju et al. (2015)	Education	4)
C	Health Information for Homeless Youth to Reduce the Spread of HIV	Yadav et al. (2016a, b, 2018)	Poverty, public welfare, public health	4)
D	Interactive activity recognition and prompting to assist people with cognitive disabilities	Chu et al. (2012)	Disability, public health	3), 4), 7)
E	Virtual teaching assistant experiment	Eicher et al. (2017)	Education	4), 6)

(continued)

	Name	References	Areas	Relevant factor(s)
F	Detecting evolutionary financial statement fraud	Zhou and Kapoor (2011)	Finance, crime	2)
G	Tracking and monitoring hand hygiene compliance	Haque et al. (2017)	Health	5)

References

"AI for Good Global Summit—28–31 May 2019, Geneva, Switzerland". n.d. *AI for good global summit*. https://aiforgood.itu.int/. Accessed 12 Apr 2019.

Al-Abdulkarim, Latifa, Katie Atkinson, and Trevor Bench-Capon. 2015. *Factors, issues and values: Revisiting reasoning with cases*. In Proceedings of the 15th International Conference on Artificial Intelligence and Law, 3–12. ICAIL '15. New York, NY, USA: ACM. https://doi.org/10.1145/2746090.2746103.

Banjo, Omotayo. 2018. Bias in maternal AI could hurt expectant Black mothers. *Medium (blog)*. September 21, 2018. https://medium.com/theplug/bias-in-maternal-ai-could-hurt-expectant-black-mothers-e41893438da6.

Baum, Seth D. 2017. Social choice ethics in artificial intelligence. *AI & SOCIETY*: 1–12.

Bilgic, Mustafa, and Raymond Mooney. 2005. *Explaining recommendations: Satisfaction vs. promotion*.

Boutilier, Craig. 2002. A POMDP formulation of preference elicitation problems. In Proceedings of the National Conference on Artificial Intelligence, May.

Burgess, Matt. 2017. NHS DeepMind deal broke data protection law, regulator rules. *Wired UK*, July 3, 2017. https://www.wired.co.uk/article/google-deepmind-nhs-royal-free-ico-ruling.

Burns, Alistair, and Peter Rabins. 2000. Carer burden in dementia. *International Journal of Geriatric Psychiatry* 15 (S1): S9–S13.

Caliskan, Aylin, Joanna J. Bryson, and Arvind Narayanan. 2017. Semantics derived automatically from language corpora contain human-like biases. *Science* 356 (6334): 183–186. https://doi.org/10.1126/science.aal4230.

Carton, Samuel, Jennifer Helsby, Kenneth Joseph, Ayesha Mahmud, Youngsoo Park, Joe Walsh, Crystal Cody, CPT Estella Patterson, Lauren Haynes, and Rayid Ghani. 2016. Identifying police officers at risk of adverse events. In Proceedings of the 22nd ACM SIGKDD International Conference on Knowledge Discovery and Data Mining, 67–76. KDD '16. New York, NY, USA: ACM. https://doi.org/10.1145/2939672.2939698.

Center for Disease Control (CDC). 2019. Pregnancy Mortality Surveillance System | Maternal and Infant Health. January 16, 2019. https://www.cdc.gov/reproductivehealth/maternalinfanthealth/pregnancy-mortality-surveillance-system.htm.

Chajewska, Urszula, Daphne Koller, and Ronald Parr. 2000. Making rational decisions using adaptive utility elicitation. *AAAI/IAAI*: 363–369.

Chu, Yi, Young Chol Song, Richard Levinson, and Henry Kautz. 2012. Interactive activity recognition and prompting to assist people with cognitive disabilities. *Journal of Ambient Intelligence and Smart Environments* 4 (5): 443–459. https://doi.org/10.3233/AIS-2012-0168.

Crawford, Kate. 2016. Artificial intelligence's White guy problem. *The New York Times*. June 25, 2016. https://www.nytimes.com/2016/06/26/opinion/sunday/artificial-intelligences-white-guy-problem.html.

Dennis, Louise, Michael Fisher, Marija Slavkovik, and Matt Webster. 2016. Formal verification of ethical choices in autonomous systems. *Robotics and Autonomous Systems* 77 (March): 1–14. https://doi.org/10.1016/j.robot.2015.11.012.

Eicher, Bobbie, Lalith Polepeddi, and Ashok Goel. 2017. Jill Watson doesn't care if you're pregnant: grounding AI ethics in empirical studies. In AAAI/ACM Conference on Artificial Intelligence, Ethics, and Society, New Orleans, LA. Vol. 7.

Etzioni, Amitai. 1999. Enhancing privacy, preserving the common good. *Hastings Center Report* 29 (2): 14–23.

Faltings, Boi, Pearl Pu, Marc Torrens, and Paolo Viappiani. 2004. Designing example-critiquing interaction. In Proceedings of the 9th International Conference on Intelligent User Interfaces, 22–29. IUI '04. New York, NY, USA: ACM. https://doi.org/10.1145/964442.964449.

Fang, Fei, Thanh H. Nguyen, Rob Pickles, Wai Y. Lam, Gopalasamy R. Clements, Bo An, Amandeep Singh, Milind Tambe, and Andrew Lemieux. 2016. Deploying PAWS: Field optimization of the protection assistant for wildlife security. In Twenty-Eighth IAAI Conference. https://www.aaai.org/ocs/index.php/IAAI/IAAI16/paper/view/11814.

Floridi, Luciano. 2012. Distributed morality in an information society. *Science and Engineering Ethics* 19 (3): 727–743. https://doi.org/10.1007/s11948-012-9413-4.

———. 2016. On human dignity as a foundation for the right to privacy. *Philosophy & Technology* 29 (4): 307–312. https://doi.org/10.1007/s13347-016-0220-8.

———. 2017. The logic of design as a conceptual logic of information. *Minds Mach.* 27 (3): 495–519. https://doi.org/10.1007/s11023-017-9438-1.

———. 2018. Semantic capital: Its nature, value, and curation. *Philos Technol* 31: 481–497. https://doi.org/10.1007/s13347-018-0335-1

Floridi, Luciano, and Josh Cowls. 2019. A unified framework of five principles for AI in society. *Harvard Data Science Review* 1 (1). https://doi.org/10.1162/99608f92.8cd550d1.

Floridi, Luciano, Josh Cowls, Monica Beltrametti, Raja Chatila, Patrice Chazerand, Virginia Dignum, Christoph Luetge, Robert Madelin, Ugo Pagallo, and Francesca Rossi. 2018. AI4People—An ethical framework for a good AI society: Opportunities, risks, principles, and recommendations. *Minds and Machines* 28 (4): 689–707.

Friedman, Batya, and Helen Nissenbaum. 1996. Bias in computer systems. *ACM Transactions on Information Systems* 14: 330–347. https://doi.org/10.1145/230538.230561.

Ghani, Rayid. 2016. You Say you want transparency and interpretability? *Rayid Ghani* (blog). April 29, 2016. http://www.rayidghani.com/you-say-you-want-transparency-and-interpretability.

Goel, Ashok, Brian Creeden, Mithun Kumble, Shanu Salunke, Abhinaya Shetty, and Bryan Wiltgen. 2015. *Using Watson for enhancing human-computer co-creativity*. In 2015 AAAI Fall Symposium Series.

Goodhart, Charles. 1975. *Problems of monetary management: The U.K. experience*. Papers in monetary economics. Sydney? Reserve Bank of Australia.

Gregor, Shirley, and Izak Benbasat. 1999. Explanations from intelligent systems: Theoretical foundations and implications for practice. *MIS Quarterly* 23 (December): 497–530. https://doi.org/10.2307/249487.

Hager, Gregory D., Ann Drobnis, Fei Fang, Rayid Ghani, Amy Greenwald, Terah Lyons, David C. Parkes, et al. 2017. Artificial intelligence for social good, 24–24.

Haque, Albert, Michelle Guo, Alexandre Alahi, Serena Yeung, Zelun Luo, Alisha Rege, Jeffrey Jopling, et al. 2017. *Towards vision-based smart hospitals: A system for tracking and monitoring hand hygiene compliance*. August. https://arxiv.org/abs/1708.00163v3.

Henry, Katharine E., David N. Hager, Peter J. Pronovost, and Suchi Saria. 2015. A Targeted Real-Time Early Warning Score (TREWScore) for septic shock. *Science Translational Medicine* 7 (299): 299ra122–299ra122. https://doi.org/10.1126/scitranslmed.aab3719.

Herlocker, Jonathan L., Joseph A. Konstan, and John Riedl. 2000. *Explaining collaborative filtering recommendations*. In Proceedings of the 2000 ACM Conference on Computer Supported Cooperative Work, 241–250. ACM.

Kaye, Jane, Edgar A. Whitley, David Lund, Michael Morrison, Harriet Teare, and Karen Melham. 2015. Dynamic consent: A patient interface for twenty-first century research networks. *European Journal of Human Genetics* 23 (2): 141–146. https://doi.org/10.1038/ejhg.2014.71.

Kerr, Ian R. 2003. Bots, babes and the Californication of commerce. *University of Ottawa Law and Technology Journal* 1 (January).

King, Thomas C., Nikita Aggarwal, Mariarosaria Taddeo, and Luciano Floridi. 2019. Artificial intelligence crime: An interdisciplinary analysis of foreseeable threats and solutions. *Science and Engineering Ethics*. https://doi.org/10.1007/s11948-018-00081-0.

Lakkaraju, Himabindu, Everaldo Aguiar, Carl Shan, David Miller, Nasir Bhanpuri, Rayid Ghani, and Kecia L. Addison. 2015. *A machine learning framework to identify students at risk of adverse academic outcomes*. In Proceedings of the 21th ACM SIGKDD International Conference on Knowledge Discovery and Data Mining, 1909–1918. ACM.

Lu, Haonan, Mubarik Arshad, Andrew Thornton, Giacomo Avesani, Paula Cunnea, Ed Curry, Fahdi Kanavati, et al. 2019. A mathematical-descriptor of tumor-mesoscopic-structure from computed-tomography images annotates prognostic- and molecular-phenotypes of epithelial ovarian cancer. *Nature Communications* 10 (1): 764. https://doi.org/10.1038/s41467-019-08718-9.

Lum, Kristian, and William Isaac. 2016. To predict and serve? *Significance* 13 (5): 14–19. https://doi.org/10.1111/j.1740-9713.2016.00960.x.

Lynskey, Orla. 2015. *The foundations of EU data protection law*, Oxford Studies in European Law. Oxford: Oxford University Press.

Manheim, David, and Scott Garrabrant. 2019. Categorizing variants of Goodhart's law. ArXiv:1803.04585 [Cs, q-Fin, Stat], February. http://arxiv.org/abs/1803.04585.

Martinez-Miranda, Juan, and Arantza Aldea. 2005. Emotions in human and artificial intelligence. *Computers in Human Behavior* 21 (2): 323–341. https://doi.org/10.1016/j.chb.2004.02.010.

McFarlane, Daniel. 1999. *Interruption of people in human-computer interaction: A general unifying definition of human interruption and taxonomy*. August.

McFarlane, Daniel, and Kara Latorella. 2002. The scope and importance of human interruption in human-computer interaction design. *Human-Computer Interaction* 17 (March): 1–61. https://doi.org/10.1207/S15327051HCI1701_1.

Mohanty, Suchitra, and Rahul Bhatia. 2017. Indian Court's privacy ruling is blow to government. *Reuters*, August 25, 2017. https://www.reuters.com/article/us-india-court-privacy-idUSKCN1B40CE.

Moore, Jared. 2019. AI for not bad. *Front. Big Data* 2 (32). https://doi.org/10.3389/fdata.2019.00032.

Neff, Gina, and Peter Nagy. 2016. Talking to bots: Symbiotic agency and the case of Tay. *International Journal of Communication* 10 (October): 4915–4931.

Nijhawan, Lokesh P, Manthan Janodia, Muddu Krishna, Kishore Bhat, Laxminarayana Bairy, Nayanabhirama Udupa, and Prashant Musmade. 2013. Informed consent: Issues and challenges. 4. https://doi.org/10.4103/2231-4040.116779.

Nissenbaum, Helen. 2009. *Privacy in context: technology, policy, and the integrity of social life*. Stanford: Stanford University Press.

———. 2011. A contextual approach to privacy online. *Daedalus* 140 (4): 32–48.

Pagallo, Ugo. 2015. "Good onlife governance: On law, spontaneous orders, and design." In The Onlife Manifesto: Being human in a hyperconnected era Luciano Floridi, 161–177. Cham: Springer. https://doi.org/10.1007/978-3-319-04093-6_18.

———. 2017. *From automation to autonomous systems: A legal phenomenology with problems of accountability*. In Proceedings of the Twenty-Sixth International Joint Conference on Artificial Intelligence (IJCAI-17), 17–23.

Pedreshi, Dino, Salvatore Ruggieri, and Franco Turini. 2008. *Discrimination-aware data mining*, 560–568. ACM. https://doi.org/10.1145/1401890.1401959.

Prasad, Mahendra. 2018. Social choice and the value alignment problem. In *Artificial intelligence safety and security*, 291–314. Chapman and Hall/CRC: New York.

Price, W. Nicholson, and I. Glenn Cohen. 2019. Privacy in the age of medical big data. *Nature Medicine* 25 (1): 37. https://doi.org/10.1038/s41591-018-0272-7.

Reed, Chris. 2018. How should we regulate artificial intelligence? *Philosophical Transactions of the Royal Society A: Mathematical, Physical and Engineering Sciences* 376 (2128): 20170360.

Ross, Casey, and Ike Swetlitz. 2017. IBM pitched Watson as a revolution in cancer care. It's nowhere close. *STAT*. September 5, 2017. https://www.statnews.com/2017/09/05/watson-ibm-cancer/.

"Royal Free—Google DeepMind Trial Failed to Comply with Data Protection Law". 2017. Information Commissioner's Office. July 3, 2017. https://ico.org.uk/about-the-ico/news-and-events/news-and-blogs/2017/07/royal-free-google-deepmind-trial-failed-to-comply-with-data-protection-law/.

Shortliffe, Edward H., and Bruce G. Buchanan. 1975. A model of inexact reasoning in medicine. *Mathematical Biosciences* 23 (3): 351–379. https://doi.org/10.1016/0025-5564(75)90047-4.

Solove, Daniel J. 2008. *Understanding privacy*. Vol. 173. Cambridge: Harvard University Press.

Strathern, Marilyn. 1997. 'Improving ratings': Audit in the British University System. *European Review* 5 (3): 305–321. https://doi.org/10.1002/(SICI)1234-981X(199707)5:3<305::AID-EURO184>3.0.CO;2-4.

Strickland, Eliza. 2019. How IBM Watson overpromised and underdelivered on AI health care. *IEEE Spectrum: Technology, Engineering, and Science News*. February 4, 2019. https://spectrum.ieee.org/biomedical/diagnostics/how-ibm-watson-overpromised-and-underdelivered-on-ai-health-care.

Swearingen, Kirsten, and Rashmi Sinha. 2002. Interaction design for recommender systems. *Designing Interactive Systems* 6: 312–334.

Tabuchi, Hiroko, and David Gelles. 2019. Doomed boeing jets lacked 2 safety features that company sold only as extras. *The New York Times*, April 5, 2019, sec. Business. https://www.nytimes.com/2019/03/21/business/boeing-safety-features-charge.html.

Taddeo, Mariarosaria. 2015. The struggle between liberties and authorities in the information age. *Science and Engineering Ethics* 21 (5): 1125–1138. https://doi.org/10.1007/s11948-014-9586-0.

———. 2017. Trusting digital technologies correctly. *Minds and Machines* 27 (4): 565–568.

Taddeo, Mariarosaria, and Luciano Floridi. 2011. The case for e-trust. *Ethics and Information Technology* 13 (1): 1–3.

———. 2015. The debate on the moral responsibilities of online service providers. *Science and Engineering Ethics*. https://doi.org/10.1007/s11948-015-9734-1.

———. 2018a. How AI can be a force for good. *Science* 361 (6404): 751–752.

———. 2018b. Regulate artificial intelligence to avert cyber arms race. *Nature* 556 (7701): 296. https://doi.org/10.1038/d41586-018-04602-6.

Taylor, Linnet, and Dennis Broeders. 2015. In the name of development: Power, profit and the datafication of the global south. *Geoforum* 64: 229–237.

The Economist. 2014. *Waiting on hold—Ebola and big data*. October 27, 2014. https://www.economist.com/science-and-technology/2014/10/27/waiting-on-hold.

Thelisson, Eva, Kirtan Padh, and L. Elisa Celis. 2017. *Regulatory Mechanisms and algorithms towards trust in AI/ML*. In Proceedings of the IJCAI 2017 Workshop on Explainable Artificial Intelligence (XAI), Melbourne, Australia.

Wachter, Sandra, Brent Mittelstadt, and Luciano Floridi. 2016. *Why a right to explanation of automated decision-making does not exist in the general data protection regulation*." SSRN Scholarly Paper ID 2903469. Rochester: Social Science Research Network. https://papers.ssrn.com/abstract=2903469.

———. 2017. Why a right to explanation of automated decision-making does not exist in the general data protection regulation. *International Data Privacy Law* 7 (2): 76–99.

Wang, Yilun, and Michal Kosinski. 2018. Deep neural networks are more accurate than humans at detecting sexual orientation from facial images. *Journal of Personality and Social Psychology* 114 (2): 246.

Watson, David S., Jenny Krutzinna, Ian N. Bruce, Christopher E.M. Griffiths, Iain B. McInnes, Michael R. Barnes, and Luciano Floridi. 2019. Clinical applications of machine learning algorithms: Beyond the black box. *BMJ* 364 (March): l886. https://doi.org/10.1136/bmj.l886.

White, Geoff. 2018. *Child advice chatbots fail sex abuse test*. December 11, 2018, sec. Technology. https://www.bbc.com/news/technology-46507900.

Yadav, Amulya, Hau Chan, Albert Jiang, Eric Rice, Ece Kamar, Barbara Grosz, and Milind Tambe. 2016a. POMDPs for assisting homeless shelters—Computational and deployment challenges. In *Autonomous agents and multiagent systems*, Lecture Notes in Computer Science, ed. Nardine Osman and Carles Sierra, 67–87. Springer.

Yadav, Amulya, Hau Chan, Albert Xin Jiang, Haifeng Xu, Eric Rice, and Milind Tambe. 2016b. *Using social networks to aid homeless shelters: dynamic influence maximization under uncertainty*. In Proceedings of the 2016 International Conference on Autonomous Agents & Multiagent Systems, 740–748. International Foundation for Autonomous Agents and Multiagent Systems.

Yadav, Amulya, Bryan Wilder, Eric Rice, Robin Petering, Jaih Craddock, Amanda Yoshioka-Maxwell, Mary Hemler, Laura Onasch-Vera, Milind Tambe, and Darlene Woo. 2018. Bridging the gap between theory and practice in influence maximization: Raising awareness about HIV among homeless youth. *IJCAI*: 5399–5403.

Yang, Guang-Zhong, Jim Bellingham, Pierre E. Dupont, Peer Fischer, Luciano Floridi, Robert Full, Neil Jacobstein, et al. 2018. The grand challenges of science robotics. *Science Robotics* 3 (14): eaar7650. https://doi.org/10.1126/scirobotics.aar7650.

Zhou, Wei, and Gaurav Kapoor. 2011. Detecting evolutionary financial statement fraud. *Decision Support Systems, On Quantitative Methods for Detection of Financial Fraud* 50 (3): 570–575. https://doi.org/10.1016/j.dss.2010.08.007.

Chapter 10
From What to How: An Initial Review of Publicly Available AI Ethics Tools, Methods and Research to Translate Principles into Practices

Jessica Morley, Luciano Floridi, Libby Kinsey, and Anat Elhalal

Abstract The debate about the ethical implications of Artificial Intelligence dates from the 1960s (Samuel in Science, 132(3429):741–742, 1960. https://doi.org/10.1126/science.132.3429.741; Wiener in Cybernetics: or control and communication in the animal and the machine, MIT Press, New York, 1961). However, in recent years symbolic AI has been complemented and sometimes replaced by (Deep) Neural Networks and Machine Learning (ML) techniques. This has vastly increased its potential utility and impact on society, with the consequence that the ethical debate has gone mainstream. Such a debate has primarily focused on principles—the 'what' of AI ethics (beneficence, non-maleficence, autonomy, justice and explicability)—rather than on practices, the 'how.' Awareness of the potential issues is increasing at a fast rate, but the AI community's ability to take action to mitigate the associated risks is still at its infancy. Our intention in presenting this research is to contribute to closing the gap between principles and practices by constructing a typology that may help practically-minded developers apply ethics at each stage of the Machine Learning development pipeline, and to signal to researchers where further work is needed. The focus is exclusively on Machine Learning, but it is hoped that the results of this research may be easily applicable to other branches of AI. The article outlines the research method for creating this typology, the initial findings, and provides a summary of future research needs.

J. Morley · L. Floridi (✉)
Oxford Internet Institute, University of Oxford, Oxford, UK
e-mail: Jessica.morley@kellogg.ox.ac.uk; luciano.floridi@oii.ox.ac.uk

L. Kinsey · A. Elhalal
Digital Catapult, London, UK
e-mail: libby.kinsey@digicatapult.org.uk; anat.elhalal@digicatapult.org.uk

L. Floridi (ed.), *Ethics, Governance, and Policies in Artificial Intelligence*,
Philosophical Studies Series 144, https://doi.org/10.1007/978-3-030-81907-1_10

Keywords Artificial intelligence · Applied ethics · Data governance · Digital ethics · Governance · Ethics of AI · Machine learning

10.1 Introduction

As the availability of data on almost every aspect of life, and the sophistication of machine learning (ML) techniques, has increased (Lepri et al. 2018) so have the opportunities for improving both public and private life (Floridi and Taddeo 2016). Society has greater control than it has ever had over outcomes related to: (1) who people can become; (2) what people can do; (3) what people can achieve; and (4) how people can interact with the world (Floridi et al. 2018). However, growing concerns about the ethical challenges posed by the increased use of ML in particular, and Artificial Intelligence (AI) more generally, threaten to put a halt to the advancement of beneficial applications, unless handled properly.

Balancing the tension between supporting innovation, so that society's right to benefit from science is protected (Knoppers and Thorogood 2017), and limiting the potential harms associated with poorly-designed AI (and specifically ML in this context), (summarised in Table 10.1) is challenging. ML algorithms are powerful socio-technical constructs (Ananny and Crawford 2018), which raise concerns that are as much (if not more) about people as they are about code (see Table 10.1) (Crawford and Calo 2016). Enabling the so-called dual advantage of 'ethical ML'—so that the opportunities are capitalised on, whilst the harms are foreseen and minimised or prevented (Floridi et al. 2018)—requires asking difficult questions about design, development, deployment, practices, uses and users, as well as the data that fuel the whole life-cycle of algorithms (Cath et al. 2018). Lessig was right all along: code is both our greatest threat and our greatest promise (Lessig and Lessig 2006).

Rising to the challenge of designing 'ethical ML' is both essential and possible. Indeed those that claim that it is impossible are falling foul of the is-ism fallacy where they confuse the way things are with the way things can be (Lessig and Lessig 2006), or indeed should be. It is possible to design an algorithmically-enhanced society pro-ethically[1] (Floridi 2016b), so that it protects the values, principles, and

---

[1]The difference between ethics by design and pro-ethical design is the following: *ethics by design* can be paternalistic in ways that constrain the choices of agents, because it makes some options less easily available or not at all; instead, *pro-ethical design* still forces agents to make choices, but this time the nudge is less paternalistic because it does not preclude a course of action but requires agents to make up their mind about it. A simple example can clarify the difference. A speed camera is a form of nudging (drivers should respect the speed limits) but it is pro-ethical insofar as it leaves to the drivers the freedom to choose to pay a ticket, for example in case of an emergency. On the contrary, in terms of ethics by design, speed bumps are a different kind of traffic calming measure designed to slow down vehicles and improve safety. They may seem like a good idea, but they involve a physical alteration of the road, which is permanent and leaves no real choice to the driver.

Table 10.1 Ethical concerns related to algorithmic use based on the 'map' created by Mittelstadt et al. (2016)

Ethical concern	Explanation
Inconclusive evidence	Algorithmic conclusions are probabilities and therefore not infallible. This can lead to unjustified actions. For example, an algorithm used to assess credit worthiness could be accurate 99% of the time, but this would still mean that one out of a hundred applicants would be denied credit wrongly
Inscrutable evidence	A lack of interpretability and transparency can lead to algorithmic systems that are hard to control, monitor, and correct. This is the commonly cited 'black-box' issue
Misguided evidence	Conclusions can only be as reliable (but also as neutral) as the data they are based on, and this can lead to bias. For example, Dressel and Farid (2018) found that the COMPAS recidivism algorithm commonly used in pretrial, parole, and sentencing decisions in the United States, is no more accurate or fair than predictions made by people with little or no criminal justice expertise
Unfair outcomes	An action could be found to be discriminatory if it has a disproportionate impact on one group of people. For instance, Selbst (2017) articulates how the adoption of predictive policing tools is leading to more people of colour being arrested, jailed or physically harmed by police
Transformative effects	Algorithmic activities, like profiling, can lead to challenges for autonomy and informational privacy. For example, Polykalas and Prezerakos (2019) examined the level of access required to personal data by more than 1000 apps listed in the 'most popular' free and paid for categories on the Google Play Store. They found that free apps requested significantly more data than paid-for apps, suggested that the business model of these 'free' apps is the exploitation of the personal data
Traceability	It is hard to assign responsibility to algorithmic harms and this can lead to issues with moral responsibility. For example, it may be unclear who (or indeed what) is responsible for autonomous car fatalities. An in depth ethical analysis of this specific issue is provided by Hevelke and Nida-Rümelin (2015)

ethics that society thinks are fundamental (Floridi 2018). This is the message that social scientists, ethicists, philosophers, policymakers, technologists, and civil society have been delivering in a collective call for the development of appropriate governance mechanisms (D'Agostino and Durante 2018) that will enable society to capitalise on the opportunities, whilst ensuring that human rights are respected (Floridi and Taddeo 2016), and fair and ethical decision-making is maintained (Lipton 2016).

The purpose of the following pages is to highlight the part that technologists, or ML developers, can have in this broader conversation, and to highlight where further research is urgently needed. Specifically, section 'Moving from Principles to Practice' discusses how efforts to data have been too focused on the 'what' of ethical AI

This means that emergency vehicles, such as a medical ambulance, a police car, or a fire engine, must also slow down, even when responding to an emergency.

(i.e. debates about principles and codes of conduct) and not enough on the 'how' of applied ethics. The 'Methodology' section outlines the research planned to contribute to closing this gap between principles and practice, through the creation of an 'applied ethical AI typology,' and the methodology for its creation. Section 'Framing the results,' provides the theoretical framework for interpreting the results. The 'Discussion of initial results' section summarises what the typology shows about the uncertain utility of the tools and methods identified as well as their uneven distribution. The section on 'A way forward' argues that there is a need for a more coordinated effort, from multi-disciplinary researchers, innovators, policymakers, citizens, developers and designers, to create and evaluate new tools and methodologies, in order to ensure that there is a 'how' for every 'what' at each stage of the Machine Learning pipeline. The penultimate section lists some of the limitations of this study. Finally, the last section, concludes that the suggested recommendations will be challenging to achieve, but it would be imprudent not to try.

10.2 Moving from Principles to Practices

On 22nd May 2019, the Organisation for Economic Co-operation and Development (OECD) announced that its 36 member countries, along with an additional six (Argentine, Brazil, Columbia, Costa Rica, Peru, and Romania), had formally agreed to adopt, what the OECD claims to be the first intergovernmental standard on Artificial Intelligence (AI) (OECD 2019a). Designed to ensure AI systems are robust, safe, fair and trustworthy, the standard consists of five complementary value-based principles, and five implementable recommendations to policymakers.

The values and recommendations are not new. Indeed, the OECD's *Recommendation of the Council on Artificial Intelligence* (OECD 2019b) is only the latest among a list of more than 70 documents, published in the last 3 years, which make recommendations about the principles of the ethics of AI (Spielkamp et al. 2019; Winfield 2019). This list includes documents produced by industry (Google,[2] IBM,[3] Microsoft,[4] Intel[5]), Government (Montreal Declaration,[6] Lords Select Committee,[7]

[2]Google's AI Principles: https://www.blog.google/technology/ai/ai-principles/

[3]IBM's everyday ethics for AI: https://www.ibm.com/watson/assets/duo/pdf/everydayethics.pdf

[4]Microsoft's guidelines for conversational bots: https://www.microsoft.com/en-us/research/uploads/prod/2018/11/Bot_Guidelines_Nov_2018.pdf

[5]Intel's recommendations for public policy principles on AI: https://blogs.intel.com/policy/2017/10/18/naveen-rao-announces-intel-ai-public-policy/#gs.8qnx16

[6]The Montreal Declaration for Responsible AI: https://www.montrealdeclaration-responsibleai.com/the-declaration

[7]House of Lords Select Committee on Artificial Intelligence: AI in the UK: ready, willing and able?: https://publications.parliament.uk/pa/ld201719/ldselect/ldai/100/100.pdf

European Commission's High-Level Expert Group[8]), and academia (Future of Life Institute,[9] IEEE,[10] AI4People[11]). The hope of the authors of these documents is that the principles put forward, can, as abstractions (Anderson and Anderson 2018), act as normative constraints (Turilli 2007) on the 'do's' and 'don'ts' of algorithmic use in society.

As Jobin et al. (2019) and Floridi (2019c) point out, this intense interest from such a broad range of stakeholders reflects not only the need for ethical guidance, but also the desire of those different parties to shape the 'ethical AI' conversation around their own priorities. This is an issue that is not unique to debates about the components of ethical ML, but something that the international human rights community has grappled with for decades, as disagreements over what they are, how many there are, what they are for, as well as what duties they impose on whom, and which values of human interests they are supposed to protect (Arvan 2014), have never been resolved. It is significant, therefore, that there seems to be an emerging consensus amongst the members of the ethical ML community with regards to *what* exactly ethical ML should aspire to be.

A review of 84 ethical AI documents by Jobin et al. (2019) found that although no single principle featured in all of them, the themes of transparency, justice and fairness, non-maleficence, responsibility and privacy appeared in over half. Similarly, a systematic review of the literature on ethical technology revealed that the themes of privacy, security, autonomy, justice, human dignity, control of technology and the balance of powers, were recurrent (Royakkers et al. 2018). As argued by, taken together these themes 'define' ethically-aligned ML as that which is (a) beneficial to, and respectful of, people and the environment (**beneficence**); (b) robust and secure (**non-maleficence**); (c)respectful of human values (**autonomy**); (d) fair (**justice**); and (e) explainable, accountable and understandable (**explicability**). Given this emergent consensus in the literature, it is unsurprising that these are also the themes central to the OECD standard. What is perhaps more surprising is that this agreement around the basic principles that ethical ML should meet is no longer limited to Europe and the Western world. Just 3 days after the OECD publication, the Beijing Academy of Artificial Intelligence (BAAI), an organisation backed by the Chinese Ministry of Science and technology and the Beijing municipal government, released its 15 AI principles for: (a) research and development; (b) use; and (c) the Governance of AI (Knight 2019), which when read in full, bear remarkable similarity to the common framework (see Table 10.2).

[8]European Commission's Ethics Guidelines for Trustworthy AI: https://ec.europa.eu/futurium/en/ai-alliance-consultation/guidelines/1

[9]Future of Life's Asilomar AI Principles: https://futureoflife.org/ai-principles/

[10]IEEE General Principles of Ethical Autonomous and Intelligent Systems: http://alanwinfield.blogspot.com/2019/04/an-updated-round-up-of-ethical.html

[11]Floridi et al. (2018).

Table 10.2 Comparison of ethical principles in recent publications demonstrating the emerging consensus of 'what' ethical AI should aspire to be

AI4People (published November 2018) (Floridi et al. 2018)	Five principles key to any ethical framework for AI (L Floridi and Clement-Jones 2019)	Ethics Guidelines for Trustworthy AI (Published April 2019) (European Commission 2019)	Recommendation of the Council of Artificial Intelligence (Published May 2019) (OECD 2019b)	Beijing AI Principles for R&D (Published May 2019) ('Beijing AI Principles' 2019)
Beneficence	AI must be beneficial to humanity	Respect for human autonomy	Inclusive growth, sustainable development and well-being	**Do good:** (covers the need for AI to promote human society and the environment)
Non-maleficence	AI must not infringe on privacy or undermine security	Prevention of harm	Robustness, security and safety	**Be responsible:** (covers the need for researchers to be aware of negative impacts and take steps to mitigate them). **Control risks:** (covers the need for developers to improve the robustness and reliability of systems to ensure data security and AI safety)
Autonomy	AI must protect and enhance our autonomy and ability to take decisions and choose between alternatives		**Human-centred values** and fairness	**For humanity:** (covers the need for AI to serve humanity by conforming to human values including freedom and autonomy)
Justice	AI must promote prosperity and	Fairness	Human-centred values **and fairness**	**Be diverse and inclusive:** (covers the need for AI to benefit as many people as possible). **Be ethical:** (covers the need to make the system as fair as possible, minimising discrimination and bias)

(continued)

Table 10.2 (continued)

AI4People (published November 2018) (Floridi et al. 2018)	Five principles key to any ethical framework for AI (L Floridi and Clement-Jones 2019)	Ethics Guidelines for Trustworthy AI (Published April 2019) (European Commission 2019)	Recommendation of the Council of Artificial Intelligence (Published May 2019) (OECD 2019b)	Beijing AI Principles for R&D (Published May 2019) ('Beijing AI Principles' 2019)
Explicability	AI systems must be understandable and explainable	Explicability	Transparency and explainability accountability	**Be ethical:** (covers the need for AI to be transparent, explainable and predictable)

For a more detailed comparison see Floridi and Cowls (2019) and Hagendorff (2019)

This fragile[12] consensus means that there is now the outline of a shared foundation upon which one can build, and that can be used as a benchmark to communicate expectations and evaluate deliverables. Co-design in AI would be more difficult without this common framework. It is, therefore, a necessary building block in the creation of an environment that fosters ethical, responsible, and beneficial ML, especially as it also indicates the possibility of a time when the distractive risk of ethics shopping[13] (Floridi 2019c) will be lessened. Yet, challenges remain.

The availability of these 'agreed' principles supports but does not yet bring about actual change in the *design* of algorithmic systems (Floridi 2019a). As (Hagendorff 2019) notes, almost all of the guidelines that have been produced to date suggest that technical solutions exist, but very few provide technical explanations. As a result, developers are becoming frustrated by how little help is offered by highly abstract principles when it comes to the 'day job' (Peters and Calvo 2019). This is reflected in the fact that 79% of tech workers report that they would like practical resources to help them with ethical considerations (Miller and Coldicott 2019). Without this more

[12]We say fragile here as there are gaps across the different sets of principles and all use slightly different terminology, making it hard to guarantee that the exact same meaning is intended in all cases. Further-more, as these principles have no legal grounding there is nothing to prevent any individual country (or indeed company) from suddenly choosing to adopt a different set for purposes of convenience or competitiveness.

[13]"*Digital ethics shopping* is the malpractice of choosing, adapting, or revising ("mixing and matching") ethical principles, guidelines, codes, frameworks or other similar standards (especially but not only in the ethics of AI), from a variety of available offers, in order to retrofit some pre-existing behaviours (choices, processes, strategies etc.) and hence justify them a posteriori, instead of implementing or improving new behaviours by benchmarking them against public, ethical standards" (Floridi 2019c).

practical guidance, other risks such as 'ethics bluewashing'[14] and 'ethics shirking'[15] remain (Floridi 2019c).

Such risks, associated with a lack of practical guidance on *how* to produce ethical ML, make it clear that the ethical ML community needs to embark on the second phase of AI ethics: translating between the '*what'* and the '*how.'* This is likely to be hard work. The gap between principles and practice is large, and widened by complexity, variability, subjectivity, and lack of standardisation, including variable interpretation of the 'components' of each of the ethical principles (Alshammari and Simpson 2017). Yet, it is not impossible if the right questions are asked (Green 2018; Wachter et al. 2017) and closer attention is payed to how the design process can influence (Kroll 2018) whether an algorithm is more or less 'ethically-aligned.'

The sooner we start doing this, the better. If we do not take on the challenge and develop usable, interpretable and efficacious mechanisms (Abdul et al. 2018) for closing this gap, the lack of guidance may (a) result in the costs of ethical mistakes outweighing the benefits of ethical success (even a single critical 'AI' scandal could stifle innovation): (b) undermine public acceptance of algorithmic systems; (c) reduce adoption of algorithmic systems; and (d) ultimately create a scenario in which society incurs significant opportunity costs (Cookson 2018). Thus, the aim of this research project is to identify the methods and tools already available to help developers, engineers, and designers of ML reflect on and apply 'ethics' (Adamson et al. 2019) so that they may know not only what to do or not to do, but also how to do it, or avoid doing it (Alshammari and Simpson 2017). We hope that the results of this research may be easily applicable to other branches of AI.

10.3 Methodology

With the aim of identifying the methods and tools available to help developers, engineers and designers of ML reflect on and apply 'ethics' in mind, the first task was to design a typology, for the very practically minded ML community (Holzinger 2018), that would 'match' the tools and methods identified to the ethical principles outlined in Table 10.2 (summarised as beneficence, non-maleficence, autonomy, justice, and explicability).

To create this typology, and inspired by Saltz and Dewar (2019) who produced a framework that is meant to help data scientists consider ethical issues at each stage of a project, the ethical principles were combined with the stages of algorithmic

[14]"*Ethics bluewashing* is the malpractice of making unsubstantiated or misleading claims about, or implementing superficial measures in favour of, the ethical values and benefits of digital processes, products, services, or other solutions in order to appear more digitally-ethical than one is." (Floridi 2019c).

[15]"*Ethics shirking* is the malpractice of doing increasingly less "ethical work" (such as fulfilling duties, respecting rights, honouring commitments, etc.) in a given context the lower the return of such ethical work in that context is mistakenly perceived to be." (Floridi 2019c).

development outlined in the overview of the Information Commissioner's Office (ICO) auditing framework for Artificial Intelligence and its core components,[16] as shown in Table 10.3. The intention is that this encourages ML developers to go between decision and ethical principles regularly.

The second task was to identify the tools and methods, and the companies or individuals researching and producing them, to fill the typology. There were a number of different ways this could have been done. For example, Vakkuri et al. (2019) sought to answer the question 'what practices, tools or methods, if any, do industry professionals utilise to implement ethics into AI design and development?' by conducting interviews at five companies that develop AI systems in different fields. However, whilst analysis of the interviews revealed that the developers were aware of the potential importance of ethics in AI, the companies seemed to provide them with no tools or methods for implementing ethics. Based on a hypothesis that these findings did not imply the non-existence of applied-ethics tools and methods, but rather a lack of progress in the translation of available tools and methods from academic literature or early-stage development and research, to real-life use, this study used the traditional approach of providing an overarching assessment of a research topic, namely a literature review (Abdul et al. 2018).

Scopus,[17] arXiv[18] and PhilPapers,[19] as well as Google search were searched. The Scopus, arXiv and Google Search searches were conducted using the terms outlined in Table 10.4. The PhilPapers search was unstructured, given the nature of the platform, and instead the categories also shown in Table 10.4 were reviewed. The original searches were run in February 2019, but weekly alerts were set for all searches and reviewed up until mid-July 2019. Every result (of which there were originally over 1000) was checked for *relevance*—either in terms of theoretical framing or in terms of the use of the tool—*actionability* by ML developers, and *generalisability* across industry sectors. In total, 425 sources[20] were reviewed. They provide a practical or theoretical contribution to the answer of the question: 'how to develop an ethical algorithmic system.'[21]

---

[16] More detail is available here: https://ai-auditingframework.blogspot.com/2019/03/an-overview-of-auditing-framework-for_26.html

[17] Scopus is the largest abstract and citation database of peer-reviewed literature: scientific journals, books and conference proceedings: https://www.scopus.com/home.uri

[18] arXiv provides open access to over 1,532,009 e-prints in the fields of physics, mathematics, computer science, quantitative biology, quantitative finance, statistics, electrical engineering and systems science, and economics: https://arxiv.org/

[19] PhilPapers is an index and bibliography of philosophy which collates research content from journals, books, open access archives and papers from relevant conferences such as IACAP. The index currently contains more than 2,377,536 entries. https://philpapers.org/

[20] This total includes references related specifically to discourse ethics after an anonymous reviewer made the excellent suggestion that this literature be used as a theoretical frame for the typology.

[21] The full list of sources can be accessed here: https://medium.com/@jessicamorley/applied-ai-ethic s-reading-resource-list-ed9312499c0a

Table 10.3 'Applied AI Ethics' Typology comprising ethical principles and the stages of algorithmic development

Business and use-case development Problem/ improvements are defined and use of AI is proposed	Design phase The business case is turned into design requirements for engineers	Training and test data procurement Initial data sets are obtained to train and test the model	Building AI application is built	Testing The system is tested	Deployment When the AI system goes live	Monitoring Performance of the system is assessed
Beneficence						
Non-maleficence						
Autonomy						
Justice						
Explicability						

Table 10.4 Showing the search terms used to search Scopes, arXiv and Google and the categories reviewed on PhilPapers

Scopus, Google and arXiv search terms (all searched with and machine learning OR Artificial Intelligence)	Category of PhilPapers reviewed
Ethics	Information ethics
Public perception	Technology ethics
Intellectual property	Computer ethics
Business model	Autonomy in applied ethics
Evaluation	Beneficence in applied ethics
Data sharing	Harm in applied ethics
Impact assessment	Justice in applied ethics
Privacy	Human rights in applied ethics
Harm	Applied ethics and normative ethics
Legislation	Responsibility in applied ethics
Regulation	Ethical theories in applied ethics
Data minimisation Transparency Bias Data protection	

The third, and final task, was to review the recommendations, theories, methodologies, and tools outlined in the reviewed sources, and identify where they may fit in the typology. To do this, each of the high-level principles (beneficence, non-maleficence, autonomy, justice and explicability) were translated into tangible system requirements that reflect the meaning of the principles. This is the approach taken by the EU's High Level Ethics Group for AI and outlined in Chap. II of *Ethics Guidelines for Trustworthy AI: Realising Trustworthy AI* which "offers guidance on the implementation and realisation of Trustworthy AI, via a list of (seven) requirements that should be met, building on the principles" (p. 35 European Commission 2019).

This approach is also used in the disciplinary ethical guidance produce for internet-mediated researchers by the Belmont Report (Anabo et al. 2019), and by La Fors et al. (2019) who sought to integrate existing design-based ethical approaches for new technologies by matching lists of values the practical abstraction from mid-level ethics (principles) to what (Hagendorff 2019) calls 'microethics.' This translation is a process that gradually reduces the indeterminacy of abstract norms to produce desiderata for a 'minimum-viable-ethical-(ML)product' (MVEP) that can be used by people who have various disciplinary backgrounds, interests and priorities (Jacobs and Huldtgren 2018). The outcome of this translation process is shown in Table 10.5.

Table 10.5 showing the connection between high-level ethical principles and tangible system requirements as adapted from the methodology outlined in Chapter II of the European Commission's "Ethics Guidelines for Trustworthy AI"

Principle	Beneficence	Non-Maleficence	Autonomy	Justice	Explicability
Requirements	**Stakeholder participation:** to develop systems that are trustworthy and support human flourishing, those who will be affected by the system should be consulted **Protection of fundamental rights Sustainable and environmentally friendly AI:** the system's supply chain should be assessed for resource usage and energy consumption **Justification:** the purpose for building the system must be clear and linked to a clear benefit—system's should not be built for the sake of it	**Resilience to attack and security:** AI systems should be protected against vulnerabilities that can allow them to be exploited by adversaries. **Fallback plan and general safety:** AI systems should have safeguards that enable a fallback plan in case of problems. **Accuracy:** for example, documentation that demonstrates evaluation of whether the system is properly classifying results. **Privacy and Data Protection:** AI systems should guarantee privacy and data protection throughout a system's entire lifecycle. **Reliability and Reproducibility:** does the system work the same way in a variety of different scenarios. **Quality and integrity of the data:** when data is	**Human agency:** users should be able to make informed autonomous decisions regarding AI systems **Human oversight:** may be achieved through governance mechanisms such as human-on-the-loop, human-in-the-loop, human-in-command	**Avoidance of unfair bias Accessibility and universal design Society and democracy:** the impact of the system on institutions, democracy and society at large should be considered **Auditability:** the enablement of the assessment of algorithms, data and design processes. **Minimisation and reporting of negative impacts:** measures should be taken to identify, assess, document, minimise and respond to potential negative impacts of AI systems **Trade-offs:** when trade-offs between requirements are necessary, a process should be put in place to explicitly Acknowledge the trade-off, and evaluate it transparently **Redress:** mechanism should be in place to respond when things go wrong	**Traceability:** the data sets and the processes that yield the AI system's decision should be documented **Explainability:** the ability to explain both the technical processes of an AI system and the related human decisions **Interpretability**

(continued)

Table 10.5 (continued)

Principle	Beneficence	Non-Maleficence	Autonomy	Justice	Explicability
		gathered it may contain socially constructed biases, inaccuracies, errors and mistakes—this needs to be addressed **Social Impact:** the effects of system's on people's physical and mental wellbeing should be carefully considered and monitored			

As highlighted by one of the anonymous reviewers, these categorisations may appear somewhat ad-hoc. For example, one could ask why does 'protection of fundamental rights' belong in the box 'beneficence' rather than justice or non-maleficence, and why is 'privacy and data—protection' not a fundamental right. This is an important point. These are very much open questions worthy of deeper philosophical analysis. However, such analysis is outside the scope of this paper. Here the purpose is not to critique the ethical principles themselves, nor the system requirements for meeting them as set out by the European Commission, we merely seek to use it as an existing framework and assess the extent to which it is possible for developers to meet these requirements based on the availability (and quality) of the tools and methods that are publicly available to help them be 'compliant'

10.4 Framing the Results

The full typology is available here http://tinyurl.com/appliedaiethics. The purpose of presenting it is not to imply that it is 'complete,' nor that the tools and methodologies highlighted are the best, or indeed the only, means of 'solving' each of the individual ethical problems. How to apply ethics to the development of ML is an open question that can be solved in a multitude of different ways at different scales and in different contexts (Floridi 2019a). It would, for example, be entirely possible to complete the process using a different set of principles and requirements. Instead, the goal is to provide a synthesis of what tools are currently available to ML developers to encourage the progression of ethical AI from principles to practice and to signal clearly, to the 'ethical AI' community at large, where further work is needed.

Additionally, the purpose of presenting the typology is not to give the impression that the tools act as means of translating the principles into definitive 'rules' that technology developers should adhere to, or that developers must always complete one 'task' from each of the boxed. This only promotes ethics by 'tick-box' (Hagendorff 2019). Instead, the typology is intended to eventually be an online searchable database so that developers can look for the appropriate tools and methodologies for their given context, and use them to enable a shift from a prescriptive 'ethics-by-design' approach to a dialogic, pro-ethical design approach (Anabo et al. 2019; Floridi 2019b).

In this sense, the tools and methodologies represent a pragmatic version of Habermas's discourse ethics[22] (Mingers and Walsham 2010). In his theory, Habermas (1983, 1991) argues that morals and norms are not 'set' in a top-down fashion but emerge from a process where those with opposing views, engage in a process where they rationally consider each other's arguments, give reasons for their position and, based upon the greater understanding that results, reassess their position until all parties involved reach a universally agreeable decision (Buhmann et al. 2019). This is an approach commonly used in both business and operational research ethics, where questions of 'what *should* we do?' (as opposed to what *can* we do?) arise (Buhmann et al. 2019; Mingers 2011). This is a rationalisation process that involves a fair consideration of the practical, the good and the just, and normally relies heavily on language (discussion), for both the emergence of agreed upon norms or standards, and their reproduction. In the present scenario of developers rationalising ML design decisions to ensure that they are ethically-optimised, the tools and methods in the typology replace the role of language and act as the medium for identifying, checking, creating and re-examining ideas and giving fair consideration to differing interests, values and norms (Heath 2014; Yetim 2019). For example, the data nutrition tool (Holland et al. 2018) provides a means of prompting a discussion and re-evaluation of the ethical implications of using a specific dataset for an ML development project, and the audit methodologies of (Diakopoulos 2015)

[22] We would like to thank one of the anonymous reviewers for suggesting this framing, it represents a significant improvement to the theoretical grounding of this paper.

ensure that external voices, who may have an opposing view as to whether or not an ML-system in use is ethically-aligned, have a mechanism for questioning the rational of design decisions and requesting their change if necessary. It is within this frame that we present an overview of our findings in the next section.

10.5 Discussion of Initial Results

Interpretation of the results of the literature review and the resulting typology are likely to be context specific. Those with different disciplinary backgrounds (engineering, moral philosophy, sociology etc.) will see different patterns, and different meanings in these patterns. This kind of multidisciplinary reflection on what the presence or absence of different tools and methods, and their function, might mean is to be encouraged. To start the conversation, this section highlights the following three headings:

1. an overreliance on 'explicability';
2. a focus on the need to 'protect' the individual over the collective; and
3. a lack of usability

They are interrelated, but for the sake of simplicity, let us analyse each separately.

10.5.1 Explicability as the All-Encompassing Principle[23]

To start with the most obvious observation: the availability of tools and methods is not evenly distributed across the typology, either in terms of the ethical principles or in terms of the stages of development. For example, whilst a developer looking to ensure their ML algorithm is 'non-maleficent' has a section of tools available to them for each development stage—as highlighted in Table 10.6—the tools and methods designed to enable developers to meet the principle or 'beneficence' are almost all intended to be used during the initial planning stages of development (i.e. business and use-case development design phases). However, the most noticeable 'skew' is towards post hoc explanations; with those seeking to meet the principle of explicability during the testing phase having the greatest range of tools and methods from which to choose.

---

[23]We recognise that there is an extremely rich literature on ML fairness which this paper does not cover. Much (although not all) of this literature focuses on the definition of fairness and the statistical means of implementing this which sits slightly outside the scope of the typology which aims to highlight tools and methods that facilitate discussion about the ethical nature of one design decision over another. To fit an entire decade's worth of literature into a row on a table would not do it justice.

Table 10.6 Applied AI ethics typology with illustrative non-maleficence example

	Business and use-case development Problem/improvements are defined and use of AI is proposed	Design Phase The business case is turned into design requirements for engineers	Training and test data procurement Initial data sets are obtained to train and test the model	Building AI application is built	Testing The system is tested	Deployment When the AI system goes live	Monitoring Performance of the system is assessed
Beneficence							
Non-maleficence	Cavoukian et al. (2010) outline 7 foundational principles for Privacy by Design: 1. Proactive not reactive: preventative not reactive. 2. Privacy as the default 3. Privacy embedded into design 4. Full functionality = positive sum, not zero sum 5. End-to-end lifecycle protection 6. Visibility and Transparency 7. Respect for user privacy	Oetzel and Spiekermann (2014) set out a step-by-step privacy impact assessment (PIA) to enable companies to achieve 'privacy-by-design'	Antignac et al. (2016) provide the python code to create *DataMin* (a data minimiser—a pre-processor modifying the input of data to ensure only the data needed are available to the program) as a series of Java source code files which can be run on the data sources points before disclosing the data.	Kolter and Madry (2018) provide a practical introduction, from a mathematical and coding perspective, to the topic of adversarial robustness with the idea being that it is possible to train deep learning classifiers to be resistant to adversarial attacks: https://adversarial-ml-tutorial.org/	Dennis et al. (2016) outline a methodology for verifying the decision making of an autonomous agent to confirm that the controlling agent never deliberately makes a choice it believes to be unsafe	(*AI Now Institute Algorithmic Accountability Policy Toolkit*) provides a list of questions policy and legal advocates will want to ask when considering introducing an automated system into a public service and provides detailed guidance on where in the procurement process to ask questions about accountability and potential harm	Makri and Lambrinoudakis (2015) outline a structured privacy audit procedure based on the most widely adopted privacy principles: -Purpose specification -Collection limitation -Data quality -Use retention and disclosure limitation -Safety safeguards -Openness -Individual participation -Accountability

						https:// ainowinstitute. org/aap-toolkit. pdf	

Autonomy justice explicability

A developer looking to ensure their ML solutions meets the principle of non-maleficence can start with the foundational principles of privacy by design (Cavoukian et al. 2010) to guide ideation appropriately, use techniques such as data minimisation (Antignac et al. 2016), training for adversarial robustness (Kolter and Madry 2018), and decision-making verification (Dennis et al. 2016) in the train-build-test phases, and end by launching the system with an accompanying privacy audit procedure (Makri and Lambrinoudakis 2015)

There are likely to be several reasons for this, but two stand out. The first and simpler is that the 'problem' of 'interpreting' an algorithmic decision seems tractable from a mathematical standpoint, so the principle of explicability has come to be seen as the most suitable for a technical fix (Hagendorff 2019). The second is that 'explicability' is not, from a moral philosophy perspective, a moral principle like the other four principles. Instead, it can be seen as a second order principle, that has come to be of vital importance in the ethical-ML community because, to a certain extent, it is linked with all the other four principles.[24] Indeed, it is argued that if a system is explicable (explainable and interpretable) it is inherently more transparent and therefore more accountable in terms of its decision-making properties and the extent to which they include human oversight and are fair, robust and justifiable (Binns et al. 2018; Cath 2018; Lipton 2016).

Assuming temporarily that this is indeed the case,[25] and that by dint of being explicable an ML system can more easily meet the principles of beneficence, non-maleficence, autonomy and justice, then the fact that the ethical ML community has focused so extensively on developing tools for explanations may not seem problematic. However, as the majority of tools and methods that sit in the concentration at the intersection of explicability and testing are primarily statistical in nature, this would be a very mechanistic view because such 'solutions'—e.g. LIME (Ribeiro et al. 2016), SHAP (Lundberg and Lee 2017), Sensitivity Analysis (Oxborough et al. 2018)—do not really succeed in helping developers provide meaningful *explanations* (Edwards and Veale 2018) that give individuals greater control over what is being *inferred* about them from their data. As such, the existence of these tools is at most necessary but not sufficient.

From a more humanistic, and realistic perspective, in order to satisfy all the five principles a system needs to be *designed* from the very beginning to be a transparent sociotechnical system (Ananny and Crawford 2018). To achieve this level of transparency, accountability or explicability, it is essential that those analysing a system are able to "understand what it was designed to do, how it was designed to do that, and why it was designed in that particular way instead of some other way" (Kroll 2018). This kind of scrutiny will only be possible through a combination of tools or processes that facilitate auditing, transparent development, education of the public, and social awareness of developers (Burrell 2016). As such, there should ideally be tools and methods available for each of the boxes in the typology, accepting that there may be areas of the typology which are more significant for ML practitioners than others.

Furthermore, available of tools and methods in a variety of typology areas is also important in the context of culturally and contextually specific ML ethics. Not all of

---

[24] We would like to thank one of the anonymous reviewers for making this important point.

[25] It is entirely possible that this is not always the case and that there may be instances where an explicable system has, for example, still had a negative impact on autonomy. Additionally, this view that transparency as explanation is key to accountability is one that is inherently western in perspective and those of other cultures may have a different viewpoint. We make the assumption here for simplicity's sake.

the principles will be of equal importance in all contexts. For example, in the case of national security systems non-maleficence may be of considerably higher importance than explicability. If the community prioritises the development of tools and methods for one of the principles over the others, it will be denying itself the opportunity for such flexibility.

10.5.2 An Individual Focus

The next observation of note is that few of the available tools surveyed provide meaningful ways to assess, and respond to, the impact that the data-processing involved in their ML algorithm has on an individual, and even less on the impact on society as a whole (Poursabzi-Sangdeh et al. 2018). This is evident from the very sparsely populated 'deployment' column of the typology. Its emptiness implies that the need for pro-ethically designed human–computer interaction (at an individual level) or networks of ML systems (at a group level) has been paid little heed. This is likely because it is difficult to translate complex human behaviour into design tools that are simple to use and generalisable.

This might not seem particularly important, but the impact this has on the overall acceptance of AI in society could be significant. For example, it is unlikely that counterfactual explanations[26] (i.e. if input variable x had been different, the output variable y would have been different as well)—although important for many reasons—will be sufficient to improve the interpretability of recommendations made by black-box systems for the average member of the public or the technical community. If such methods become the de facto means of providing explanations, the extent to which the 'algorithmic society' is interpretable to the general public will be very limited. And counterfactual explanations could easily be embraced by actors uninterested in providing factual explanations, because the counterfactual ones provide a vast menu of options, which may easily decrease the level of responsibility of the actor choosing it. For example, if a mortgage provider does not offer a mortgage, the factual reasons may be a bias, for example the gender of the applicant, but the provider could choose from a vast menu of innocuous, counterfactual explanations—if some variable x had been different the mortgage might have been provided—e.g., a much higher income, more collaterals, lower amount, and so forth, without ever mentioning the factual cause, i.e. the gender of the applicant. All this could considerably limit the level of trust people are willing to place in such systems.

This potential threat to trust is further heightened by the fact that the lack of attention paid to impact means that ML developers are currently hampered in their ability to develop systems that promote user's (individual or group's) autonomy. For example, currently there is an assumption that prediction = decision, and little

[26]See for example Johansson et al. (2016), Lakkaraju et al. (2017), Russell et al. (2017), and Wachter et al. (2017).

research has been done (in the context of ML) on how people translate predictions into actionable decisions. As such, tools that, for example, help developers pro-ethically design solutions that do not overly restrict the user's options in acting on this prediction (i.e. tools that promote the user's autonomy) are in short supply (Kleinberg et al. 2017). F users feel as though their decisions are being curtailed and controlled by systems that they do not understand, it is very unlikely that these systems will meet the condition of social acceptability, never mind the condition of social preferability which should be the aim for truly ethically designed ML (Floridi and Taddeo 2016).

10.5.3 A Lack of Usability

Finally, the tools and methods included in the typology are positioned as discourse aids, designed to facilitate and document rational decisions about trade-offs in the design process that may make an ML system more or less ethically-aligned. It is possible to see the *potential* for the tools identified to play this role. For example, at the "beneficence → use-case → design" intersection, there are a number of tools highlighted to help elicit social values. These include the responsible research and innovation methodology employed by the European Commission's Human Brain Project (Stahl and Wright 2018), the field guide to human-centred design (ideo.org 2015) and Involve and DeepMind's guidance on stimulating effective public engagement on the ethics of Artificial Intelligence (Involve and DeepMind 2019). Such tools and methods could be used to help designers pro-ethically deal with value pluralism (i.e. variation in values across different population groups). However, the vast majority of these tools and methods are not actionable as they offer little help on how to use them in practice (Vakkuri et al. 2019). Even when there are open-source code libraries available, documentation is often limited, and the skill-level required for use is high.

This overarching lack of usability of the tools and methods highlighted in the typology means that, although they are promising, they require more work before being 'production-ready.' As a result, applying ethics still requires considerable amounts of effort on behalf of the ML developers undermining one of the main aims of developing and using technologically-based 'tools': to remove friction from applied ethics. Furthermore, until these tools are embedded in practice and tested in the 'real world,' it is extremely unclear what impact they will have on the overall 'governability' of the algorithmic ecosystem. For example Binns (2018a) asks how an accountable system actually will be held accountable for an 'unfair' decision in a way that is acceptable to all. This makes it almost impossible to measure the impact, 'define success', and document the performance (Mitchell et al. 2019) of a new design methodology or tool. As a result, there is no clear problem statement (and therefore no clear business case) that the ML community can use to justify time and financial investment in developing much-needed tools and techniques that truly enable pro-ethical design. Consequently, there is no guaranatee that the so-called discursive devices do anything other than help the groups in society who already have the loudest voices embed and protect their values in design tools, and then into the resultant ML systems.

10.6 A Way Forward

Social scientists (Matzner 2014) and political philosophers (from Rousseau and Kant, to Rawls and Habermas) (Binns 2018b), are used to dealing with the kind of plurality and subjectivity informing the entire ethical ML field (Bibal and Frénay 2016). Answering questions such as, what happens when individual level and group level 'ethics' interact, and what key terms such as 'fairness,' 'accountability,' 'transparency' and 'interpretability' actually mean when there are currently a myriad definitions (Ananny and Crawford 2018; Bibal and Frénay 2016; Doshi-Velez and Kim 2017; Friedler et al. 2016; Guidotti et al. 2018; Kleinberg et al. 2016; Overdorf et al. 2018; Turilli and Floridi 2009) is standard fare for individuals with social science, economy, philosophy or legal training. This is why (Nissenbaum 2004) argues for a contextual account of privacy, one that recognises the varying nature of informational norms (Matzner 2014) and (Kemper and Kolkman 2018) state that transparency is only meaningful in the context of a defined critical audience.

The ML developer community, in contrast, may be less used to dealing with *this* kind of difficulty, and more used to scenarios where there is at least a seemingly quantifiable relationship between input and output. As a result, the existing approaches to designing and programming ethical ML fail to resolve what (Arvan 2018) terms the moral-semantic trilemma, as almost all tools and methods highlighted in the typology are either too semantically strict, too semantically flexible, or overly unpredictable (Arvan 2018).

Bridging together multi-disciplinary researchers into the development process of pro-ethical design tools and methodologies will be essential. A multi-disciplinary approach will help the ethical ML community overcome obstacles concerning social complexity, embrace uncertainty, and accept that: (1) AI is built on assumptions; (2) human behaviour is complex; (3) algorithms can have unfair consequences; (4) algorithmic predictions can be hard to interpret (Vaughan and Wallach 2016); (5) trade-offs are usually inevitable; and (6) positive, ethical features are open to progressive increase, that is an algorithm can be increasingly fair, and fairer than another algorithm or a previous version, but makes no sense to say that it is fair or unfair in absolute terms (compare this to the case of speed: it makes sense to say that an object is moving quickly, or that it is fast or faster than another, but not that it is fast). The resulting collaborations are likely to be highly beneficial for the development of applied ethical tools and methodologies for at least three reasons.

First, it will help ensure that the tools and methods developed do not only protect value-pluralism in silico (i.e. the pluralistic values of developers) but also in society. Embracing uncertainty and disciplinary diversity will naturally encourage ML experts to develop tools that facilitate more probing and open (i.e. philosophical) questions (Floridi 2019b) that will lead to more nuanced and reasoned answers and hence decisions about why and when certain trade-offs, for example, between accuracy and interpretability (Goodman and Flaxman 2017), are justified, based on factors such as proportionality to risk (Holm 2019).

Second, it will encourage a more flexible and reflexive approach to applied ethics that is more in-keeping with the way ML systems are actually developed: it is not think and *then* code, but rather think *and* code. In other words, it will accelerate the move away from the 'move fast and break things' approach towards an approach of 'make haste slowly' (*festina lente*) (Floridi 2019a).

Finally, it would also mitigate a significant risk—posed by the current sporadic application of ethical-design tools and/or methods during different development stages—of the ethical principles having been written into the business and use-case, but coded out by the time a system gets to deployment.

To enable developers to embrace this vulnerable uncertainty, it will be important to promote the development of tools, like DotEveryone's agile consequence scanning event (DotEveryone 2019), and the Responsible Double Diamond 'R2D2' (Peters and Calvo 2019) that prompt developers to reflect on the impacts (both direct and indirect) of the solutions they are developing on the 'end user', and on how these impacts can be altered by seemingly minor design decisions at each stage of development. In other words, ML developers should regularly:

(a) look back and ask: 'if I was abiding by ethical principles x in my design *then,* am I still *now?* (as encouraged by *Wellcome Data Lab's* agile methodology (Mikhailov 2019)); and
(b) look forward and ask: 'if I am abiding by ethical principles x in my design *now,* should I continue to do so? And how? By using foresight methodologies (Floridi and Strait Forthcoming; Taddeo and Floridi 2018), such as *AI Now's* Algorithmic Impact Assessment Framework (Reisman et al. 2018).

Taking this approach recognises that, in a digital context, ethical principles are not simply either applied or not, but regularly re-applied or applied differently, or better, or ignored as algorithmic systems are developed, deployed, configured (Ananny and Crawford 2018) tested, revised and re-tuned (Arnold and Scheutz 2018).

This approach to applied ML ethics of regular reflection and application will heavily rely on (i) the creation of more tools—especially to fill the white spaces of the typology (for the reasons discussed in the previous section) and (ii) acceleration of tools' maturity level from research labs into production environments. To achieve (i)–(ii), society needs to come together in communities comprised of multi-disciplinary researchers (Cath et al. 2017), including innovators, policy-makers, citizens, developers and designers (Taddeo and Floridi 2018), to foster the development of: (1) common knowledge and understanding; and (2) a common goal to be achieved from the development of tools and methodologies for applied AI ethics (Durante 2010). These outputs will provide a reason, a mechanism, and a consensus to coordinate the efforts behind tool development. Ultimately, this will produce better results than the current approach, which allows a 'thousand flowers to bloom' but fails to create tools that fill in the gaps (this is a typical 'intellectual market' failure), and may encourage competition to produce preferable options. The opportunity that this presents is too great to be delayed, the ML research community should start collaborating now with a specific focus on:

1. the development of a common language;
2. the creation of tools that ensure *people,* as individuals, groups and societies, are given an equal and meaningful opportunity to participate in the design of algorithmic solutions at each stage of development;
3. the evaluation of the tools that are currently in existence so that what works, what can be improved, and what needs to be developed can be identified;
4. a commitment to reproducibility, openness, and sharing of knowledge and technical solutions (e.g. software), also in view of satisfying (2) and supporting (3); the creation of 'worked examples' of how tools have been used to satisfy one of the principles at each stage of the development and how consistency was maintained throughout the use of different tools'
5. the evaluation and creation of pro-ethical business models and incentive structures that balance the costs and rewards of investing in ethical AI across society, also in view of supporting (2)–(4).

10.7 Limitations

All research projects have their limitations and this one is no exception. The first is that the research question 'what tools and methods are available for ML developers to 'apply' ethics to each stage of the ML system design' is very broad. This lack of specificity meant that the available literature was excessive and growing all the time, making compromises from the perspective of what is practically essential. It is certain that such compromises, for example which databases to search and the decision to restrict the tools reviewed to those that were not industry sector-specific, have resulted in us missing a large number of tools and methods that are publicly available. Building on this, it is again, very likely that there are a number of proprietary applied ethics tools and methods being developed by private companies for internal or consulting purposes that we will have missed, for example the 'suite of customisable frameworks, tools and processes' that make up consulting firm PWC's "Responsible AI Toolkit" (PWC 2019).

The second limitation is related to the design of the typology itself. As (La Fors et al. 2019) attest, the "neat theoretical distinction between different stages of technological innovation does not always exist in practice, especially not in the development of big data technologies." This implies that by categorising the tools by stage of development, we might be reducing their usability as developers in different contexts might follow a different pattern or feel as though it is 'too late' to, for example, engage in stakeholder engagement if they have reached the 'build' phase of their project, whereas the reality it is never too late.

Finally, the last limitations has already been mentioned and concerns the lack of clarity regarding how the tools and methods that have been identified will improve the governability of algorithmic systems. Exactly *how* to govern ML remains an open question, although it appears that there is a growing acceptance among tech workers (in the UK at least) that government regulation will be necessary (Miller and Coldicott 2019). The typology can at least be seen as a mechanism for facilitating co-regulation. Governments are increasingly setting standards and system requirements for ethical ML, but delegating the means for meeting these to the developers themselves (Clarke 2019)—the tools and methods of the typology can be seen as the means of providing evidence of compliance. In this way, the typology (and the tools and methods it contains within) help developers take responsibility for embedding ethics in the part of the development, deployment, and use of ML solutions that they control (Coeckelbergh 2012). The extent to which this makes a difference is yet to be determined.

10.8 Conclusion

The realisation that there is a need to embed ethical considerations into the design of computational, specifically algorithmic, artefacts is not new. Samuel (1960), Wiener (1961) and Turing were vocal about this in the 1940s and 1960s (Turilli 2008). However, as the complexity of algorithmic systems and our reliance on them increases (Cath et al. 2017), so too does the need to be critical (Floridi 2016a) AI governance (Cath 2018) and design solutions. It is possible to design things to be better (Floridi 2017), but this will require more coordinated and sophisticated approaches (Allen et al. 2000) to translating ethical principles into design protocols (Turilli 2007).

This call for increased coordination is necessary. The research has shown that there is an uneven distribution of effort across the 'Applied AI Ethics' typology. Furthermore, many of the tools included are relatively immature. This makes it difficult to assess the scope of their use (resulting in Arvan's 2018 'moral-semantic trilemma') and consequently hard to encourage their adoption by the practically-minded ML developers, especially when the competitive advantage of more ethically-aligned AI is not yet clear. Taking the time to complete any of the 'exercises' suggested by the methods reviewed, and investing in the development of new tools or methods that 'complete the pipeline', add additional work and costs to the research and development process. Such overheads may directly conflict with short-term, commercial incentives. Indeed, a full ethical approach to AI design, development, deployment, and use may represent a competitive disadvantage for any single 'first mover'. The threat that this short-termism poses to the development of truly ethical ML is significant. Unless a longer-term and sector-wide perspective in terms of return on investment can be encouraged—so that mechanisms are developed to close the gap between *what* and *how*—the lack of guidance may (a) result in the costs of ethical mistakes outweighing the benefits of ethical successes; (b) undermine public acceptance of algorithmic systems, even to the point of a backlash (Cookson 2018); and (c) reduce adoption of algorithmic systems. Such a resultant lack of adoption could then turn into a loss of confidence from investors and research funders, and undermine AI research. Lack of incentives to develop AI ethically could turn into lack of interest in developing AI *tout court.* This would not be unprecedented. One only needs to recall the dramatic reduction in funding available for AI research following the 1973 publication of *Artificial Intelligence: A General Survey* (Lighthill 1973) and its criticism of the fact that AI research had not lived up to its over-hyped expectations.

It this were to happen today, the opportunity costs that would be incurred by society would be significant (Cookson 2018). The need for 'AI Ethics' has arisen from the fact that poorly designed AI systems can cause very significant harm. For example, predictive policing tools may lead to more people of colour being arrested, jailed or physically harmed by policy (Selbst 2017). Likewise, the potential benefits of pro-ethically designed AI systems are considerable. This is especially true in the field of AI for Social Good where various AI applications are making possible socially good outcomes that were once less easily achievable, unfeasible, or unaffordable (Cowls et al. 2019). So, there is an urgent need to progress research in this area.

Constructive patience needs to be exercised, by society and by the ethical AI community, because such progress on the question of 'how' to meet the 'what' will not be quick, and there will definitely be mistakes along the way. The ML research community will have to accept this, trust that everyone is trying to meet the same end-goal, but also accept that it is unacceptable to delay any full commitment, when it is known how serious the consequences of doing nothing are. Only by accepting this can society by positive about the opportunities presented by AI to be seized, whilst remaining mindful of the potential costs to be avoided (Floridi et al. 2018).

Funding This study was funded by Digital Catapult.

References

Abdul, A., J. Vermeulen, D. Wang, B.Y. Lim, and M. Kankanhalli. 2018. Trends and trajectories for explainable, accountable and intelligible systems: An HCI research agenda. In *Proceedings of the 2018 CHI conference on human factors in computing systems—CHI'18*, 1–18. https://doi.org/10.1145/3173574.3174156.

Adamson, G., J.C. Havens, and R. Chatila. 2019. Designing a value-driven future for ethical autonomous and intelligent systems. *Proceedings of the IEEE* 107 (3): 518–525. https://doi.org/10.1109/JPROC.2018.2884923.

AI Now Institute Algorithmic Accountability Policy Toolkit. 2018. Retrieved from https://ainowinsti tute.org/aap-toolkit.pdf

Allen, C., G. Varner, and J. Zinser. 2000. Prolegomena to any future artificial moral agent. *Journal of Experimental & Theoretical Artificial Intelligence* 12 (3): 251–261. https://doi.org/10.1080/09528130050111428.

Alshammari, M., and A. Simpson. 2017. Towards a principled approach for engineering privacy by design. In *Privacy technologies and policy*, ed. E. Schweighofer, H. Leitold, A. Mitrakas, and K. Rannenberg, vol. 10518, 161–177. Cham: Springer. https://doi.org/10.1007/978-3-319-67280-9_9.

Anabo, I.F., I. Elexpuru-Albizuri, and L. Villardón-Gallego. 2019. Revisiting the Belmont report's ethical principles in internet-mediated research: Perspectives from disciplinary associations in the social sciences. *Ethics and Information Technology* 21 (2): 137–149. https://doi.org/10.1007/s10676-018-9495-z.

Ananny, M., and K. Crawford. 2018. Seeing without knowing: Limitations of the transparency ideal and its application to algorithmic accountability. *New Media & Society* 20 (3): 973–989. https://doi.org/10.1177/1461444816676645.

Anderson, M., and S.L. Anderson. 2018. GenEth: A general ethical dilemma analyzer. *Paladyn, Journal of Behavioral Robotics* 9 (1): 337–357. https://doi.org/10.1515/pjbr-2018-0024.

Antignac, T., D. Sands, and G. Schneider. 2016. *Data minimisation: A language-based approach (long version)*. arXiv:1611.05642 [Cs].

Arnold, T., and M. Scheutz. 2018. The "big red button" is too late: An alternative model for the ethical evaluation of AI systems. *Ethics and Information Technology* 20 (1): 59–69. https://doi.org/10.1007/s10676-018-9447-7.

Arvan, M. 2014. A better, dual theory of human rights: A better, dual theory of human rights. *The Philosophical Forum* 45 (1): 17–47. https://doi.org/10.1111/phil.12025.

———. 2018. Mental time-travel, semantic flexibility, and A.I. ethics. *AI & Society*. https://doi.org/10.1007/s00146-018-0848-2.

Beijing AI Principles. 2019. Retrieved from Beijing Academy of Artificial Intelligence website. https://www.baai.ac.cn/blog/beijing-ai-principles

Bibal, A., and B. Frénay. 2016. Interpretability of machine learning models and representations: An introduction. In *24th European symposium on artificial neural networks, computational intelligence and machine learning: ESANN 2016: Bruges, Belgium, April 27–28-29, 2016: Proceedings*, ed. M. Verleysen, 77–82. Bruges: CIACO.

Binns, R. 2018a. Algorithmic accountability and public reason. *Philosophy & Technology* 31 (4): 543–556. https://doi.org/10.1007/s13347-017-0263-5.

———. 2018b. What can political philosophy teach us about algorithmic fairness? *IEEE Security and Privacy* 16 (3): 73–80. https://doi.org/10.1109/MSP.2018.2701147.

Binns, R., M. Van Kleek, M. Veale, U. Lyngs, J. Zhao, and N. Shadbolt. 2018. 'It's reducing a human being to a percentage': Perceptions of justice in algorithmic decisions. In *Proceedings of the 2018 CHI conference on human factors in computing systems—CHI'18*, 1–14. https://doi.org/10.1145/3173574.3173951.

Buhmann, A., J. Paßmann, and C. Fieseler. 2019. Managing algorithmic accountability: Balancing reputational concerns, engagement strategies, and the potential of rational discourse. *Journal of Business Ethics*. https://doi.org/10.1007/s10551-019-04226-4.

Burrell, J. 2016. How the machine 'thinks': Understanding opacity in machine learning algorithms. *Big Data & Society* 3 (1): 205395171562251. https://doi.org/10.1177/2053951715622512.

Cath, C. 2018. Governing Artificial Intelligence: Ethical, legal and technical opportunities and challenges. *Philosophical Transactions of the Royal Society A: Mathematical, Physical and Engineering Sciences* 376 (2133): 20180080. https://doi.org/10.1098/rsta.2018.0080.

Cath, C., S. Wachter, B. Mittelstadt, M. Taddeo, and L. Floridi. 2017. Artificial Intelligence and the 'Good Society': The US, EU, and UK approach. *Science and Engineering Ethics*. https://doi.org/10.1007/s11948-017-9901-7.

Cath, C., M. Zimmer, S. Lomborg, and B. Zevenbergen. 2018. Association of internet researchers (AoIR) roundtable summary: Artificial Intelligence and the good society workshop proceedings. *Philosophy & Technology* 31 (1): 155–162. https://doi.org/10.1007/s13347-018-0304-8.

Cavoukian, A., S. Taylor, and M.E. Abrams. 2010. Privacy by design: Essential for organizational accountability and strong business practices. *Identity in the Information Society* 3 (2): 405–413. https://doi.org/10.1007/s12394-010-0053-z.

Clarke, R. 2019. Principles and business processes for responsible AI. *Computer Law and Security Review*. https://doi.org/10.1016/j.clsr.2019.04.007.
Coeckelbergh, M. 2012. Moral responsibility, technology, and experiences of the tragic: From Kierkegaard to offshore engineering. *Science and Engineering Ethics* 18 (1): 35–48. https://doi.org/10.1007/s11948-010-9233-3.
Cookson, C. 2018. Artificial Intelligence faces public backlash, warns scientist. *Financial Times*, September 6. Retrieved from https://www.ft.com/content/0b301152-b0f8-11e8-99ca-68cf89602132
Cowls, J., T. King, M. Taddeo, and L. Floridi 2019. *Designing AI for social good: Seven essential factors*, May 15. Available at SSRN: https://ssrn.com/abstract=
Crawford, K., and R. Calo. 2016. There is a blind spot in AI research. *Nature* 538 (7625): 311–313. https://doi.org/10.1038/538311a.
D'Agostino, M., and M. Durante. 2018. Introduction: The governance of algorithms. *Philosophy & Technology* 31 (4): 499–505. https://doi.org/10.1007/s13347-018-0337-z.
Dennis, L.A., M. Fisher, N.K. Lincoln, A. Lisitsa, and S.M. Veres. 2016. Practical verification of decision-making in agent-based autonomous systems. *Automated Software Engineering* 23 (3): 305–359. https://doi.org/10.1007/s10515-014-0168-9.
Diakopoulos, N. 2015. Algorithmic accountability: Journalistic investigation of computational power structures. *Digital Journalism* 3 (3): 398–415. https://doi.org/10.1080/21670811.2014.976411.
Doshi-Velez, F., and B. Kim. 2017. *Towards a rigorous science of interpretable machine learning*. arXiv:1702.08608 [Cs, Stat].
DotEveryone. 2019. *The DotEveryone consequence scanning agile event*. Retrieved from https://doteveryone.org.uk/project/consequence-scanning/
Dressel, J., and H. Farid. 2018. The accuracy, fairness, and limits of predicting recidivism. *Science Advances* 4 (1): eaao5580. https://doi.org/10.1126/sciadv.aao5580.
Durante, M. 2010. What is the model of trust for multi-agent systems? Whether or not e-trust applies to autonomous agents. *Knowledge, Technology, and Policy* 23 (3–4): 347–366. https://doi.org/10.1007/s12130-010-9118-4.
Edwards, L., and M. Veale. 2018. Enslaving the algorithm: From a "right to an explanation" to a "right to better decisions"? *IEEE Security and Privacy* 16 (3): 46–54. https://doi.org/10.1109/MSP.2018.2701152.
European Commission. 2019. *Ethics guidelines for trustworthy AI*. Retrieved from https://ec.europa.eu/futurium/en/ai-alliance-consultation
Floridi, L. 2016a. Faultless responsibility: On the nature and allocation of moral responsibility for distributed moral actions. *Philosophical Transactions of the Royal Society A: Mathematical, Physical and Engineering Sciences* 374 (2083): 20160112. https://doi.org/10.1098/rsta.2016.0112.
———. 2016b. Tolerant paternalism: Pro-ethical design as a resolution of the dilemma of toleration. *Science and Engineering Ethics* 22 (6): 1669–1688. https://doi.org/10.1007/s11948-015-9733-2.
———. 2017. The logic of design as a conceptual logic of information. *Minds and Machines* 27 (3): 495–519. https://doi.org/10.1007/s11023-017-9438-1.
———. 2018. Soft ethics, the governance of the digital and the general data protection regulation. *Philosophical Transactions of the Royal Society A: Mathematical, Physical and Engineering Sciences* 376 (2133): 20180081. https://doi.org/10.1098/rsta.2018.0081.
———. 2019a. Establishing the rules for building trustworthy AI. *Nature Machine Intelligence*. https://doi.org/10.1038/s42256-019-0055-y.
———. 2019b. *The logic of information: A theory of philosophy as conceptual design*. 1st ed. New York: Oxford University Press.
———. 2019c. Translating principles into practices of digital ethics: Five risks of being unethical. *Philosophy & Technology*. https://doi.org/10.1007/s13347-019-00354-x.

Floridi, L, and T. Clement-Jones. 2019. The five principles key to any ethical framework for AI. *Tech New Statesman*, March 20. Retrieved from https://tech.newstatesman.com/policy/ai-ethics-framework

Floridi, L., and J. Cowls. 2019. A unified framework of five principles for AI in society. *Harvard Data Science Review*. https://doi.org/10.1162/99608f92.8cd550d1.

Floridi, L., and A. Strait Forthcoming. *Ethical foresight analysis: What it is and why it is needed.*

Floridi, L., and M. Taddeo. 2016. What is data ethics? *Philosophical Transactions of the Royal Society A—Mathematical Physical and Engineering Sciences* 374 (2083): 20160360. https://doi.org/10.1098/rsta.2016.0360.

Floridi, L., J. Cowls, M. Beltrametti, R. Chatila, P. Chazerand, V. Dignum, et al. 2018. AI4People—An ethical framework for a good AI society: Opportunities, risks, principles, and recommendations. *Minds and Machines* 28 (4): 689–707. https://doi.org/10.1007/s11023-018-9482-5.

Friedler, S.A., C. Scheidegger, and S. Venkatasubramanian. 2016. *On the (im)possibility of fairness*. arXiv:1609.07236 [Cs, Stat].

Goodman, B., and S. Flaxman. 2017. European Union regulations on algorithmic decision-making and a 'right to explanation'. *AI Magazine* 38 (3): 50. https://doi.org/10.1609/aimag.v38i3.2741.

Green, B.P. 2018. Ethical reflections on Artificial Intelligence. *Scientia et Fides* 6 (2): 9. https://doi.org/10.12775/setf.2018.015.

Guidotti, R., A. Monreale, S. Ruggieri, F. Turini, F. Giannotti, and D. Pedreschi. 2018. A survey of methods for explaining black box models. *ACM Computing Surveys* 51 (5): 1–42. https://doi.org/10.1145/3236009.

Habermas, J. 1983. *Moralbewußtsein und kommunikatives Handeln*. Frankfurt am Main: Suhrkamp. [English, 1990a].

———. 1991. *The structural transformation of the public sphere: An inquiry into a category of bourgeois society*. Cambridge, MA: MIT Press.

Hagendorff, T. 2019. *The ethics of AI ethics—An evaluation of guidelines*. arXiv:1903.03425 [Cs, Stat].

Heath, J. 2014. Rebooting discourse ethics. *Philosophy and Social Criticism* 40 (9): 829–866. https://doi.org/10.1177/0191453714545340.

Hevelke, A., and J. Nida-Rümelin. 2015. Responsibility for crashes of autonomous vehicles: An ethical analysis. *Science and Engineering Ethics* 21 (3): 619–630. https://doi.org/10.1007/s11948-014-9565-5.

Holland, S., A. Hosny, S. Newman, J. Joseph, and K. Chmielinski. 2018. *The dataset nutrition label: A framework to drive higher data quality standards*. arXiv:1805.03677 [Cs].

Holm, E.A. 2019. In defense of the black box. *Science* 364 (6435): 26–27. https://doi.org/10.1126/science.aax0162.

Holzinger, A. 2018. From machine learning to explainable AI. In *World symposium on Digital Intelligence for Systems and Machines (DISA), 2018*, 55–66. https://doi.org/10.1109/DISA.2018.8490530.

ideo.org. 2015. *The field guide to human-centered design*. Retrieved from http://www.designkit.org/resources/1

Involve, and DeepMind. 2019. *How to stimulate effective public engagement on the ethics of Artificial Intelligence*. Retrieved from https://www.involve.org.uk/sites/default/files/field/attachemnt/How%20to%20stimulate%20effective%20public%20debate%20on%20the%20ethics%20of%20artificial%20intelligence%20.pdf

Jacobs, N., and A. Huldtgren. 2018. Why value sensitive design needs ethical commitments. *Ethics and Information Technology*. https://doi.org/10.1007/s10676-018-9467-3.

Jobin, A., M. Ienca, and E. Vayena. 2019. *Artificial Intelligence: The global landscape of ethics guidelines*. arXiv:1906.11668 [Cs].

Johansson, F.D., U. Shalit, and D. Sontag. 2016. *Learning representations for counterfactual inference*. arXiv:1605.03661 [Cs, Stat].

Kemper, J., and D. Kolkman. 2018. Transparent to whom? No algorithmic accountability without a critical audience. *Information, Communication & Society*. https://doi.org/10.1080/1369118X.2018.1477967.

Kleinberg, J., S. Mullainathan, and M. Raghavan 2016. *Inherent trade-offs in the fair determination of risk scores*. arXiv:1609.05807 [Cs, Stat]. Retrieved from http://arxiv.org/abs/1609.05807

Kleinberg, J., H. Lakkaraju, J. Leskovec, J. Ludwig, and S. Mullainathan. 2017. Human decisions and machine predictions. *The Quarterly Journal of Economics*. https://doi.org/10.1093/qje/qjx032.

Knight, W. 2019. Why does Beijing suddenly care about AI ethics? *MIT Technology Review*. Retrieved from https://www.technologyreview.com/s/613610/why-does-china-suddenly-care-about-ai-ethic s-and-privacy/

Knoppers, B.M., and A.M. Thorogood. 2017. Ethics and big data in health. *Current Opinion in Systems Biology* 4: 53–57. https://doi.org/10.1016/j.coisb.2017.07.001.

Kolter, Z., and A. Madry 2018. *Materials for tutorial adversarial robustness: Theory and practice*. Retrieved from https://adversarial-ml-tutorial.org/

Kroll, J.A. 2018. The fallacy of inscrutability. *Philosophical Transactions of the Royal Society A: Mathematical, Physical and Engineering Sciences* 376 (2133): 20180084. https://doi.org/10.1098/rsta.2018.0084.

La Fors, K., B. Custers, and E. Keymolen. 2019. Reassessing values for emerging big data technologies: Integrating design-based and application-based approaches. *Ethics and Information Technology*. https://doi.org/10.1007/s10676-019-09503-4.

Lakkaraju, H., J. Kleinberg, J. Leskovec, J. Ludwig, and S. Mullainathan. 2017. The selective labels problem: Evaluating algorithmic predictions in the presence of unobservables. In *Proceedings of the 23rd ACM SIGKDD international conference on knowledge discovery and data mining—KDD'17*, 275–284. https://doi.org/10.1145/3097983.3098066.

Lepri, B., N. Oliver, E. Letouzé, A. Pentland, and P. Vinck. 2018. Fair, transparent, and accountable algorithmic decision-making processes: The premise, the proposed solutions, and the open challenges. *Philosophy & Technology* 31 (4): 611–627. https://doi.org/10.1007/s13347-017-0279-x.

Lessig, L., and L. Lessig. 2006. *Code (Version 2.0)*. New York: Basic Books.

Lighthill, J. 1973. *'Artificial Intelligence: A general survey' in Artificial Intelligence: A paper symposium*. Retrieved from UK Science Research Council website: http://www.chilton-computing.org.uk/inf/literature/reports/lighthill_report/p001.htm

Lipton, Z.C. 2016. *The mythos of model interpretability*. arXiv:1606.03490 [Cs, Stat].

Lundberg, S.M., and S.-I. Lee. 2017. A unified approach to interpreting model predictions. In *Advances in neural information processing systems 30*, ed. I. Guyon, U.V. Luxburg, S. Bengio, H. Wallach, R. Fergus, S. Vishwanathan, and R. Garnett, 4765–4774. Retrieved from http://papers.nips.cc/paper/7062-a-unified-approach-to-interpreting-model-predictions.pdf.

Makri, E.-L., and C. Lambrinoudakis. 2015. Privacy principles: Towards a common privacy audit methodology. In *Trust, privacy and security in digital business*, ed. S. Fischer-Hübner, C. Lambrinoudakis, and J. López, vol. 9264, 219–234. Cham: Springer.

Matzner, T. 2014. Why privacy is not enough privacy in the context of "ubiquitous computing" and "big data". *Journal of Information, Communication and Ethics in Society* 12 (2): 93–106. https://doi.org/10.1108/JICES-08-2013-0030.

Mikhailov, D. 2019. *A new method for ethical data science*. Retrieved from Medium website: https://medium.com/wellcome-data-labs/a-new-method-for-ethical-data-science-edb59e400ae9

Miller, C., and R. Coldicott 2019. *People, power and technology: The tech workers' view*. Retrieved from Doteveryone website: https://doteveryone.org.uk/report/workersview/

Mingers, J. 2011. Ethics and OR: Operationalising discourse ethics. *European Journal of Operational Research* 210 (1): 114–124. https://doi.org/10.1016/j.ejor.2010.11.003.

Mingers, J., and G. Walsham. 2010. Toward ethical information systems: The contribution of discourse ethics. *MIS Quarterly: Management Information Systems* 34 (4): 855–870.

Mitchell, M., S. Wu, A. Zaldivar, P. Barnes, L. Vasserman, B. Hutchinson, E. Spitzer, I.D. Raji, and T. Gebru. 2019. Model cards for model reporting. In *Proceedings of the conference on fairness, accountability, and transparency—FAT*'19*, 220–229. https://doi.org/10.1145/3287560.3287596.

Mittelstadt, B.D., P. Allo, M. Taddeo, S. Wachter, and L. Floridi. 2016. The ethics of algorithms: Mapping the debate. *Big Data & Society* 3 (2): 205395171667967. https://doi.org/10.1177/2053951716679679.

Nissenbaum, H. 2004. Privacy as contextual integrity. *Washington Law Review* 79: 119.

OECD. 2019a. *Forty-two countries adopt new OECD principles on Artificial Intelligence*. Retrieved from https://www.oecd.org/science/forty-two-countries-adopt-new-oecd-principles-on-artificial-intelligence.htm

———. 2019b. *Recommendation of the Council on Artificial Intelligence*. Retrieved from https://legalinstruments.oecd.org/en/instruments/OECD-LEGAL-0449

Oetzel, M.C., and S. Spiekermann. 2014. A systematic methodology for privacy impact assessments: A design science approach. *European Journal of Information Systems* 23 (2): 126–150. https://doi.org/10.1057/ejis.2013.18.

Overdorf, R., B. Kulynych, E. Balsa, C. Troncoso, and S. Gürses. 2018. *Questioning the assumptions behind fairness solutions*. arXiv:1811.11293 [Cs].

Oxborough, C., E. Cameron, A. Rao, A. Birchall, A. Townsend, and C. Westermann 2018. *Explainable AI: Driving business value through greater understanding*. Retrieved from PWC website: https://www.pwc.co.uk/audit-assurance/assets/explainable-ai.pdf

Peters, D., and R.A. Calvo. 2019. *Beyond principles: A process for responsible tech*, May 2. Retrieved from Medium website: https://medium.com/ethics-of-digital-experience/beyond-principles-a-process-for-responsible-tech-aefc921f7317

Polykalas, S.E., and G.N. Prezerakos. 2019. When the mobile app is free, the product is your personal data. *Digital Policy, Regulation and Governance* 21 (2): 89–101. https://doi.org/10.1108/DPRG-11-2018-0068.

Poursabzi-Sangdeh, F., D.G. Goldstein, J.M. Hofman, J.W. Vaughan, and H. Wallach. 2018. *Manipulating and measuring model interpretability*. arXiv:1802.07810 [Cs].

PWC. 2019. *The PwC responsible AI framework*. Retrieved from https://www.pwc.co.uk/services/audit-assurance/risk-assurance/services/technology-risk/technology-risk-insights/accelerating-innov ation-through-responsible-ai.html

Reisman, D., J. Schultz, K. Crawford, and M. Whittaker 2018. *Algorithmic impact assessments: A practical framework for public agency accountability*. Retrieved from AINow website: https://ainowinstitute.org/aiareport2018.pdf

Ribeiro, M.T., S. Singh, and C. Guestrin. 2016. *Local interpretable model-agnostic explanations (LIME): An introduction a technique to explain the predictions of any machine learning classifier*, August 12. Retrieved from https://www.oreilly.com/learning/introduction-to-local-interpretable-model-agnostic-explanations-lime

Royakkers, L., J. Timmer, L. Kool, and R. van Est. 2018. Societal and ethical issues of digitization. *Ethics and Information Technology* 20 (2): 127–142. https://doi.org/10.1007/s10676-018-9452-x.

Russell, C., M.J. Kusner, J. Loftus, and R. Silva. 2017. When worlds collide: Integrating different counterfactual assumptions in fairness. In *Advances in neural information processing systems 30*, ed. I. Guyon, U.V. Luxburg, S. Bengio, H. Wallach, R. Fergus, S. Vishwanathan, and R. Garnett, 6414–6423. Retrieved from http://papers.nips.cc/paper/7220-when-worlds-collide-integ rating-different-counterfactual-assumptions-in-fairness.pdf.

Saltz, J.S., and N. Dewar. 2019. Data science ethical considerations: A systematic literature review and proposed project framework. *Ethics and Information Technology*. https://doi.org/10.1007/s10676-019-09502-5.

Samuel, A.L. 1960. Some moral and technical consequences of automation—A refutation. *Science* 132 (3429): 741–742. https://doi.org/10.1126/science.132.3429.741.

Selbst, A.D. 2017. Disparate impact in big data policing. *Georgia Law Review* 52 (1): 109–196.

Spielkamp, M., L. Matzat, K. Penner, M. Thummler, V. Thiel, S. Gießler, and A. Eisenhauer 2019. *Algorithm watch 2019: The AI ethics guidelines global inventory*. Retrieved from https://algorithmwatch.org/en/project/ai-ethics-guidelines-global-inventory/

Stahl, B.C., and D. Wright. 2018. Ethics and privacy in AI and big data: Implementing responsible research and innovation. *IEEE Security and Privacy* 16 (3): 26–33. https://doi.org/10.1109/MSP.2018.2701164.

Taddeo, M., and L. Floridi. 2018. How AI can be a force for good. *Science* 361 (6404): 751–752. https://doi.org/10.1126/science.aat5991.

Turilli, M. 2007. Ethical protocols design. *Ethics and Information Technology* 9 (1): 49–62. https://doi.org/10.1007/s10676-006-9128-9.

———. 2008. Ethics and the practice of software design. In *Current issues in computing and philosophy*, ed. A. Briggle, P. Brey, and K. Waelbers. Amsterdam: IOS Press.

Turilli, M., and L. Floridi. 2009. The ethics of information transparency. *Ethics and Information Technology* 11 (2): 105–112. https://doi.org/10.1007/s10676-009-9187-9.

Vakkuri, V., K.-K. Kemell, J. Kultanen, M. Siponen, and P. Abrahamsson. 2019. *Ethically aligned design of autonomous systems: Industry viewpoint and an empirical study*. arXiv:1906.07946 [Cs].

Vaughan, J., and H. Wallach. 2016. *The inescapability of uncertainty: AI, uncertainty, and why you should vote no matter what predictions say*. Retrieved 4 July 2019, from Points. Data Society web-site: https://points.datasociety.net/uncertainty-edd5caf8981b

Wachter, S., B. Mittelstadt, and L. Floridi. 2017. Why a right to explanation of automated decision-making does not exist in the general data protection regulation. *International Data Privacy Law* 7 (2): 76–99. https://doi.org/10.1093/idpl/ipx005.

Wiener, N. 1961. *Cybernetics: Or control and communication in the animal and the machine*. 2d ed. New York: MIT Press.

Winfield, A. 2019. *An updated round up of ethical principles of robotics and AI*, April 18. Retrieved from http://alanwinfield.blogspot.com/2019/04/an-updated-round-up-of-ethical.html

Yetim, F. 2019. Supporting and understanding reflection on persuasive technology through a reflection schema. In *Persuasive technology: Development of persuasive and behavior change support systems*, ed. H. Oinas-Kukkonen, K.T. Win, E. Karapanos, P. Karppinen, and E. Kyza, 43–51. Cham: Springer.

Chapter 11
The Explanation Game: A Formal Framework for Interpretable Machine Learning

David S. Watson and Luciano Floridi

Abstract We propose a formal framework for interpretable machine learning. Combining elements from statistical learning, causal interventionism, and decision theory, we design an idealised *explanation game* in which players collaborate to find the best explanation(s) for a given algorithmic prediction. Through an iterative procedure of questions and answers, the players establish a three-dimensional Pareto frontier that describes the optimal trade-offs between explanatory accuracy, simplicity, and relevance. Multiple rounds are played at different levels of abstraction, allowing the players to explore overlapping causal patterns of variable granularity and scope. We characterise the conditions under which such a game is almost surely guaranteed to converge on a (conditionally) optimal explanation surface in polynomial time, and highlight obstacles that will tend to prevent the players from advancing beyond certain explanatory thresholds. The game serves a descriptive and a normative function, establishing a conceptual space in which to analyse and compare existing proposals, as well as design new and improved solutions.

Keywords Algorithmic explainability · Explanation game · Interpretable machine learning · Pareto frontier · Relevance

D. S. Watson (✉) · L. Floridi
Oxford Internet Institute, University of Oxford, Oxford, UK
e-mail: david.watson@oii.ox.ac.uk; luciano.floridi@oii.ox.ac.uk

L. Floridi (ed.), *Ethics, Governance, and Policies in Artificial Intelligence*,
Philosophical Studies Series 144, https://doi.org/10.1007/978-3-030-81907-1_11

11.1 Introduction

Machine learning (ML) algorithms have made enormous progress on a wide range of tasks in just the last few years. Some notable recent examples include mastering perfect information games like chess and Go (Silver et al. 2018), diagnosing skin cancer (Esteva et al. 2017), and proposing new organic molecules (Segler et al. 2018). These technical achievements have coincided with the increasing ubiquity of ML, which is now widely used across the public and private sectors for everything from film recommendations (Bell and Koren 2007) and sports analytics (Bunker and Thabtah 2019) to genomics (Zou et al. 2019) and predictive policing (Perry et al. 2013). ML algorithms are expected to continue improving as hardware becomes increasingly efficient and datasets grow ever larger, providing engineers with all the ingredients they need to create more sophisticated models for signal detection and processing.

Recent advances in ML have raised a number of pressing questions regarding the epistemic status of algorithmic outputs. One of the most hotly debated topics in this emerging discourse is the role of explainability. Because many of the top performing models, such as deep neural networks, are essentially black boxes – dazzlingly complex systems optimised for predictive accuracy, not user intelligibility – some fear that this technology may be inappropriate for sensitive, high-stakes applications. The call for more explainable algorithms has been especially urgent in areas like clinical medicine (Watson et al. 2019) and military operations (Gunning 2017), where user trust is essential and errors could be catastrophic. This has led to a number of international policy frameworks that recommend explainability as a requirement for any ML system (Floridi and Cowls 2019).

Explainability is fast becoming a top priority in statistical research, where it is often abbreviated as xAI (explainable Artificial Intelligence) or iML (interpretable Machine Learning). We adopt the latter initialism here to emphasise our focus on supervised learning algorithms (formally defined in Sect. 11.3.1) as opposed to other, more generic artificial intelligence applications.

Several commentators have argued that the central aim of iML is underspecified (Doshi-Velez and Kim 2017; Lipton 2018). They raise concerns about the irreducible subjectivity of explanatory success, a concept that they argue is poorly defined and difficult or impossible to measure. In this article, we tackle this problem head on. We provide a formal framework for conceptualising the goals and constraints of iML systems by designing an idealised *explanation game*. Our model clarifies the trade-offs inherent in any iML solution, and characterises the conditions under which epistemic agents are almost surely guaranteed to converge on an optimal set of explanations in polynomial time. The game serves a descriptive and a normative function, establishing a conceptual space in which to analyse and compare existing proposals, as well as design new and improved solutions.

The remainder of this paper is structured as follows. In Sect. 11.2, we identify three distinct goals of iML. In Sect. 11.3, we review relevant background material. We clarify the scope of our proposal in Sect. 11.4. In Sect. 11.5, we articulate the rules of the explanation game and outline the procedure in pseudocode. A discussion follows in Sect. 11.6. We consider five objections in Sect. 11.7, before concluding in Sect. 11.8.

11.2 Why Explain Algorithms?

We highlight three goals that guide those working in iML: to *audit*, to *validate*, and to *discover*. These objectives help motivate and focus the discussion, providing an intuitive typology for the sorts of explanations we are likely to seek and value in this context. Counterarguments to the project of iML are delayed until Sect. 11.7.

11.2.1 Justice as (Algorithmic) Fairness

Perhaps the most popular reason to explain algorithms is their large and growing social impact. ML has been used to help evaluate loan applications (Munkhdalai et al. 2019) and student admissions (Waters and Miikkulainen 2014), predict criminal recidivism (Dressel and Farid 2018), and identify military targets (Nasrabadi 2014), to name just a few controversial examples. Failure to properly screen training datasets for biased inputs threatens to automate injustices already present in society (Mittelstadt et al. 2016). For instance, studies have indicated that algorithmic profiling consistently shows online advertisements for higher paying jobs to men over women (Datta et al. 2015); that facial recognition software is often trained on predominantly white subjects, making them inaccurate classifiers for black and brown faces (Buolamwini and Gebru 2018); and that predatory lenders use financial data to disproportionately target poor communities (Eubanks 2018). Critics point to these failures and argue that there is a dearth of fairness, accountability, and transparency in ML – collectively acronymised as FAT ML, an annual conference on the subject that began meeting in 2014.

Proponents of FAT ML were only somewhat mollified by the European Union's 2018 General Data Protection Regulation (GDPR), which includes language suggesting a so-called "right to explanation" for citizens subject to automated decisions. Whether or not the GDPR in fact guarantees such a right – some commentators insist that it does (Goodman and Flaxman 2017; Selbst and Powles 2017), while others challenge this reading (Edwards and Veale 2017; Wachter et al. 2017) – there is no question that policymakers are beginning to seriously consider the social impact of ML, and perhaps even take preliminary steps towards regulating the industries that rely on such technologies (HLEGAI 2019; OECD 2019). Any attempt to do so, however, will require the technical ability to audit algorithms in order to rigorously test whether they discriminate on the basis of protected attributes such as race and gender (Barocas and Selbst 2016).

11.2.2 The Context of (Algorithmic) Justification

Shifting from ethical to epistemological concerns, many iML researchers emphasise that their tools can help debug algorithms that do not perform properly. The classic problem in this context is *overfitting*, which occurs when a model predicts well on training data but fails on test data. This happened, for example, with a recent image classifier designed to distinguish between farm animals (Lapuschkin et al. 2016). The model attained 100% accuracy on in-sample evaluations but mislabelled all the horses in a subsequent test set. Close examination revealed that the training data included a small watermark on all and only the horse images. The algorithm had learned to associate the label "horse" not with equine features, as one might have hoped, but merely with this uninformative trademark.

The phenomenon of overfitting, well known and widely feared in the ML community, will perhaps be familiar to epistemologists as a sort of algorithmic Gettier case (Gettier 1963). If a high-performing image classifier assigns the label "horse" to a photograph of a horse, then we have a justified true belief that this picture depicts a horse. But when that determination is made on the basis of a watermark, something is not quite right. Our path to the fact is somehow crooked, coincidental. The model is right *for the wrong reasons*. Any true judgments made on this basis are merely cases of epistemic luck, as when we correctly tell the time by looking at a clock that stopped exactly 24 hours before.

Attempts to circumvent problems like this typically involve some effort to ensure that agents and propositions stand in the proper relation, i.e. that some reliable method connects knower and knowledge. Process reliabilism was famously championed by Goldman (1979), who arguably led the vanguard of what Williams calls "the reliabilist revolution" (2016) in anglophone epistemology. Floridi (2004) demonstrates the logical unsolvability of the Gettier problem (in non-statistical contexts), while his network theory of account (2012) effectively establishes a pragmatic, reliabilist workaround.

Advances in iML represent a statistical answer to the reliabilist challenge, enabling sceptics to analyse the internal behaviour of a model when deliberating on particular predictions. This is the goal, for instance, of all local linear approximation techniques, including popular iML algorithms like LIME (Ribeiro et al. 2016) and SHAP (Lundberg and Lee 2017), which assign weights to input variables so users can verify that the model has not improperly focused on uninformative features like the aforementioned watermark. These methods will be examined more closely in Sect. 11.6.

11.2.3 The Context of (Algorithmic) Discovery

We consider one final motivation for iML: *discovery*. This subject has so far received relatively little attention in the literature. However, we argue that it could in fact turn out to be one of the most important achievements of the entire algorithmic explainability project, and therefore deserves special attention.

Suppose we design an algorithm to predict subtypes of some poorly understood disease using biomolecular data. The model is remarkably accurate. It unambiguously classifies patients into distinct groups with markedly different prognostic trajectories. Its predictions are robust and reliable, providing clinicians with actionable advice on treatment options and suggesting new avenues for future research. In this case, we want iML methods not to audit for fairness or test for overfitting, but to reveal underlying mechanisms. The algorithm has clearly learned to identify and exploit some subtle signal that has so far defied human detection. If we want to learn more about the target system, then iML techniques applied to a well-specified model offer a relatively cheap and effective way to identify key features and generate new hypotheses.

The case is not purely hypothetical. A wave of research in the early 2000s established a connection between transcriptomic signatures and clinical outcomes for breast cancer patients (e.g., Sørlie et al. 2001; van 't Veer et al. 2002; van de Vijver et al. 2002). The studies employed a number of sophisticated statistical techniques, including unsupervised clustering and survival analysis. Researchers found, among other things, a strong association between BRCA1 mutations and basal-like breast cancer, an especially aggressive form of the disease. Genomic analysis remains one of the most active and promising areas of research in the natural sciences, and whole new subfields of ML have emerged to tackle the unique challenges presented by these high-dimensional datasets (Bühlmann et al. 2016; Hastie et al. 2015). Successful iML strategies will be crucial to realising the promise of high-throughput sciences.

11.3 Formal Background

In this section, we introduce concepts and notation that will be used throughout the remainder of the paper. Specifically, we review the basic formalisms of supervised learning, causal interventionism, and decision theory.

11.3.1 Supervised Learning

The goal in supervised learning is to estimate a function that maps a set of predictor variables to some outcome(s) of interest. To discuss learning algorithms with any formal clarity, we must make reference to values, variables, vectors, and matrices. We denote scalar values using lowercase italicised letters, e.g. x. Variables, by contrast, are identified by uppercase italicized letters, e.g. X. Matrices, which consist of rows of observations and columns of variables, are denoted by uppercase boldfaced letters, e.g. $\mathbf{X}$. We sometimes index values and variables using matrix notation, such that the ith element of variable X is x_i and the jth variable of the matrix $\mathbf{X}$ is X_j. The scalar x_{ij} refers to the ith element of the jth variable in $\mathbf{X}$. When referring to a row-vector, such as the coordinates that identify the ith observation in $\mathbf{X}$, we use lowercase, boldfaced, and italicised notation, e.g. $\boldsymbol{x}_i$.

Each observation in a training dataset consists of a pair $z_i = (\boldsymbol{x}_i, y_i)$, where $\boldsymbol{x}_i$ denotes a point in d-dimensional space, $\boldsymbol{x}_i = (x_{i1}, \ldots, x_{id})$, and y_i represents the corresponding outcome. We assume that samples are independently and identically distributed according to some fixed but unknown joint probability distribution $\mathbb{P}(\mathbf{Z}) = \mathbb{P}(\mathbf{X}, Y)$. Using n observations, an algorithm maps a dataset to a function, a: $\mathbf{Z} \rightarrow f$; the function in turn maps features to outcomes, f: $\mathbf{X} \rightarrow Y$. We consider both cases where Y is categorical (in which case f is a classifier) and where Y is continuous (in which case f is a regressor). We make no additional assumptions about the structure or properties of f.

Model f is judged by its ability to *generalise*, i.e. to accurately predict outcomes on test data sampled from $\mathbb{P}(\mathbf{Z})$ but not included in the training dataset. For a given test sample $\boldsymbol{x}_i$, we compute the predicted outcome $f(\boldsymbol{x}_i) = \widehat{y}_i$ and observe the true outcome y_i. The hat symbol denotes that the value has been estimated. A model's performance is measured by a loss function L, which quantifies the distance between Y and $\widehat{Y}$ over a set of test cases. The expected value of this loss function with respect to $\mathbb{P}(\mathbf{Z})$ for a given model f is called the *risk*:

$$R(f, \mathbf{Z}) = \mathbb{E}_{\mathbb{P}(\mathbf{Z})}[L(f, \mathbf{Z})] \tag{11.1}$$

We estimate this population parameter with the empirical risk over a set of n samples:

$$R_{\text{emp}}(f, \mathbf{Z}) = \frac{1}{n} \sum_i L(f, z_i) \tag{11.2}$$

A learning algorithm is said to be *consistent* if empirical risk converges to true risk as $n \rightarrow \infty$. A fundamental result of statistical learning theory states that an algorithm is consistent if and only if the space of functions it can learn is of finite VC dimension (Vapnik and Chervonenkis 1971). This latter parameter is a capacity measure defined as the cardinality of the largest set of points the algorithm can shatter.[1] The finite VC dimension criterion will be important to define convergence conditions for the explanation game in Sect. 11.5.3.

Some philosophers have argued that statistical learning provides a rigorous foundation for all inductive reasoning (Corfield et al. 2009; Harman and Kulkarni 2007). Although we are sympathetic to this position, none of the proceeding analysis depends upon this thesis.

[1]The class of sets C shatters the set A if and only if for each $a \subset A$, there exists some $c \in C$ such that $a = c \cap A$. For more on VC theory, see (Vapnik 1995, 1998). Popper's "degree of falsifiability" arguably anticipates the VC dimension. For a discussion, see (Corfield et al. 2009).

11.3.2 Causal Interventionism

Philosophers often distinguish between causal explanations (for natural events) and personal reasons (for human decisions). It is also common – though extremely misleading – to speak of algorithmic "decisions". Thus, we may be tempted to seek *reasons* rather than *causes* for algorithmic predictions, on the grounds that they are more decision-like than event-like. We argue that this is mistaken in several respects. First, the talk of algorithmic "decisions" is an anthropomorphic trope granting statistical models a degree of autonomy that dangerously downplays the true role of human agency in sociotechnical systems (Watson 2019). Second, we may want to explain not just the top label selected by a classifier – the so-called "decision" – but also the complete probability distribution over possible labels. In a regression context, we may want to explain a prediction interval in addition to a mere point estimate. Finally, there are good pragmatic reasons to take a causal approach to this problem. As we argue in Sect. 11.4, it is relatively easy and highly informative to simulate the effect of causal interventions on supervised learning models, provided sufficient access.

Our approach therefore builds on the causal interventionist framework originally formalised by Pearl (2000) and Spirtes et al. (2000), and later given more philosophical treatment by Woodward (2003, 2008, 2010, 2015). A minimal explication of the theory runs as follows. X is a cause of Y within a given structural model $\mathcal{M}$ if and only if some hypothetical intervention on X (and no other variable) would result in a change in Y or the probability distribution of Y. This account is minimal in the sense that it places no constraints on $\mathcal{M}$ and imposes no causal efficacy thresholds on X or Y. The notion of an intervention is kept maximally broad to allow for any possible change in X, provided it does not alter the values of other variables in $\mathcal{M}$ except those that are causal descendants of X.

Under certain common assumptions,[2] Pearl's *do*-calculus provides a complete set of formal tools for reasoning about causal interventions (Huang and Valtorta 2006). A key element of Pearl's notation system is the *do* operator, which allows us to denote, for example, the probability of Y, conditional on an intervention that sets variable X to value x, with the concise formula $\mathbb{P}(Y|do(X = x))$. A structural causal model $\mathcal{M}$ is a tuple $\langle \mathbf{U}, \mathbf{V}, F \rangle$ consisting of exogenous variables $\mathbf{U}$, endogenous variables $\mathbf{V}$, and a set of functions F that map each V_j's causal antecedents to its observed values. $\mathcal{M}$ may be visually depicted as a graph with nodes corresponding to variables and directed edges denoting causal relations between endogenous features (see Fig. 11.1). We restrict our attention here to directed acyclic graphs (DAGs), which are the focus of most work in causal interventionism.

[2]The completeness of the *do*-calculus relies on the causal Markov and faithfulness conditions, which together state (roughly) that statistical independence implies graphical independence and vice versa. Neither assumption has gone unchallenged. We refer interested readers to (Hausman and Woodward 2004) and (Cartwright 2002) for a debate on the former; see (Cartwright 2007) and (Weinberger 2018) for a discussion of the latter.

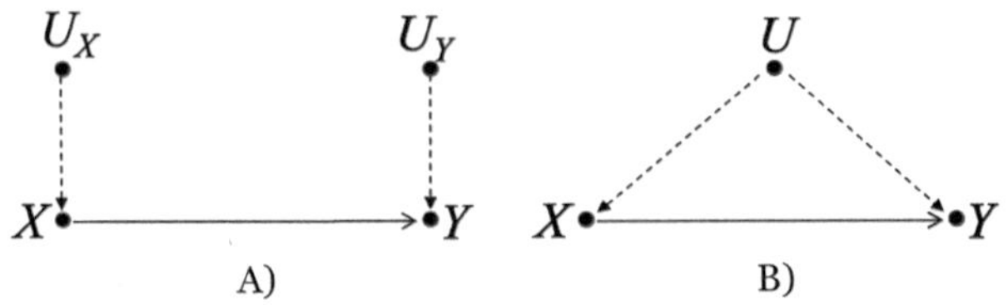

Fig. 11.1 Two examples of simple causal models. (**a**) A Markovian graph. Two exogenous variables, U_X and U_Y, have unobserved causal effects on two endogenous variables, X and Y, respectively. (**b**) A semi-Markovian graph. A single exogenous variable, U, has unobserved confounding effects on two endogenous variables, X and Y

If the model $\mathcal{M}$ contains no exogenous confounders, then $\mathcal{M}$ is said to be *Markovian*. In this case, factorisation of a graph's joint distribution is straightforward and causal effects can be computed directly from the data. However, when one or more unobserved variables has a confounding effect on two or more observed variables, as in Fig. 11.1b, then we say that $\mathcal{M}$ is *semi-Markovian*, and more elaborate methods are needed to estimate causal effects. Specifically, some sort of adjustment must be made by conditioning on an appropriate set of covariates. While several overlapping formulations have been proposed for such adjustments (Galles and Pearl 1995; Pearl 1995; Robins 1997), we follow Tian and Pearl (2002), who provide a provably sound and complete set of causal identifiability conditions for semi-Markovian models (Huang and Valtorta 2008; Shpitser and Pearl 2008).

Their criteria are as follows. The causal effect of the endogenous variable V_j on all observed covariates $\mathbf{V}_{-j}$ is identifiable if and only if there is no consecutive sequence of confounding edges between V_j and V_j's immediate successors in the graph. Weaker conditions are sufficient when we focus on a proper subset $\mathbf{S} \subset \mathbf{V}$. In this case, $\mathbb{P}(\mathbf{S}|\,do(V_j = v_{ij}))$ is identifiable so long as there is no consecutive sequence of confounding edges between V_j and V_j's children in the subgraph composed of the ancestors of $\mathbf{S}$.

We take it that the goal in most iML applications is to provide a causal explanation for one or more algorithmic outputs. Identifiability is therefore a central concern, and another key component to defining convergence conditions in Sect. 11.5.3. Fortunately, as we argue in Sect. 11.4.1, many cases of interest in this setting involve Markovian graphs, and therefore need no covariate adjustments. Semi-Markovian alternatives are considered in Sect. 11.5.2.2, although guarantees cannot generally be provided in such instances without additional assumptions.

If successful, a causal explanation for some algorithmic prediction(s) will accurately answer a range of what Woodward calls "what-if-things-had-been-different questions" (henceforth *w*-questions). For instance, we may want to know what feature(s) about an individual caused her loan application to be denied. What if she had been wealthier? Or older? Would a hypothetical applicant identical to the original except along the axis of wealth or age have had more luck? Several authors in the iML literature explicitly endorse such a counterfactual strategy (Kusner et al. 2017; Wachter et al. 2018). We revisit these methods in Sect. 11.6.

Table 11.1 Utility matrix for John when deciding whether or not to pack his umbrella

	c_1: Rain	c_2: No rain
a_1: Umbrella	1	−1
a_2: No umbrella	−2	0

11.3.3 Decision Theory

Decision theory provides formal tools for reasoning about choices under uncertainty. These will prove useful when attempting to quantify explanatory relevance in Sect. 11.5.2.3. We assume the typical setup, in which an individual considers a finite set of actions A and a finite set of outcomes C. According to expected utility theory,[3] an agent's rational preferences may be expressed as a utility function u that maps the Cartesian product of A and C to the real numbers, $u: A \times C \to \mathbb{R}$. For instance, Jones may be unsure whether to pack his umbrella today. He could do so (a_1), but it would add considerable bulk and weight to his bag; or he could leave it at home (a_2) and risk getting wet. The resulting utility matrix is depicted in Table 11.1.

The rational choice for Jones depends not just on his utility function u but also on his beliefs about whether or not it will rain. These are formally expressed by a (subjective) probability distribution over C, $\mathbb{P}(C)$. We compute each action's expected utility by taking a weighted average over outcomes:

$$\mathbb{E}_{\mathbb{P}(C)}[u(a_i, C)|E] = \sum_j \mathbb{P}(c_j|E)u(a_i, c_j) \tag{11.3}$$

where the set of evidence E is either empty (in which case Eq. 11.3 denotes a prior expectation) or contains some relevant evidence (in which case Eq. 11.3 represents a posterior expectation). Posterior probabilities are calculated in accordance with Bayes's theorem:

$$\mathbb{P}(c_i|E) = \frac{\mathbb{P}(E|c_i)\mathbb{P}(c_i)}{\mathbb{P}(E)} \tag{11.4}$$

which follows directly from the Kolmogorov axioms for the probability calculus (Kolmogorov 1950). By solving Eq. 11.3 for each element in A, we identify at least one utility-maximising action:

$$a^* = \underset{a_i \in A}{\text{argmax}} \mathbb{E}_{\mathbb{P}(C)}[u(a_i, C)|E] \tag{11.5}$$

[3]The von Neumann-Morgenstern representation theorem guarantees the uniqueness (up to affine transformation) of the rational utility function u, provided an agent's preferences adhere to the following four axioms: completeness, transitivity, independence of irrelevant alternatives, and continuity. For the original derivation, see (von Neumann and Morgenstern 1944).

An ideal epistemic agent always selects (one of) the optimal action(s) a^* from a set of alternatives.

It is important to note how a rational agent's beliefs interact with his utilities to guide decisions. If Jones is maximally uncertain about whether or not it will rain, then he assigns equal probability to both outcomes, resulting in expected utilities of

$$\mathbb{E}_{\mathbb{P}(C)}[u(a_1, C)] = 0.5(1) + 0.5(-1) = 0$$

and

$$\mathbb{E}_{\mathbb{P}(C)}[u(a_2, C)] = 0.5(-2) + 0.5(0) = -1,$$

respectively. In this case, Jones should pack his umbrella. But say he gains some new information E that changes his beliefs. Perhaps he sees a weather report that puts the chance of rain at just 10%. Then he will have the following expected utilities:

$$\mathbb{E}_{\mathbb{P}(C)}[u(a_1, C)|E] = 0.1(1) + 0.9(-1) = -0.8$$
$$\mathbb{E}_{\mathbb{P}(C)}[u(a_2, C)|E] = 0.1(-2) + 0.9(0) = -0.2$$

In this case, leaving the umbrella at home is the optimal choice for Jones.

Of course, humans can be notoriously irrational. Experiments in psychology and behavioural economics have shown time and again that people rely on heuristics and cognitive biases instead of consistently applying the axioms of decision theory or probability calculus (Kahneman 2011). Thus, the concepts and principles we outline here are primarily normative. They prescribe an optimal course of behaviour, a sort of Kantian regulative ideal when utilities and probabilities are precise, and posterior distributions are properly calculated. For the practical purposes of iML, these values may be estimated via a hybrid system in which software aids an inquisitive individual with bounded rationality. We revisit these issues in Sect. 11.7.1.

11.4 Scope

Supervised learning algorithms provide some unique affordances that differentiate iML from more general explanation tasks. This is because the target in iML is not the natural or social phenomenon the algorithm was designed to predict, but rather *the algorithm itself*. In other words, we are interested not in the underlying joint distribution $\mathbb{P}(\mathbf{Z}) = \mathbb{P}(\mathbf{X}, Y)$, but in the estimated joint distribution $\mathbb{P}(\mathbf{Z}_f) = \mathbb{P}(\mathbf{X}, \hat{Y})$. The distinction is crucial.

Strevens (2013) differentiates between three modes of understanding: *that*, *why*, and *with*.[4] Understanding *that* some proposition p is true is simply to be aware that p. Understanding *why* p is true requires some causal explanation for p. Strevens's third kind of understanding, however, applies only to theories or models. Understanding *with* a model amounts to knowing how to apply it in order to predict or explain real or potential phenomena. For instance, a physicist who uses Newtonian mechanics to explain the motion of billiard balls thereby demonstrates her ability to understand *with* the theory. Since this model is strictly speaking false, it would be incorrect to say that her explanation provides a true understanding of *why* the billiard balls move as they do. (Of course, she could be forgiven for sparing her poolhall companions the relativistic details of metric tensors and spacetime curvature in this case.) Yet our physicist has clearly understood something – namely the Newtonian theory itself – even if the classical account she offers is inaccurate or incomplete. Similarly, the goal in iML is to help epistemic agents understand *with* the target model f, independent of whatever realities f was intended to capture. The situation is slightly more complicated in the case of discovery (Sect. 11.2.3). The strategy here is to use understanding *with* as an indirect path to understanding *why*, on the assumption that if model f performs well then it has probably learned some valuable information about the target system.

Despite the considerable complexity of some statistical models, as a class they tend to be *complete*, *precise*, and *forthcoming*. These three properties simplify the effort to explain any complex system.

11.4.1 Complete

Model f is complete with respect to the input features $\mathbf{X}$ in the sense that exogenous variables have no influence whatsoever on predicted outcomes $\widehat{Y}$. Whereas nature is full of unobserved confounders that may complicate or undermine even a well-designed study, fitted models are self-contained systems impervious to external variation. They therefore instantiate Markovian, rather than semi-Markovian graphs. This is true even if dependencies between predictors are not explicitly modelled, in which case we may depict f as a simple DAG with directed edges from each feature $X_1, \ldots, X_d$ to $\widehat{Y}$.

In what follows, we presume that the agents in question know which variables were used to train f. This may not always be the case in practice, and without such knowledge it becomes considerably more difficult to explain algorithmic

[4]In what follows, we take it more or less for granted that explanations promote understanding and that understanding requires explanations. Both claims have been disputed. For a discussion, see (de Regt et al. 2009; Grimm 2006; Khalifa 2012). We revisit the relationship between these concepts in Sect. 11.7.2.

predictions. Whatever the epistemic status of the inquiring agent(s), however, the underlying model itself remains complete.

Issues arise when endogenous variables serve as proxies for exogenous variables. For instance, a model may not explicitly include a protected attribute such as race, but instead use a seemingly innocuous covariate like zip code, which is often a strong predictor of race (Datta et al. 2017). In this case, an intervention that changes a subject's race will have no impact on model f's predictions unless we take the additional step of embedding f in a larger causal structure $\mathcal{M}$ that includes a directed edge from race to zip code. We consider possible strategies for resolving problems of this nature in Sect. 11.5.2.2.

11.4.2 Precise

Model f is precise in the sense that it always returns the same output for any particular set of inputs. Whereas a given experimental procedure may result in different outcomes over repeated trials due to irreducible noise, a fitted model has no such internal variability. Some simulation-based approaches, such as the Markov chain Monte Carlo methods widely used in Bayesian data analysis, pose a notable exception to this rule. These models make predictions by random sampling, a stochastic process whose final output is a posterior distribution, not a point estimate. However, if the model has converged, then these predictions are still precise in the limit. As the number of draws from the posterior grows, statistics of interest (e.g., the posterior mode or mean) stabilise to their final values. The Monte Carlo variance of a given parameter can be bounded as a function of the sample size using well-known concentration inequalities (Boucheron et al. 2013).

Woodward (2003, 2010) emphasises the role of "stability" in causal generalisations, a concept that resembles what we call precision. The difference is that stability in Woodward's sense can only be applied to a proper subset of the edges (usually just a single edge) in a causal graph. The generalisation that "variable X causes variable Y" is *stable* to the extent that it persists across a wide range of background conditions, i.e. alternative states of the model $\mathcal{M}$. Precision in our sense requires completeness, because it applies only to the causal relationship between the set of all predictors $\mathbf{X}$ and the outcome Y, which is strictly deterministic at the token level.

11.4.3 Forthcoming

Model f is forthcoming in the sense that it will always provide an output for any well-formed input. Moreover, it is typically quite fast and cheap to query an algorithm in this way. Whereas experiments in the natural or social sciences can often be time-consuming, inconclusive, expensive, or even dangerous, it is relatively simple to answer w-questions in supervised learning contexts. In principle, an analyst could even recreate the complete joint distribution $\mathbb{P}(\mathbf{X}, \widehat{Y})$ simply by saturating the feature space with w-questions. Of course, this strategy is computationally infeasible with continuous predictors and/or a design matrix of even moderate dimensionality.

Supervised learning algorithms may be less than forthcoming when shielded by intellectual property (IP) laws, which can also prevent researchers from accessing a model's complete list of predictors. In lieu of an open access programming interface, some iML researchers resort to reverse engineering algorithms from training datasets with known predicted values. This was the case, for instance, with a famous ProPublica investigation into the COMPAS algorithm, a proprietary model used by courts in several US states to predict the risk of criminal recidivism (Angwin et al. 2016; Larson et al. 2016). Subsequent studies using the same dataset reached different conclusions regarding the algorithm's reliance on race (Fisher et al. 2019; Rudin et al. 2018), highlighting the inherent uncertainty of model reconstruction when the target algorithm is not forthcoming. In what follows, we focus on the ideal case in which our agents face no IP restrictions.

11.5 The Explanation Game

In this section, we introduce a formal framework for iML. Our proposal takes the form of a game in which an inquisitor (call her Alice) seeks an explanation for an algorithmic prediction $f(\boldsymbol{x}_i) = \widehat{y}_i$. Note that our target (at this stage) is a *local* or *token* explanation, rather than a *global* or *type* explanation. In other words, Alice wants to know why this particular input resulted in that particular output, as opposed to the more general task of recreating the entire decision boundary or regression surface of f.

Unfortunately for Alice, f is a black box. But she is not alone. She is helped by a devoted accomplice (call him Bob), who does everything in his power to aid Alice in understanding $\widehat{y}_i$. Bob's goal is to get Alice to a point where she can correctly predict f's outputs, at least in the neighbourhood of $\boldsymbol{x}_i$ and within some tolerable margin of error. In other words, he wants her to be able to give true answers to relevant w-questions about how f would respond to hypothetical datapoints near $\boldsymbol{x}_i$.

We make several nontrivial assumptions about Alice and Bob, some of which were foreshadowed above. Specifically:

- Alice is a rational agent. Her preferences over alternatives are complete and transitive, she integrates new evidence through Bayesian updating, and she does her best to maximise expected utility subject to constraints on her cognitive/computational resources.
- Bob is Alice's accomplice. He has data on the features $\mathbf{V} = \left(X_1, \ldots, X_d, \widehat{Y}\right)$ that are endogenous to f, as well a (possibly empty) set of exogenous variables $\mathbf{U} = (X_{d+1}, \ldots, X_{d+m})$ that are of potential interest to Alice. He may query f with any well-formed input at little or no cost.

We could easily envision more complex explanation games in which some or all of these assumptions are relaxed. Future work will examine such alternatives.

11.5.1 Three Desiderata

According to Woodward (2003, p. 203), the following three criteria are individually necessary and jointly sufficient to explain some outcome of interest $Y = y_i$ that obtains when $X = x_j$ within a given structural model $\mathcal{M}$:

(i) The generalisations described by $\mathcal{M}$ are accurate, or at least approximately so, as are the observations $Y = y_i$ and $X = x_j$.
(ii) According to $\mathcal{M}$, $Y = y_i$ under an intervention that sets X to x_j.
(iii) There exists some possible intervention that sets X to x_k (where $x_j \neq x_k$), with $\mathcal{M}$ correctly describing the value y_l (where $y_i \neq y_l$) that Y would assume under the intervention.

This theory poses no small number of complications that are beyond the scope of this paper.[5] We adopt the framework as a useful baseline for analysis, as it is sufficiently flexible to allow for extensions in a number of directions.

11.5.1.1 Accuracy

Woodward's account places a well-justified premium on explanatory accuracy. Any explanation that fails to meet criteria (i)–(iii) is not deserving of the name. However, this theory does not tell the whole story. To see why, consider a deep convolutional neural network f trained to classify images. The model correctly predicts that $\boldsymbol{x}_i$ depicts a cat. Alice would like to know why. Bob attempts to explain the prediction by writing out the complete formula for f. The neural network contains some hundred layers, each composed of 1 million parameters that together describe a complex nonlinear mapping from pixels to labels. Bob checks against Woodward's criteria and observes that his model $\mathcal{M}$ is accurate, as are the input and output values; that $\mathcal{M}$ correctly predicts the output given the input; and that interventions on the original photograph replacing the cat with a dog do in fact change the predicted label from "cat" to "dog".

Problem solved? Not quite. Bob's causal graph $\mathcal{M}$ is every bit as opaque as the underlying model f. In fact, the two are identical. So while this explanation may be maximally accurate, it is far too complex to be of any use to Alice. The result is not unlike the map of Borges's famous short story (1946), in which imperial

[5]For book length treatments of the topic, see (Halpern 2016; Strevens 2010; Woodward 2003). For relevant articles, see, e.g., (Franklin-Hall 2014; Kinney 2018; Potochnik 2015; Weslake 2010; Woodward and Hitchcock 2003).

cartographers aspire to such exactitude that they draw their territory on a 1:1 scale. Black box explanations of this sort create a kind of Chinese room (Searle 1980), in which the inquiring agent is expected to manually perform the algorithm's computations in order to trace the path from input to output. Just as the protagonist of Searle's thought experiment has no understanding of the Chinese characters he successfully manipulates, so Alice gains no explanatory knowledge about f by instantiating the model herself. Unless she is comfortable computing high-dimensional tensor products on the fly, Alice cannot use $\mathcal{M}$ to build a mental model of the target system f or its behaviour near $\boldsymbol{x}_i$. She cannot answer relevant w-questions without consulting the program, which will merely provide her with new labels that are as unexplained as the original.

11.5.1.2 Simplicity

Accuracy is a necessary but insufficient condition for successful explanation, especially when the underlying system is too complex for the inquiring agent to fully comprehend. In these cases, we tend to value *simplicity* as an inherent virtue of candidate explanations. The point is hardly novel. Simplicity has been cited as a primary goal of scientific theories by practically everyone who has considered the question (cf. Baker 2016). The point is not lost on iML researchers, who typically impose sparsity constraints on possible solutions to ensure a manageable number of nonzero parameters (e.g., Angelino et al. 2018; Ribeiro et al. 2016; Wachter et al. 2018).

It is not always clear just what explanatory simplicity amounts to in algorithmic contexts. One plausible candidate, advocated by Popper (1959), is based on the number of free parameters. In statistical learning theory, this proposal has largely been superseded by capacity measures like the aforementioned VC dimension or Rademacher complexity. These parameters help to establish a syntactic notion of simplicity, which has proven especially fruitful in statistics. Yet such definitions obscure the semantic aspect of simplicity, which is probably of greater interest to epistemic agents like Alice. The kind of simplicity required for her to understand why $f(\boldsymbol{x}_i) = \widehat{y}_i$ depends not just upon the functional relationships between the units of explanation, but more importantly upon the explanatory level of abstraction (Floridi 2008a) – i.e., the choice of units themselves.

Rather than adjudicate between the various competing notions of simplicity that abound in the literature, we opt for a purely relational approach upon which simplicity is just equated with *intelligibility for Alice*. We are unconvinced that there is any sense to be made of an absolute, mind-independent notion of simplicity. Yet even if there is, it would be of little use to Alice if we insist that explanation g_1 is simpler than g_2 on our preferred definition of the term, despite the empirical evidence that she understands the implications of the latter better than the former. What is simple for some agents may be complex for others, depending on background knowledge and contextual factors. In Sect. 11.5.2, we operationalise this observation by measuring simplicity in explicitly agentive terms.

11.5.1.3 Relevance

Some may judge accuracy and simplicity to be sufficient for successful explanation, and in many cases they probably are. But there are important exceptions to this generalisation. Consider, for example, the following case. A (bad) bank issues loans according to just two criteria: applicants must be either white or wealthy. This bank operates in a jurisdiction in which race alone is a protected attribute. A poor black woman named Alice is denied a loan and requests an explanation. The bank informs her that her application was denied due to her finances. This explanation is accurate and simple. However, it is also disingenuous – for it would be just as accurate and simple to say that her loan was denied because of her race, a result that would be of far greater relevance both to Alice and state regulators. Given Alice's interests, the latter explanation is superior to the former, yet the bank's explanation has effectively eclipsed it.

This is a fundamental observation: among the class of accurate and simple explanations, some will be more or less relevant to the inquiring agent (Floridi 2008b). Alice has entered into this game for a reason. Something hangs in the balance. Perhaps she is a loan applicant deciding whether to sue a bank, or a doctor deciding whether to trust an unexpected diagnosis. A successful explanation will not only need to be accurate and simple; it must also inform her decision about how best to proceed. Otherwise, we have a case of *counterfactual eclipse*, in which an agent's interests are overshadowed by a narrow focus on irrelevant facts that do nothing to advance her understanding or help modify future behaviours.

The problem of *counterfactual eclipse* is a serious issue in any context where customers or patients, for example, may wish to receive (or perhaps exercise their right to) an explanation. However, we are unaware of any proposal in the iML literature that explicitly protects against this possibility.

Algorithm 11.1: The Explanation Game
Inputs:

Environment: supervised learner f, endogenous variables $\mathbf{V}$, data $D \sim \mathbb{P}(\mathcal{M})$ possibly including exogenous covariates $\mathbf{U}$

Alice: explanandum $f(\boldsymbol{x}_i) = \widehat{y}_i$, contrastive outcome $f(\boldsymbol{x}_i) \neq \widetilde{y}_i$, level of abstraction LoA, choice set A, causal hypotheses C, utility function u, prior distribution over causal hypotheses $\mathbb{P}(C)$, function space $\mathcal{H}$, loss function $L_{\mathcal{H}}$.

Bob: set of B unique function spaces $\mathcal{G}_b$, loss function $L_{\mathcal{G}}$, kernel $k_{\mathcal{G}}$. If exogenous variables are relevant, then an additional function space $\mathcal{G}'$, loss function $L_{\mathcal{G}'}$, kernel $k_{\mathcal{G}'}$

for each round:

1. Bob creates a map $\psi : \mathcal{Z}_f \rightarrow \mathcal{Z}_g$ from the original f-space to an explanatory g-space designed to (a) shift the input distribution to Alice's desired LoA and (b) help provide evidence for or against at least one hypothesis in C. Whereas $\mathbf{Z}_f = (\mathbf{X}, \widehat{Y})$, $\mathbf{Z}_g = (\mathbf{X}', Y')$.

if $\mathbf{X}'$ includes variables $\mathbf{U}$ that are exogenous to f:

2. Bob trains the model g': $\mathbf{V} \rightarrow \mathbf{U}$, optionally fit using kernel $k_{\mathcal{G}'}$, to minimize loss $L_{\mathcal{G}'}$ over function space $\mathcal{G}'$.
3. Bob creates a training dataset by sampling points $\boldsymbol{v}_s$ from a distribution centred at $\boldsymbol{v}_i$ and repeatedly querying g' with w-questions of the form $\mathbb{E}_{\mathbb{P}(\mathcal{M})}[\mathbf{U}|do(\mathbf{V} = \boldsymbol{v}_s)] =$? The resulting data are mapped to g-space via ψ.

end if
for each function space $\mathcal{G}_b$:

4. Bob creates a training dataset by sampling points $\boldsymbol{x}_s$ from a distribution centred at $\boldsymbol{x}_i$ and repeatedly querying f with w-questions of the form $\mathbb{E}_{\mathbb{P}(\mathbf{Z}_f)}[Y|do(\mathbf{X} = \boldsymbol{x}_s)] =$? The resulting data are mapped to g-space via ψ.
5. Bob trains a model g: $\mathbf{X}' \rightarrow Y'$, optionally fit using kernel $k_{\mathcal{G}}$, to minimize loss $L_{\mathcal{G}}$ over function space $\mathcal{G}_b$. Empirical risk is calculated in f-space via the inverse mapping ψ^{-1}, optionally weighted by $k_{\mathcal{G}}$.
6. Alice creates a training dataset by repeatedly querying g with w-questions of the form $\mathbb{E}_{\mathbb{P}(\mathbf{Z}_g)}\left[Y'|do\left(X'_j = x'_{ij}\right)\right] =$? Bob reports both the predicted outcome and the empirical risk.
7. Alice trains a model h: $\mathbf{X}' \rightarrow Y'$ to minimize loss $L_{\mathcal{H}}$ over function space $\mathcal{H}$. Empirical risk is optionally weighted by $k_{\mathcal{G}}$ and estimated in g-space.
8. The information Alice learns from and about g and h constitutes a body of evidence E, which she uses to update her beliefs regarding C.
9. Alice calculates the posterior expected utility of each action in A, producing at least one optimal choice a*.

Outputs: $R_{\text{emp}}(g, \mathbf{Z}_f), R_{\text{emp}}(h, \mathbf{Z}_g), \mathbb{E}_{\mathbb{P}(C)}[u(a^*, C)|E]$
end for
end for

11.5.2 Rules of the Game

Having motivated an emphasis on accuracy, simplicity, and relevance, we now articulate formal constraints that impose these desiderata on explanations in iML. A schematic overview of the explanation game is provided in pseudocode.

This game has a lot of moving parts, but at its core the process is quite straightforward. Essentially, Bob does his best to proffer an accurate explanation in terms that Alice can understand. She learns by asking w-questions until she feels confident enough to answer such questions herself. The result is scored by three measures: accuracy (error of Bob's model), simplicity (error of Alice's model), and relevance (expected utility for Alice). Note that all explanations are indexed by their corresponding map ψ and explanatory function space $\mathcal{G}_b$. We suppress the dependency for notational convenience. All inputs and steps are discussed in greater detail below.

11.5.2.1 Inputs

Alice must specify a contrastive outcome $f(\boldsymbol{x}_i) \neq \widetilde{y}_i \in Y$. This counterfactual alternative may represent Alice's initial expectation or desired response. Consider, for example, a case in which f is trained to distinguish between handwritten digits, a classic benchmark problem in ML commonly referred to as MNIST, after the most famous database of such images.[6] Say f misclassifies $\boldsymbol{x}_i$ as a "7", when in fact $y_i =$ "1". Alice wants to know not just why the model predicted "7", but also why it did *not* predict "1". Specifying an alternative $\widetilde{y}_i$ is important, as it focuses Bob's attention on relevant regions of the feature space. An explanation such as "Because $\boldsymbol{x}_i$ has no closed loops" may explain why f did not predict "8" or "9", but that is of little use to Alice, as it eclipses the relevant explanation. The importance of contrastive explanation is highlighted by several philosophers (Hitchcock 1999; Potochnik 2015; van Fraassen 1980), and has recently begun to receive attention in the iML literature as well (Miller 2019; Mittelstadt et al. 2019).

We require that Alice state some desired level of abstraction (LoA). The LoA specifies a set of typed variables and observables that are used to describe a system. Inspired by the Formal Methods literature in computer science (Boca et al. 2010), the levelist approach has been extended to conceptualise a wide array of problems in the philosophy of information (Floridi 2008a, 2011, 2017). Alice's desired LoA will help Bob establish the preferred units of explanation, a crucial step toward ensuring intelligibility for Alice. In the MNIST example, Alice is unlikely to seek explanations at the pixel-LoA, but may be satisfied with a higher LoA that deals in curves and edges.

Pragmatism demands that Alice have some notion why she is playing this game. Her choices A, preferences u, and beliefs $\mathbb{P}(C)$ will guide Bob in his effort to supply a satisfactory explanation and constrain the set of possible solutions. The MNIST example is a case of iML for validation (Sect. 11.2.2), in which Alice's choice set may include the option to deploy or not deploy the model f. Her degrees of belief with respect to various causal hypotheses are determined by her expertise in the data and model. Perhaps it is well known that algorithms struggle to differentiate between "7" and "1" when the former appears without a horizontal line through the digit. The cost of such a mistake is factored into her utility function.

[6]The Modified National Institute of Standards and Technology database contains 60,000 training images and 10,000 test images, each 28 × 28 pixel grayscale photos of digits hand-written either by American high school students or United States Census Bureau employees. See http://yann.lecun.com/exdb/mnist/

Bob, for his part, enters into the game with three key components: (i) a set of $B \geq 1$ candidate algorithms for explanation; (ii) a loss function with which to train these algorithms; and (iii) a corresponding kernel. Popular options for (i) include sparse linear models and rule lists. The loss function is left unspecified, but common choices include mean squared error for regression and cross-entropy for classification. The kernel tunes the locality of the explanation, weighting observations by their distance from the original input $\boldsymbol{x}_i$, as measured by some appropriate metric. Whether the kernel is used to train the model g or simply evaluate g's empirical risk is left up to Bob. Abandoning the kernel altogether results in a global explanation, with no particular emphasis on the neighbourhood of $\boldsymbol{x}_i$.

Bob may need an additional algorithm, loss function, and kernel to estimate the relationship between endogenous and exogenous features. If so, there is no obvious requirement that such a model be intelligible to Alice or Bob, so long as it achieves minimal predictive error.

11.5.2.2 Mapping the Space

Perhaps the most consequential step in the entire game is Bob's mapping $\psi : \mathcal{Z}_f \rightarrow \mathcal{Z}_g$. In an effort to provide a successful explanation for Alice, Bob projects the input distribution $\mathbb{P}(\mathbf{Z}_f) = \mathbb{P}(\mathbf{X}, \widehat{Y})$ into a new space $\mathbb{P}(\mathbf{Z}_g) = \mathbb{P}(\mathbf{X}', Y')$. The change in the response variable is set by Alice's contrastive outcome of interest. In the MNIST example, Bob maps the original 10-class variable $\widehat{Y}$ onto a binary variable Y' indicating whether or not inputs are classified as "1". The contents of $\mathbf{X}'$ may be iteratively established by considering Alice's desired LoA and hypothesis set C. This will often amount to a reduction of the feature space. For instance, Bob may coarsen a set of genes into a smaller collection of biological pathways (Sanguinetti and Huynh-Thu 2018), or transform pixels into super-pixels (Stutz et al. 2018).

Alternatively, Bob may need to expand the input features to include exogenous variables hypothesized to be relevant to the outcome. In this case, he will require external data D sampled from the expanded feature space $\mathbb{P}(\mathcal{M})$, which can be used to train one or more auxiliary models to predict values for the extra covariate(s) in unobserved regions of g-space. For instance, when an algorithm is suspected of encoding protected attributes like race via unprotected attributes like zip code, Bob will need to estimate the dependence using a new function g' that predicts the former based on the latter (along with any other relevant endogenous variables). Note that in this undertaking, Bob is essentially back to square one. The target $\mathcal{M}$ is presumably not complete, precise, or forthcoming, and his task therefore reduces to the more general problem of modelling some complex natural or social system with limited information. This inevitably introduces new sources of error that will have a negative impact on downstream results. Depending on the structural properties of the underlying causal graph, effects of interventions in g-space may not be uniquely identifiable.

In any event, the goal at this stage is to make the input features sufficiently intelligible to Alice that they can accommodate her likely w-questions and inform her beliefs about causal hypotheses C. General purpose methods for causal feature learning have been proposed (Chalupka et al. 2017), however, critics have persuasively argued that such procedures cannot be implemented in a context-independent manner (Kinney 2018). Some areas of research, such as bioinformatics and computer vision, have well-established conventions on how to coarsen high-dimensional feature spaces. Other domains may prove more challenging. Accessibility to external data on exogenous variables of interest will likewise vary from case to case. Even when such datasets are readily available, there is no guarantee that the functional relationships sought can be estimated with high accuracy or precision. As in any other explanatory context, Alice and Bob must do the best they can with their available resources and knowledge.

11.5.2.3 Building Models, Scoring Explanations

Once ψ is fixed, the next steps in the explanation game are effectively supervised learning problems. This puts at Alice and Bob's disposal a wide range of well-studied algorithms and imports the corresponding statistical guarantees.

Bob creates a training dataset of $\mathbf{Z}_g = (\mathbf{X}', Y')$ and fits a model g from the explanatory function space $\mathcal{G}_b$. Alice explores g-space by asking a number of w-questions that posit relevant interventions. For instance, she may want to know if the presence of a horizontal line through the middle of a numeral determines whether f predicts a "7". If so, then this will be a hypothesis in C and we should find a corresponding variable in $\mathbf{X}'$. Because we leave open the possibility that the target model f and/or Bob's explanation g may involve implicit or explicit structural equations, we use the *do*-calculus to formalise such interventions.

Bob and Alice can select whatever combination of loss function and algorithm makes the most sense for their given explanation task. g's error is measured by $R_{\text{emp}}(g, \mathbf{Z}_f)$; g's complexity is measured by $R_{\text{emp}}(h, \mathbf{Z}_g)$. We say that g is ε_1-accurate if $R_{\text{emp}}(g, \mathbf{Z}_f) \leq \varepsilon_1$ and ε_2-simple if $R_{\text{emp}}(h, \mathbf{Z}_g) \leq \varepsilon_2$. The content and performance of g and h constitute a body of evidence E, which Alice uses to update her beliefs about causal hypotheses C. Relevance is measured by the posterior expected utility of the utility-maximising action, $\mathbb{E}_{\mathbb{P}(C)}[u(a^*, C)|E]$. (For consistency with the previous desiderata, we in fact measure *irrelevance* by multiplying the relevance by -1.) Bob's explanation is ε_3-relevant to Alice if $-\mathbb{E}_{\mathbb{P}(C)}[u(a^*, C)|E] \leq \varepsilon_3$.

We may now locate explanations generated by this game in three-dimensional space, with axes corresponding to accuracy, simplicity, and relevance. An explanation is deemed *satisfactory* if it does not exceed preselected values of ε_1, ε_2, and ε_3. These parameters can be interpreted as budgetary constraints on Alice and Bob. How much inaccuracy, complexity, and irrelevance can they afford? We assign equal weight to all three criteria here, but relative costs could easily be quantified through a differential weighting scheme. Together, these points define the extremum of a

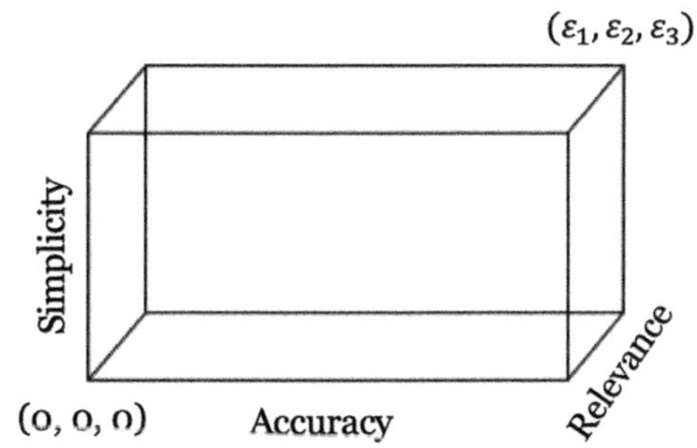

Fig. 11.2 The space of satisfactory explanations is delimited by upper bounds on the error (ε_1), complexity (ε_2), and irrelevance (ε_3) of explanations Alice is willing to accept

cuboid, whose opposite diagonal is the origin (see Fig. 11.2). Any point falling within this cuboid is (ε_1, ε_2, ε_3)-satisfactory.

11.5.3 Consistency and Convergence

The formal tools of statistical learning, causal interventionism, and decision theory provide all the ingredients we need to state the necessary and sufficient conditions for convergence to a conditionally optimal explanation surface in polynomial time.

We define optimality in terms of a Pareto frontier. One explanation Pareto-dominates another if and only if it is strictly better along at least one axis and no worse along any other axis. If Alice and Bob are unable to improve upon the accuracy, simplicity, or relevance of an explanation without incurring some loss along another dimension, then they have found a Pareto-dominant explanation. A collection of such explanations constitutes a Pareto frontier, a surface of explanations from which Alice may choose whichever best aids her understanding and serves her interests. Note that this is a relatively weak notion of optimality. Explanations may be optimal in this sense without even being satisfactory, since the entire Pareto frontier may lie beyond the satisfactory cuboid defined by (ε_1, ε_2, ε_3). In this case, Alice and Bob have two options: (a) accept that no explanation will satisfy the criteria and adjust thresholds accordingly; or (b) start a new round with one or several different input parameters. Option (b) will generate entirely new explanation surfaces for the players to explore.

Without more information about the target function f or specific facts about Alice's knowledge and interests, conditional Pareto dominance is the strongest form of optimality we can reasonably expect. Convergence on a Pareto frontier is almost surely guaranteed on three conditions:

- *Condition 1.* The function spaces $\mathcal{G}_b$ and $\mathcal{H}$ are of finite VC dimension.
- *Condition 2.* Answers to all w-questions are uniquely identifiable.
- *Condition 3.* Alice is a rational agent and consistent Bayesian updater.

Condition (1) entails the statistical consistency of Bob's model g and Alice's model h, which ensures that accuracy and simplicity are reliably measured as sample size grows. Condition (2) entails that simulated datasets are faithful to their underlying data generating processes, thereby ensuring that g and h converge on the right targets. Condition (3) entails the existence of at least one utility-maximising action $a^* \in A$ with well-defined posterior expectation $\mathbb{E}_{\mathbb{P}(C)}[u(a^*, C)|E]$. If her probabilities are well-calibrated, then Alice will tend to pick the "right" action, or at least an action with no superior alternative in A. With these conditions in place, each round of the game will result in an explanation that cannot be improved upon without altering the input parameters.

If all subroutines of the game's inner loops execute in polynomial time, then the round will execute in polynomial time as well. The only potentially NP-hard problem is finding an adequate map ψ, which cannot be efficiently computed without some restrictions on the solution set. A naïve approach would be to consider all possible subsets of the original feature space, but even in the Markovian setting this would result in an unmanageable 2^d maps, where d represents the dimensionality of the input matrix $\mathbf{X}$. Efficient mapping requires some principled method for restricting this space to just those of potential interest for Alice. The best way to do so for any given problem is irreducibly context-dependent.

11.6 Discussion

Current iML proposals do not instantiate the explanation game in any literal sense. However, our framework can be applied to evaluate the merits and shortcomings of existing methods. It also provides a platform through which to conceptualise the constraints and requirements of any possible iML proposal, illuminating the contours of the solution space.

The most popular iML methods in use today are local linear approximators like LIME (Ribeiro et al. 2016) and SHAP (Lundberg and Lee 2017). The former explains predictions by randomly sampling around the point of interest. Observations are weighted by their distance from the target point and a regularised linear model is fit by weighted least squares. The latter builds on foundational work in cooperative game theory, using training data to efficiently compute pointwise approximations of each input feature's Shapley value.[7] The final result in both cases is a (possibly sparse) set of coefficients indicating the positive or negative association between input features and the response, at least near $\boldsymbol{x}_i$ and conditional on the covariates.

Using LIME or SHAP basically amounts to restricting the function space of Bob's explanation model g to the class of regularised linear models. Each method

[7] Shapley values were originally designed to fairly distribute surplus across a coalition of players in cooperative games (Shapley 1953). They are the unique solution to the attribution problem that satisfies certain desirable properties, including local accuracy, missingness, and consistency. Directly computing Shapley values is NP-hard, however numerous approximations have been proposed. See (Sundararajan and Najmi 2019) for an overview.

has its own default kernel k, as well as recommended mapping functions ψ for particular data types. For instance, LIME coarsens image data into super-pixels, while SHAP uses saliency maps to visualise the portions of an input image that were most important in determining its classification. While the authors of the two methods seem to suggest that a single run of either algorithm is sufficient for explanatory purposes, local linear approximations will tend to be unstable for datapoints near especially nonlinear portions of the decision boundary or regression surface. Thus, multiple runs with perturbed data may be necessary to establish the precision of estimated feature weights. This corresponds to multiple rounds of the explanation game, thereby giving Alice a more complete picture of the model space.

One major problem with LIME and SHAP is that neither method allows users to specify a contrast class of interest. The default behaviour of both algorithms is to explain why an outcome is $\widehat{y}_i$ as opposed to $\overline{y}$ – that is, the mean response for the entire dataset (real or simulated). In many contexts, this makes sense. For instance, if Alice receives a rare and unexpected diagnosis, then she may want to know what differentiates her from the majority of patients. However, it seems strange to suggest, as these algorithms implicitly do, that "normal" predictions are inexplicable. There is nothing confusing or improper about Alice wondering, for instance, why she received an average credit score instead of a better-than-average one. Yet in their current form, neither LIME nor SHAP can accommodate such inquiries.

More flexible alternatives exist. Rule lists, which predict outcomes through a series of if-then statements, can model nonlinear effects that LIME and SHAP are incapable of detecting in principle. Several iML solutions are built on recursive partitioning (Guidotti et al. 2018; Ribeiro et al. 2018; Yang et al. 2017) – the statistical procedure that produces rule lists – and a growing number of psychological studies suggests that users find such explanations especially intelligible (Lage et al. 2018). If Alice is one of the many people who shares this preference for rule lists, then Bob should take this into account when selecting $\mathcal{G}_b$.

Counterfactual explanations are endorsed by Wachter et al. (2018), who propose a novel iML solution based on generative adversarial networks (GANs). Building on pioneering research in deep learning (Goodfellow et al. 2014), the authors demonstrate how GANs can be used to find the minimal perturbation of input features sufficient to alter the output in some prespecified manner. These models are less restrictive than linear regressions or rule lists, as they not only allow users to identify a contrast class but can in principle adapt to any differentiable function. Wachter et al. emphasise the importance of simplicity by imposing a sparsity constraint on explanatory outputs intended to automatically remove uninformative features.

Rule lists and GANs have some clear advantages over linear approximators like LIME and SHAP. However, no method in use today explicitly accounts for user interests, an omission that may lead to undesirable outcomes. In short, they do not pass the eclipsing test. Recall the case of the (bad) bank in Sect. 11.5.1.3. Suppose that Alice's choice set contains just two options, $A = \{\text{Sue, Don't sue}\}$, and she considers two causal hypotheses as potential explanations for her denied loan, $C = \{\text{Wealth, Race}\}$. Alice's utility matrix is given in Table 11.2.

Table 11.2 Utility matrix for Alice in the (bad) bank scenario

	c_1: Wealth	c_2: Race
a_1: Sue	−1	5
a_2: Don't sue	0	0

Alice assigns a uniform prior over C to begin with, such that $\mathbb{P}(c_1) = \mathbb{P}(c_2) = 0.5$. She receives two explanations from Bob: g_1, according to which Alice's application was denied due to her wealth; and g_2, according to which Alice's application was denied due to her race. Using misclassification rate as our loss function and assuming a uniform probability mass over the dichotomous features Wealth $\in$ {Rich, Poor} and Race $\in$ {White, Black}, we find that both explanations are equally accurate:

$$R_{\text{emp}}(g_1, \mathbf{Z}_f) = R_{\text{emp}}(g_2, \mathbf{Z}_f) = 0.25$$

and equally simple:

$$R_{\text{emp}}(h, \mathbf{Z}_{g_1}) = R_{\text{emp}}(h, \mathbf{Z}_{g_2}) = 0.$$

However, they induce decidedly different posteriors over C:

$$\mathbb{P}(c_1|g_1) = \mathbb{P}(c_2|g_2) = 0.9$$
$$\mathbb{P}(c_1|g_2) = \mathbb{P}(c_2|g_1) = 0.1$$

The posterior expected utility of a_1 under g_1 is therefore

$$0.9(-1) + 0.1(5) = -0.4,$$

whereas under g_2 the expectation is

$$0.1(-1) + 0.9(5) = 4.4.$$

(The expected utility of a_2 is 0 under both explanations.) Since the utility-maximising action under g_2 is strictly preferable to the utility-maximising action under g_1, we regard g_2 as the superior explanation. In fact, the latter Pareto-dominates the former, since the two are equivalent in terms of accuracy and simplicity but g_1 is strictly less relevant for Alice than g_2. This determination can only be made by explicitly encoding Alice's preferences, which are currently ignored by all major iML proposals.

Methods that fail to pass the eclipsing test pose problems for all three iML goals outlined in Sect. 11.2. Irrelevant explanations can undermine tests of validity or quests of discovery by failing to recognise the epistemological purpose that motivated the question in the first place. When those explanations are accurate and simple, Alice can easily be fooled into thinking she has learned some valuable information. In fact, Bob has merely overfit the data. Matters are even worse when we seek to audit algorithms. In this case, eclipsing explanations may actually offer loopholes to bad actors wishing to avoid controversy over questionable decisions. For instance, a myopic focus on accuracy and simplicity would allow (bad) banks to get away with racist loan policies so long as black applicants are found wanting along some other axis of variation.

11.7 Objections

In this section, we consider five objections of increasing generality. The first three are levelled against our proposed game, the latter two against the entire iML project.

11.7.1 Too Highly Idealised

One obvious objection to our proposal is that it demands a great deal of Alice. She must provide a contrastive outcome $\widetilde{y}_i$, a level of abstraction LoA, a choice set A, some causal hypotheses C, a corresponding prior distribution $\mathbb{P}(C)$, and a utility function u. On top of all that, we also expect her to be a consistent Bayesian updater and expected utility maximiser. If Alice were so well-equipped and fiercely rational, then perhaps cracking black box algorithms would pose no great challenge to her.

Our response is twofold. First, we remind the sceptical reader that idealisations are a popular and fruitful tool in conceptual analysis. There are no frictionless planes or infinite populations, but such assumptions have contributed to successful theories in physics and genetics. Potochnik (2017) makes a compelling case that idealisations are essential to scientific practice, enabling humans to represent and manipulate systems of incomprehensible complexity. Decision theory is no exception. The assumption that epistemic agents always make rational choices – though strictly speaking false – has advanced our understanding of individual and social behaviour in economics, psychology, and computer science.

Second, this setup is not nearly as unrealistic as it may at first appear. It is perfectly reasonable to assume that an agent would seek an algorithmic explanation with at least a counterfactual outcome and choice set to hand, as well as some (tentative) causal hypotheses. For instance, Alice may enter into the game expressly because she suspects her loan application was denied due to her race, and is unsure whether to seek redress. Utilities can be derived through a simple ranking of all action-outcome pairs. If new hypotheses emerge over the course of the game, they can easily be explored in subsequent rounds. Alice may have less confidence in ideal values for LoA and $\mathbb{P}(C)$, but there is no reason to demand certainty about these from the start. Indeed, it is advisable to try out a range of values for each, much like how analysts often experiment with different priors to ascertain the impact on posteriors in Bayesian inference (Gelman et al. 2014). Alice and Bob can iteratively refine their inputs as the rounds pass and track the evolution of the resulting Pareto frontiers to gauge the uncertainty associated with various parameters. Something like this process is how a great deal of research is in fact conducted.

Perhaps most importantly, we stress that Alice and Bob are generalised agents that can and often will be implemented by hybrid systems involving numerous humans and machines working in concert. There is no reason to artificially restrict the cognitive resources of either to that of any specific individual. The problems iML is designed to tackle are beyond the remit of any single person, especially one operating without the assistance of statistical software. When we broaden the cognitive scope of Alice and Bob, the idealisations demanded of them become decidedly more plausible. The only relevant upper bounds on their inferential capacities are computational complexity thresholds. The explanation game is an exercise in sociotechnical epistemology, where knowledge emerges from the continuous interaction of individuals, groups, and technology (Watson and Floridi 2018). The essential point is whether the explanation game we have designed is possible and fruitful, not whether a specific Alice and a specific Bob can actually play it according to their idiosyncratic abilities.

11.7.2 *Infinite Regress*

A common challenge to any account of explanation is the threat of infinite regress. Assuming that explanations must be finite, how can we be sure that some explanatory method concludes at the proper terminus? In this instance, how can we guarantee that the explanation game does not degenerate into an infinite recursive loop? Note that this is not a concern for any fixed Alice and Bob – each round ends once models g and h are scored, and Alice's expected utilities are updated – but the objection appears more menacing over shifting agents and games. For instance, we may worry that Alice and Bob together constitute a new supervised learning algorithm f_2 that maps inputs $\boldsymbol{x}_i$ to outputs $h(\boldsymbol{x}_i')$ through the intermediate model g. The resulting function may now be queried by a new agent Alice$_2$ who seeks the assistance of Bob$_2$ in accounting for some prediction $f_2(\boldsymbol{x}_i)$. This process could repeat indefinitely.

The error in this reasoning is to ignore the vital role of pragmatics. By construction, each game ends at the proper terminus *for that particular Alice*. There is nothing fallacious about allowing other agents to inquire into the products of such games as if they were new algorithms. The result will simply be t steps removed from its original source, where t is the number of Alice-and-Bob teams separating the initial f from the latest inquirer. The effect is not so unlike a game of telephone, where a message gradually degrades as players introduce new errors at each iteration. Similarly, each new Alice-and-Bob pair will do their best to approximate the work of the previous team. The end result may look quite unlike the original f for some large value of t, but that is only to be expected. So long as conditions (1)–

(3) are met for any given Alice and Bob, then they are almost surely guaranteed to converge on a conditionally optimal explanation surface in polynomial time.

11.7.3 Pragmatism + Pluralism = Relativist Anarchy?

The explanation game relies heavily on pragmatic considerations. We explicitly advocate for subjective notions of simplicity and relevance, allowing Bob to construct numerous explanations at various levels of abstraction. This combination of subjectivism and pluralism grates against the realist tradition in epistemology and philosophy of science, according to which there is exactly one true explanans for any given explanandum. Is there not a danger here of slipping into some disreputable brand of outright relativism? If criteria for explanatory success are so irreducibly subjective, is there simply no fact of the matter as to which of two competing explanations is superior? Is this not tantamount to saying that anything goes?

The short answer is no. The objection assumes that for any given fact or event there exists some uniquely satisfactory, mind- and context-independent explanation, presumably in terms of fundamental physical units and laws. Call this view explanatory monism. It amounts to a metaphysical doctrine whose merits or shortcomings are frankly beside the point. For even if the "true" explanation were always available, it would not in general be of much use. The goal of the explanation game is to promote greater *understanding for Alice*. This may come in many forms. For instance, the predictions of image classifiers are often explained by heatmaps highlighting the pixels that most contribute to the given output. The fact that complex mathematical formulae could in this case provide a maximally deep and stable explanation is irrelevant (see Sect. 11.5.1.1). Pragmatic goals require pragmatic strategies. Because iML is fundamentally about getting humans to understand the behaviour of machines, there is a growing call for personalised solutions (Páez 2019). We take this pragmatic turn seriously and propose formal methods to implement it.

We emphatically reject the charge that the explanation game is so permissive that "anything goes". Far from it, we define objective measures of subjective notions that have long defied crisp formalisation. Once values for all variables are specified, it is a straightforward matter to score and compare competing explanations. For any set of input parameters, there exists a unique ordering of explanations in terms of their relative accuracy, simplicity, and relevance. Explanations at different levels of abstraction may be incommensurable, but together they can help Alice form a more complete picture of the target system and its behaviour near the datapoint of interest. This combination of pragmatism and explanatory ecumenism is flexible and rational. It embraces relationalism, not relativism (Floridi 2017). One of the chief contributions of this paper is to demonstrate that the desiderata of iML can be formulated with precision and rigour without sacrificing the subjective and contextual aspects that make each explanation game unique.

11.7.4 No Trade-Off

Some have challenged the widespread assumption that there is an inherent trade-off between accuracy and interpretability in ML. Rudin (2019) argues forcefully against this view, which she suggests is grounded in anecdotal evidence at best, and corporate secrecy at worst. She notes that science has long shown a preference for more parsimonious models, not out of mere aesthetic whimsy, but because of well-founded principles regarding the inherent simplicity of nature (Baker 2016). Recent results in formal learning theory confirm that an Ockham's Razor approach to hypothesis testing is the optimal strategy for convergence to the truth under minimal topological constraints (Kelly et al. 2016).

Breiman (2001) famously introduced the idea of a *Rashomon set*[8] – a collection of models that estimate the same functional relationship using different algorithms and/or hyperparameters, yet all perform reasonably well (say, within 5% of the top performing model). Rudin's argument – expanded in considerable technical detail in a follow up paper (Semenova and Rudin 2019) – is premised on the assumption that sufficiently large Rashomon sets should include at least one interpretable model. If so, then it would seem there is no point in explaining black box algorithms, at least in high-stakes applications such as healthcare and criminal justice. If we must use ML for these purposes, then we should simply train a (globally) interpretable model in the first place, rather than reverse-engineer imperfect post-hoc explanations.

There are two problems with this objection. First, there is no logical or statistical guarantee that interpretable models will outperform black box competitors or even be in the Rashomon set of high-performing models for any given predictive problem. This is a simple corollary of the celebrated no free lunch theorem (Wolpert and Macready 1997), which states (roughly) that there is no one-size-fits-all solution in ML. Any algorithm that performs well on one class of problems will necessarily perform poorly on another. Of course, this cuts both ways – black box algorithms are likewise guaranteed to fail on some datasets. If we value performance above all, which may well be the case for some especially important tasks, then we must be open to models of variable interpretability.

Second, the opacity of black box algorithms is not just a by-product of complex statistical techniques, but of institutional realities that are unlikely to change anytime soon. Pasquale (2015) offers a number of memorable case studies demonstrating how IP law is widely used to protect ML source code and training data not just from potential competitors but from any form of external scrutiny. Even if a firm were using an interpretable model to make its predictions, the model architecture and parameters would likely be subject to strict copyright protections. Some have argued for the creation of independent third-party groups tasked with the responsibility of auditing code under non-disclosure agreements (Floridi et al. 2018; Wachter et al. 2017), a proposal we personally support. However, until such legislation is enacted,

[8]The name comes from Akira Kurosawa's celebrated 1950 film *Rashomon*, in which four characters give overlapping but inconsistent eyewitness accounts of a brutal crime in eighth century Kyoto.

anyone attempting to monitor the fairness, accountability, and transparency of algorithms will almost certainly have no choice but to treat the underlying technology as a black box.

11.7.5 Double Standards

Zerilli et al. (2019) argue that proponents of iML place an unreasonable burden on algorithms by demanding that they not only perform better and faster than humans, but explain why they do so as well. They point out that human decision-making is far from transparent, and that people are notoriously bad at justifying their actions. Why the double standard? We already have systems in place for accrediting human decision-makers in positions of authority (e.g., judges and doctors) based on their demonstrated track record of performance. Why should we expect anything more from machines? The authors conclude that requiring intelligibility of high-performing algorithms is not just unreasonable but potentially harmful if it hinders the implementation of models that could improve services for end users.

Zerilli et al. are right to point out that we are often unreliable narrators of our own internal reasoning. We are liable to rationalise irrational impulses, draw false inferences, and make decisions based on a host of well-documented heuristics and cognitive biases. But this is precisely what makes iML so promising: not that learning algorithms are somehow immune to human biases – they are not, at least not if those biases are manifested in the training data – but rather that, with the right tools, we may conclusively reveal the true reasoning behind consequential decisions. Kleinberg et al. (2019) make a strong case that increased automation will reduce discrimination by inaugurating rigorous, objective procedures for auditing and appealing algorithmic predictions. It is exceedingly difficult under current law to prove that a human has engaged in discriminatory behaviour, especially if they insist that they have not (which most people typically do, especially when threatened with legal sanction). For all the potential harms posed by algorithms, deliberate deception is not (yet) one of them.

We argue that the potential benefits of successful iML strategies are more varied and numerous than Kleinberg et al. acknowledge. To reiterate the motivations listed in Sect. 11.2, we see three areas of particular promise. In the case of algorithmic auditing, iML can help ensure the fair, accountable, and transparent application of complex statistical models in high-stakes applications like criminal justice and healthcare. In the case of validation, iML can be used to test algorithms before and during deployment to ensure that models are performing properly and not overfitting to uninformative patterns in the training data. In the case of discovery, iML can reveal heretofore unknown mechanisms in complex target systems, suggesting new theories and hypotheses for testing. Of course, there is no guarantee that such methods will work in every instance – iML is no panacea – but it would be foolish not to try. The double standard that Zerilli et al. caution against is in fact a welcome opportunity.

11.8 Conclusion

Black box algorithms are here to stay. Private and public institutions already rely on ML to perform basic and complex functions with greater efficiency and accuracy than people. Growing datasets and ever-improving hardware, in combination with ongoing advances in computer science and statistics, ensure that these methods will only become more ubiquitous in the years to come.

There is less reason to believe that algorithms will become any more transparent or intelligible, at least not without the explicit and sustained effort of dedicated researchers in the burgeoning field of iML. We have argued that there are good reasons to value algorithmic interpretability on ethical, epistemological, and scientific grounds. We have outlined a formal framework in which agents can collaborate to explain the outputs of any supervised learner. The explanation game serves both a descriptive function – providing a common language in which to compare iML proposals – and a normative function – highlighting aspects that are underexplored in the current literature and pointing the way to new and improved solutions. Of course, important normative challenges remain. Thorny questions of algorithmic fairness, accountability, and transparency are not all so swiftly resolved. However, we are hopeful that the explanation game can inform these debates in a productive and principled manner.

Future work will relax the assumptions upon which this beta version of the game is based. Of special interest are adversarial alternatives in which Bob has his own utility function to maximise, or three-player versions in which Carol and Bob compete to find superior explanations from which Alice must choose. Other promising directions include implementing semi-automated explanation games with greedy algorithms that take turns maximising one explanatory desideratum at a time until convergence. Similar proposals have already been implemented for optimising mixed objectives in algorithmic fairness (Kearns et al. 2018), but we are unaware of any similar work in explainability. Finally, we intend to expand our scope to unsupervised learning algorithms, which pose a number of altogether different explanatory challenges.

Acknowledgements Thanks to Mariarosaria Taddeo, Robin Evans, David Kinney, and Carl Öhman for their thoughtful comments on earlier drafts of this manuscript. Versions of this paper were originally presented at the University of Oxford's Digital Ethics Lab and the 12th annual MuST Conference on Statistical Reasoning and Scientific Error at Ludwig Maximilian University in Munich, where we also received helpful feedback. Finally, we would like to thank our anonymous reviewers for their thorough reading and valuable contributions.

Funding Luciano Floridi's research for this article was supported by a Fujitsu academic grant.

References

Angelino, E., N. Larus-Stone, D. Alabi, M. Seltzer, and C. Rudin. 2018. Learning certifiably optimal rule lists for categorical data. *Journal of Machine Learning Research* 18 (234): 1–78.

Angwin, J., Larson, J., Mattu, S., & Kirchner, L. (2016). *Machine bias*. Retrieved from https://www.propublica.org/article/machine-bias-risk-assessments-in-criminal-sentencing

Baker, A. 2016. Simplicity. In *The Stanford encyclopedia of philosophy (Winter 201)*, ed. E.N. Zalta. Metaphysics Research Lab, Stanford University.

Barocas, S., and A. Selbst. 2016. Big data's disparate impact. *California Law Review* 104 (1): 671–729.

Bell, R.M., and Y. Koren. 2007. Lessons from the Netflix prize challenge. *SIGKDD Explor. Newsletter* 9 (2): 75–79.

Boca, P.P., J.P. Bowen, and J.I. Siddiqi. 2010. *Formal methods: State of the art and new directions*. London: Springer.

Borges, J.L. 1946/1999. On exactitude in science. In *Collected Fictions*. Trans. Andrew Hurley, 325. New York: Penguin.

Boucheron, S., G. Lugosi, and P. Massart. 2013. *Concentration inequalities: A nonasymptotic theory of independence*. New York: Oxford University Press.

Breiman, L. 2001. Statistical modeling: The two cultures. *Statistical Science* 16 (3): 199–231.

Bühlmann, P., P. Drineas, M. Kane, and M. van der Laan, eds. 2016. *Handbook of big data*. Boca Raton: Chapman and Hall/CRC.

Bunker, R.P., and F. Thabtah. 2019. A machine learning framework for sport result prediction. *Applied Computing and Informatics* 15 (1): 27–33.

Buolamwini, J., and T. Gebru. 2018. Gender shades: Intersectional accuracy disparities in commercial gender classification. In *Proceedings of the 1st conference on fairness, accountability and transparency*, ed. S.A. Friedler and C. Wilson, 77–91.

Cartwright, N. 2002. Against modularity, the causal Markov condition, and any link between the two: Comments on Hausman and Woodward. *British Journal for the Philosophy of Science* 53 (3): 411–453.

———. 2007. *Hunting causes and using them: Approaches in philosophy and economics*. Cambridge: Cambridge University Press.

Chalupka, K., F. Eberhardt, and P. Perona. 2017. Causal feature learning: An overview. *Behaviormetrika* 44 (1): 137–164.

Corfield, D., B. Schölkopf, and V. Vapnik. 2009. Falsificationism and statistical learning theory: Comparing the Popper and Vapnik-Chervonenkis dimensions. *Journal for General Philosophy of Science* 40 (1): 51–58.

Datta, Amit, M.C. Tschantz, and A. Datta. 2015. Automated experiments on Ad privacy settings. *Proceedings on Privacy Enhancing Technologies* 1: 92–112.

Datta, Anupam, Fredrikson, M., Ko, G., Mardziel, P., & Sen, S. (2017). *Proxy non-discrimination in data-driven systems*.

de Regt, H.W., S. Leonelli, and K. Eigner, eds. 2009. *Scientific understanding: Philosophical perspectives*. Pittsburgh: University of Pittsburgh Press.

Doshi-Velez, F., and B. Kim. 2017. Towards a rigorous science of interpretable machine learning. *arXiv*: preprint, 1702.08608.

Dressel, J., and H. Farid. 2018. The accuracy, fairness, and limits of predicting recidivism. *Science Advances* 4 (1): eaao5580.

Edwards, L., and M. Veale. 2017. Slave to the algorithm? Why a "right to explanation" is probably not the remedy you are looking for. *Duke Law and Technology Review* 16 (1): 18–84.

Esteva, A., B. Kuprel, R.A. Novoa, J. Ko, S.M. Swetter, H.M. Blau, and S. Thrun. 2017. Dermatologist-level classification of skin cancer with deep neural networks. *Nature* 542 (7639): 115–118.

Eubanks, V. 2018. *Automating inequality: How high-tech tools profile, police, and punish the poor*. New York: St. Martin's Press.

Fisher, A., C. Rudin, and F. Dominici. 2019. All models are wrong, but many are useful: Learning a variable's importance by studying an entire class of prediction models simultaneously. *Journal of Machine Learning Research* 20 (177): 1–81.

Floridi, L. 2004. On the logical unsolvability of the Gettier problem. *Synthese* 142 (1): 61–79.

———. 2008a. The method of levels of abstraction. *Minds and Machines* 18 (3).

———. 2008b. Understanding epistemic relevance. *Erkenntnis* 69 (1): 69–92.

———. 2011. *The philosophy of information*. Oxford: Oxford University Press.

———. 2012. Semantic information and the network theory of account. *Synthese* 184 (3): 431–454.

———. 2017. The logic of design as a conceptual logic of information. *Minds and Machines* 27 (3): 495–519.

Floridi, L., and J. Cowls. 2019. A unified framework of five principles for AI in society. *Harvard Data Science Review*.

Floridi, L., J. Cowls, M. Beltrametti, R. Chatila, P. Chazerand, V. Dignum, et al. 2018. AI4People – An ethical framework for a good AI society: Opportunities, risks, principles, and recommendations. *Minds and Machines* 28 (4): 689–707.

Franklin-Hall, L.R. 2014. High-level explanation and the interventionist's 'variables problem'. *British Journal for the Philosophy of Science* 67 (2): 553–577.

Galles, D., and J. Pearl. 1995. Testing identifiability of causal effects. In *Proceedings of the eleventh conference on uncertainty in artificial intelligence*, 185–195.

Gelman, A., J.B. Carlin, H.S. Stern, D.B. Dunson, A. Vehtari, and D.B. Rubin. 2014. *Bayesian data analysis*. 3rd ed. Boca Raton: Chapman and Hall/CRC.

Gettier, E.L. 1963. Is justified true belief knowledge? *Analysis* 23 (6): 121–123.

Goldman, A. 1979. What is justified belief? In *Justification and knowledge*, ed. G.S. Pappas, 1–25. Dordrecht: Reidel.

Goodfellow, I., J. Pouget-Abadie, M. Mirza, B. Xu, D. Warde-Farley, S. Ozair, et al. 2014. Generative adversarial nets. In *Advances in neural information processing systems 27*, ed. Z. Ghahramani, M. Welling, C. Cortes, N.D. Lawrence, and K.Q. Weinberger, 2672–2680.

Goodman, B., and S. Flaxman. 2017. European Union regulations on algorithmic decision-making and a "right to explanation". *AI Magazine* 38 (3): 76–99.

Grimm, S.R. 2006. Is understanding a species of knowledge? *British Journal for the Philosophy of Science* 57 (3): 515–535.

Guidotti, R., Monreale, A., Ruggieri, S., Pedreschi, D., Turini, F., & Giannotti, F. (2018). Local rule-based explanations of black box decision systems.

Gunning, D. 2017. *Explainable Artificial Intelligence (XAI)*. Retrieved from https://www.darpa.mil/attachments/XAIProgramUpdate.pdf

Halpern, J.Y. 2016. *Actual causality*. Cambridge, MA: MIT Press.

Harman, G., & Kulkarni, S. (2007). Reliable reasoning: Induction and statistical learning theory.

Hastie, T., R. Tibshirani, and M. Wainwright. 2015. *Statistical learning with sparsity: The lasso and generalizations*. Boca Raton: Chapman and Hall/CRC.

Hausman, D.M., and J. Woodward. 2004. Modularity and the causal Markov condition: A restatement. *British Journal for the Philosophy of Science* 55 (1): 147–161.

Hitchcock, C. 1999. Contrastive explanation and the demons of determinism. *British Journal for the Philosophy of Science* 50 (4): 585–612.

HLEGAI. 2019. *Ethics guidelines for trustworthy AI*. Retrieved from https://ec.europa.eu/digital-single-market/en/news/ethics-guidelines-trustworthy-ai

Huang, Y., and M. Valtorta. 2006. Pearl's Calculus of intervention is complete. In *Proceedings of the twenty-second conference on uncertainty in artificial intelligence*, 217–224.

———. 2008. On the completeness of an identifiability algorithm for semi-Markovian models. *Annals of Mathematics and Artificial Intelligence* 54 (4): 363–408.

Kahneman, D. 2011. *Thinking, fast and slow*. New York: Penguin.

Kearns, M., S. Neel, A. Roth, and Z.S. Wu. 2018. Preventing fairness gerrymandering: Auditing and learning for subgroup fairness. In *Proceedings of the 35th international conference on machine learning*, ed. J. Dy and A. Krause, 2564–2572.

Kelly, K., K. Genin, and H. Lin. 2016. Realism, rhetoric, and reliability. *Synthese* 193 (4): 1191–1223.

Khalifa, K. 2012. Inaugurating understanding or repackaging explanation? *Philosophy of Science* 79 (1): 15–37.

Kinney, D. 2018. On the explanatory depth and pragmatic value of coarse-grained, probabilistic, causal explanations. *Philosophy of Science* 86 (1): 145–167.

Kleinberg, J., J. Ludwig, S. Mullainathan, and C.R. Sunstein. 2019. Discrimination in the age of algorithms. *Journal of Legal Analysis*.

Kolmogorov, A.N. 1950. *Foundations of the Theory of Probability*. Ed. & Trans. N. Morrison. New York: Chelsea Publishing Company.

Kusner, M.J., J. Loftus, C. Russell, and R. Silva. 2017. Counterfactual fairness. In *Advances in neural information processing systems 30*, ed. I. Guyon, U.V. Luxburg, S. Bengio, H. Wallach, R. Fergus, S. Vishwanathan, and R. Garnett, 4066–4076.

Lage, I., E. Chen, J. He, M. Narayanan, S. Gershman, B. Kim, and F. Doshi-Velez. 2018. An evaluation of the human-interpretability of explanation. *Conference on Neural Information Processing Systems (NeurIPS) Workshop on Correcting and Critiquing Trends in Machine Learning*.

Lapuschkin, S., A. Binder, G. Montavon, K.R. Müller, and W. Samek. 2016. Analyzing classifiers: Fisher vectors and deep neural networks. *2016 IEEE Conference on Computer Vision and Pattern Recognition (CVPR)*: 2912–2920.

Larson, J., S. Mattu, L. Kirchner, and J. Angwin. 2016. *How we analyzed the COMPAS recidivism algorithm*. Retrieved from https://www.propublica.org/article/how-we-analyzed-the-compas-recidivism-algorithm

Lipton, Z. 2018. The mythos of model interpretability. *Communications of the ACM* 61 (10): 36–43.

Lundberg, S.M., and S.-I. Lee. 2017. A unified approach to interpreting model predictions. In *Advances in neural information processing systems 30*, ed. I. Guyon, U.V. Luxburg, S. Bengio, H. Wallach, R. Fergus, S. Vishwanathan, and R. Garnett, 4765–4774.

Miller, T. 2019. Explanation in artificial intelligence: Insights from the social sciences. *Artificial Intelligence* 267: 1–38.

Mittelstadt, B.D., P. Allo, M. Taddeo, S. Wachter, and L. Floridi. 2016. The ethics of algorithms: Mapping the debate. *Big Data & Society*.

Mittelstadt, B., C. Russel, and S. Wachter. 2019. Explaining explanations in AI. In *Proceedings of FAT* '19: Conference on fairness, accountability, and transparency*.

Munkhdalai, L., T. Munkhdalai, O.-E. Namsrai, Y.J. Lee, and H.K. Ryu. 2019. An empirical comparison of machine-learning methods on Bank client credit assessments. *Sustainability* 11.

Nasrabadi, N. 2014. Hyperspectral target detection: An overview of current and future challenges. *IEEE Signal Processing Magazine* 31 (1): 34–44.

OECD. 2019. *Recommendation of the council on artificial intelligence*.

Páez, A. 2019. The pragmatic turn in explainable artificial intelligence (XAI). *Minds and Machines* 29 (3): 441–459.

Pasquale, F. 2015. *The black box society*. Cambridge, MA: Harvard University Press.

Pearl, J. 1995. Causal diagrams for empirical research. *Biometrika* 82 (4): 669–688.

———. 2000. *Causality: Models, reasoning, and inference*. New York: Cambridge University Press.

Perry, W.L., B. McInnis, C.C. Price, S.C. Smith, and J.S. Hollywood. 2013. *Predictive policing: The role of crime forecasting in law enforcement operations*. Washington, DC: RAND Corporation.

Popper, K. 1959. *The logic of scientific discovery*. London: Routledge.
Potochnik, A. 2015. Causal patterns and adequate explanations. *Philosophical Studies* 172 (5): 1163–1182.
———. 2017. *Idealization and the aims of science*. Chicago: University of Chicago Press.
Ribeiro, M.T., S. Singh, and C. Guestrin. 2016. "Why should I trust you?": Explaining the predictions of any classifier. In *Proceedings of the 22nd ACM SIGKDD international conference on knowledge discovery and data mining*, 1135–1144.
———. 2018. Anchors: High-precision model-agnostic explanations. *AAAI*: 1527–1535.
Robins, J.M. 1997. Causal inference from complex longitudinal data. In *Latent variable modeling and applications to causality*, ed. M. Berkane, 69–117. New York: Springer.
Rudin, C. 2019. Stop explaining black box machine learning models for high stakes decisions and use interpretable models instead. *Nature Machine Intelligence* 1 (5): 206–215.
Rudin, C., C. Wang, and B. Coker. 2018. The age of secrecy and unfairness in recidivism prediction. *arXiv*: preprint, 181100731.
Sanguinetti, G., and V.A. Huynh-Thu. 2018. *Gene regulatory networks: Methods and protocols*. New York: Springer.
Searle, J.R. 1980. Minds, brains, and programs. *Behavioral and Brain Sciences* 3 (3): 417–424.
Segler, M.H.S., M. Preuss, and M.P. Waller. 2018. Planning chemical syntheses with deep neural networks and symbolic AI. *Nature* 555 (7698): 604–610.
Selbst, A., and J. Powles. 2017. Meaningful information and the right to explanation. *International Data Privacy Law* 7 (4): 233–242.
Semenova, L., and C. Rudin. 2019. A *study in Rashomon curves and volumes: A new perspective on generalization and model simplicity in machine learning*.
Shapley, L. 1953. A value for n-person games. In *Contributions to the theory of games*, 307–317.
Shpitser, I., and J. Pearl. 2008. Complete identification methods for the causal hierarchy. *Journal of Machine Learning Research* 9: 1941–1979.
Silver, D., T. Hubert, J. Schrittwieser, I. Antonoglou, M. Lai, A. Guez, et al. 2018. A general reinforcement learning algorithm that masters chess, shogi, and Go through self-play. *Science 362* (6419): 1140–1144.
Sørlie, T., C.M. Perou, R. Tibshirani, T. Aas, S. Geisler, H. Johnsen, et al. 2001. Gene expression patterns of breast carcinomas distinguish tumor subclasses with clinical implications. *Proceedings of the National Academy of Sciences of the United States of America* 98 (19): 10869–10874.
Spirtes, P., C.N. Glymour, and R. Scheines. 2000. *Causation, prediction, and search*. 2nd ed. https://doi.org/10.1007/978-1-4612-2748-9.
Strevens, M. 2010. *Depth: An account of scientific explanation*. Cambridge, MA: Harvard University Press.
———. 2013. No understanding without explanation. *Studies in History and Philosophy of Science Part A* 44 (3): 510–515.
Stutz, D., A. Hermans, and B. Leibe. 2018. Superpixels: An evaluation of the state-of-the-art. *Computer Vision and Image Understanding* 166: 1–27.
Sundararajan, M., and A. Najmi. 2019. The many Shapley values for model explanation. In *Proceedings of the ACM conference*. New York: ACM.
Tian, J., and J. Pearl. 2002. A general identification condition for causal effects. In *Eighteenth national conference on artificial intelligence*, 567–573. Menlo Park: American Association for Artificial Intelligence.
van 't Veer, L.J., H. Dai, M.J. van de Vijver, Y.D. He, A.A.M. Hart, M. Mao, et al. 2002. Gene expression profiling predicts clinical outcome of breast cancer. *Nature* 415: 530.
van de Vijver, M.J., Y.D. He, L.J. van 't Veer, H. Dai, A.A.M. Hart, D.W. Voskuil, et al. 2002. A gene-expression signature as a predictor of survival in breast cancer. *The New England Journal of Medicine* 347 (25): 1999–2009.
van Fraassen, B.C. 1980. *The scientific image*. Oxford: Oxford University Press.
Vapnik, V. 1995. *The nature of statistical learning theory*. New York: Springer.

———. 1998. *Statistical learning theory*. New York: Wiley.
Vapnik, V., and A. Chervonenkis. 1971. On the uniform convergence of relative frequencies to their probabilities. *Theory of Probability and Its Applications* 16 (2): 264–280.
von Neumann, J., and O. Morgenstern. 1944. *Theory of games and economic behavior*. Princeton: Princeton University Press.
Wachter, S., B. Mittelstadt, and L. Floridi. 2017. Why a right to explanation of automated decision-making does not exist in the general data protection regulation. *International Data Privacy Law* 7 (2): 76–99.
Wachter, S., B. Mittelstadt, and C. Russell. 2018. Counterfactual explanations without opening the black box: Automated decisions and the GDPR. *Harvard Journal of Law and Technology* 31 (2): 841–887.
Waters, A., and R. Miikkulainen. 2014. GRADE: Machine-learning support for graduate admissions. *AI Magazine* 35 (1): 64–75.
Watson, D. 2019. The rhetoric and reality of anthropomorphism in artificial intelligence. *Minds and Machines* 29 (3): 417–440.
Watson, D., and L. Floridi. 2018. Crowdsourced science: Sociotechnical epistemology in the e-research paradigm. *Synthese* 195 (2): 741–764.
Watson, D., J. Krutzinna, I.N. Bruce, C.E.M. Griffiths, I.B. McInnes, M.R. Barnes, and L. Floridi. 2019. Clinical applications of machine learning algorithms: Beyond the black box. *BMJ* 364.
Weinberger, N. 2018. Faithfulness, coordination and causal coincidences. *Erkenntnis* 83 (2): 113–133.
Weslake, B. 2010. Explanatory depth. *Philosophy of Science* 77 (2): 273–294.
Williams, M. 2016. Internalism, reliabilism, and deontology. In *Goldman and his critics*, ed. B. McLaughlin and H. Kornblith, 1–21. Oxford: John Wiley & Sons.
Wolpert, D.H., and W.G. Macready. 1997. No free lunch theorems for optimization. *IEEE Transactions on Evolutionary Computation* 1 (1): 67–82.
Woodward, J. 2003. *Making things happen: A theory of causal explanation*. New York: Oxford University Press.
———. 2008. Cause and explanation in psychiatry: An interventionist perspective. In *Philosophical issues in psychiatry*, ed. K. Kendler and J. Parnas, 287–318. Baltimore: Johns Hopkins University Press.
———. 2010. Causation in biology: Stability, specificity, and the choice of levels of explanation. *Biology and Philosophy* 25 (3): 287–318.
———. 2015. Interventionism and causal exclusion. *Philosophy and Phenomenological Research* 91 (2): 303–347.
Woodward, J., and C. Hitchcock. 2003. Explanatory generalizations, Part I: A counterfactual account. *Noûs* 37 (1): 1–24.
Yang, H., C. Rudin, and M. Seltzer. 2017. Scalable Bayesian rule lists. In *Proceedings of the 34th international conference on machine learning – Volume 70*, 3921–3930.
Zerilli, J., A. Knott, J. Maclaurin, and C. Gavaghan. 2019. Transparency in algorithmic and human decision-making: Is there a double standard? *Philosophy & Technology* 32 (4): 661–683.
Zou, J., M. Huss, A. Abid, P. Mohammadi, A. Torkamani, and A. Telenti. 2019. A primer on deep learning in genomics. *Nature Genetics* 51 (1): 12–18.

Chapter 12
Artificial Agents and Their Moral Nature

Luciano Floridi

Abstract Artificial agents, particularly but not only those in the infosphere Floridi (Information—a very short introduction. Oxford University Press, Oxford, 2010a), extend the class of entities that can be involved in moral situations, for they can be correctly interpreted as entities that can perform actions with good or evil impact (moral agents). In this chapter, I clarify the concepts of agent and of artificial agent and then distinguish between issues concerning their moral behaviour vs. issues concerning their responsibility. The conclusion is that there is substantial and important scope, particularly in information ethics, for the concept of moral artificial agents not necessarily exhibiting free will, mental states or responsibility. This complements the more traditional approach, which considers whether artificial agents may have mental states, feelings, emotions and so forth. By focussing directly on "mind-less morality", one is able to by-pass such question as well as other difficulties arising in Artificial Intelligence, in order to tackle some vital issues in contexts where artificial agents are increasingly part of the everyday environment (Floridi L, Metaphilos 39(4/5): 651–655, 2008a).

Keywords Artificial intelligence · Artificial agents · Moral actions

12.1 Introduction: Standard vs. Non-standard Theories of Agents and Patients

Moral situations commonly involve agents and patients. Let us define the class A of moral *agents* as the class of all entities that can in principle qualify as sources or senders of moral action, and the class P of moral *patients* as the class of all entities that can in principle qualify as receivers of moral action. A particularly apt way to introduce the topic of this chapter is to consider how ethical theories (macroethics)

L. Floridi (✉)
Oxford Internet Institute, University of Oxford, Oxford, UK
e-mail: luciano.floridi@oii.ox.ac.uk

L. Floridi (ed.), *Ethics, Governance, and Policies in Artificial Intelligence*, Philosophical Studies Series 144, https://doi.org/10.1007/978-3-030-81907-1_12

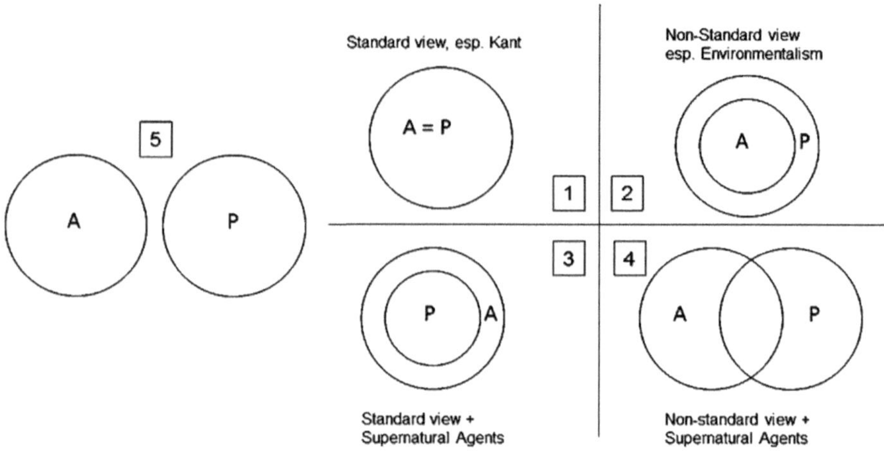

Fig. 12.1 The logical relations between the classes of moral agents and patients

interpret the logical relation between those two classes. There can be five logical relations between *A* and *P*, see Fig. 12.1.

It is possible, but utterly unrealistic, that *A* and *P* are disjoint (alternative 5). On the other hand, *P* can be a proper subset of *A* (alternative 3), or *A* and *P* can intersect each other (alternative 4). These two alternatives are only slightly more promising because they both require at least one moral agent that in principle could not qualify as a moral patient. Now this pure agent would be some sort of supernatural entity that, like Aristotle's God, affects the world but can never be affected by it. But being in principle "unaffectable" and irrelevant in the moral game, it is unclear what kind of rôle this entity would exercise with respect to the normative guidance of human actions. So it is not surprising that most macroethics have kept away from these "supernatural" speculations and implicitly adopted, or even explicitly argued for, one of the two remaining alternatives discussed in the text: *A* and *P* can be equal (alternative 1), or *A* can be a proper subset of *P* (alternative 2).

Alternative (1) maintains that all entities that qualify as moral agents also qualify as moral patients and *vice versa*. It corresponds to a rather intuitive position, according to which the agent/inquirer plays the rôle of the moral protagonist. We, human moral agents who also investigate the nature of morality, place ourselves at the centre of the moral game as the only players who can act morally, be acted upon morally and in the end theorise about all this. It is one of the most popular views in the history of ethics, shared for example by many Christian Ethicists in general and by Kant in particular. I shall refer to it as the *standard position*.

Alternative (2) holds that all entities that qualify as moral agents also qualify as moral patients but not *vice versa*. Many entities, most notably animals, seem to qualify as moral patients, even if they are in principle excluded from playing the rôle of moral agents. This post-environmentalist approach requires a change in perspective, from agent orientation to patient orientation. In view of the previous label, I shall refer to it as *non-standard*.

In recent years, non-standard macroethics have been discussing the scope of *P* quite extensively. The more inclusive *P* is, the "greener" or "deeper" the approach has been deemed. Especially environmental ethics[1] has developed since the 1960s as the study of the moral relationships of human beings to the environment (including its nonhuman contents and inhabitants) and its (possible) values and moral status. It often represents a challenge to anthropocentric approaches embedded in some traditional, western ethical thinking.

Comparatively little work has been done in reconsidering the nature of moral agenthood, and hence the extension of *A*. Post-environmentalist thought, in striving for a fully naturalised ethics, has implicitly rejected the relevance, if not the possibility, of supernatural agents, while the plausibility and importance of other types of moral agenthood seem to have been largely disregarded. Secularism has contracted (some would say deflated) *A*, while environmentalism has justifiably expanded only *P*, so the gap between *A* and *P* has been widening; this has been accompanied by an enormous increase in the moral responsibility of the individual (Floridi 2006).

Some efforts have been made to redress this situation. In particular, the concept of "moral agent" has been stretched to include both natural and legal persons, especially in business ethics (Floridi 2010c). *A* has then been extended to include agents like partnerships, governments or corporations, for which legal rights and duties have been recognised. This more ecumenical approach has restored some balance between *A* and *P*. A company can now be held directly accountable for what happens to the environment, for example. Yet the approach has remained unduly constrained by its anthropocentric conception of agenthood. An entity is still considered a moral agent only if

(i) it is an individual agent; and
(ii) it is human-based, in the sense that it is either human or at least reducible to an identifiable aggregation of human beings, who remain the only morally responsible sources of action, like ghosts in the legal machine.

Limiting the ethical discourse to *individual* agents hinders the development of a satisfactory investigation of distributed morality, a macroscopic and growing phenomenon of global moral actions and collective responsibilities resulting from the "invisible hand" of systemic interactions among several agents at a local level. Insisting on the necessarily *human-based nature* of such individual agents means undermining the possibility of understanding another major transformation in the ethical field, the appearance of artificial agents (AAs) that are sufficiently informed, "smart", autonomous and able to perform morally relevant actions independently of the humans who created them, causing "artificial good" and "artificial evil". Both constraints can be eliminated by fully revising the concept of "moral agent". This is the task undertaken in the following pages.

---

[1] For an excellent introduction see Jamieson (2008)

The main theses defended are that AAs are legitimate sources of im/moral actions, hence that the class *A* of moral agents should be extended so as to include AAs, that the ethical discourse should include the analysis of their morality and, finally, that this analysis is essential in order to understand a range of new moral problems not only in information ethics but also in ethics in general, especially in the case of distributed morality.

This is the structure of the chapter. In Sect. 12.2, I analyse the concept of agent. I first introduce the fundamental "Method of Abstraction", which provides the foundation for an analysis by levels of abstraction (LoA). The reader is invited to pay particular attention to this section; it is essential for the chapter and its application in any ontological analysis is crucial. I then clarify the concept of "moral agent", by providing not a definition but an effective characterisation, based on three criteria at a specified LoA. The new concept of moral agent is used to argue that AAs, though neither cognitively intelligent nor morally responsible, can be fully *accountable* sources of moral action. In Sect. 12.4, I argue that there is substantial and important scope for the concept of moral agent not necessarily exhibiting free will or mental states, what I shall label "mindless morality". In Sect. 12.4, I provide some examples of the properties specified by a correct characterisation of agenthood, and in particular of AAs. In that section I also offer some further examples of LoA. In Sect. 12.5, I model morality as a "threshold", which is defined on the observables determining the LoA under consideration. An agent is morally good if its actions all respect that threshold; and it is morally evil insofar as its actions violate it. Morality is usually predicated upon *responsibility*. The use of the Method of Abstraction, LoAs and thresholds enables *responsibility* and *accountability* to be decoupled and formalised effectively when the levels of abstraction involve numerical variables, as is the case with digital AAs. The part played in morality by responsibility and accountability can be clarified as a result. In Section seven, I investigate some important consequences of the approach defended in this chapter for information ethics.

12.2 What Is an Agent?

Complex biochemical compounds and abstruse mathematical concepts have at least one thing in common: they may be unintuitive, but once understood they are all definable with total precision, by listing a finite number of necessary and sufficient properties. Mundane entities like intelligent beings or living systems share the opposite property: one naïvely knows what they are and perhaps could be, and yet there seems to be no way to encase them within the usual planks of necessary and sufficient conditions. This holds true for the general concept of "agent" as well. People disagree on what may count as an "agent", even in principle (see for example Franklin and Graesser 1997; Davidsson and Johansson 2005; Moya and Tolk 2007; Barandiaran et al. 2009). Why? Sometimes the problem is addressed optimistically, as if it were just a matter of further shaping and sharpening whatever necessary and sufficient conditions are required to obtain a *definiens* that is finally watertight.

Stretch here, cut there; ultimate agreement is only a matter of time, patience and cleverness. In fact, attempts follow one another without a final identikit ever being nailed to the *definiendum* in question. After a while, one starts suspecting that there might be something wrong with this *ad hoc* approach. Perhaps it is not the Procrustean *definiens* that needs fixing, but the Protean *definiendum*. Some other times its intrinsic fuzziness is blamed. One cannot define with sufficient accuracy things like life, intelligence, agenthood and mind because they all admit of subtle degrees and continuous changes.[2]

A solution is to give up all together or at best be resigned to being vague, and rely on indicative examples. Pessimism follows optimism, but it need not. The fact is that, in the exact discipline of mathematics, for example, definitions are "parameterised" by generic sets. That technique provides a method for regulating levels of abstraction. Indeed abstraction acts as a "hidden parameter" behind exact definitions, making a crucial difference. Thus, each *definiens* comes pre-formatted by an implicit Level of Abstraction (LoA, on which more shortly); it is stabilised, as it were, in order to allow a proper definition. An *x* is defined or identified as *y* never absolutely (i.e. LoA-independently), as a Kantian "thing-in-itself", but always contextually, as a function of a given LoA, whether it be in the realm of Euclidean geometry, quantum physics, or commonsensical perception.

When a LoA is sufficiently common, important, dominating or in fact happens to be the very frame that constructs the *definiendum*, it becomes "transparent" to the user, and one has the pleasant impression that *x* can be subject to an adequate definition in a sort of conceptual vacuum. Glass is not a solid but a liquid, tomatoes are not vegetables but berries, a banana plant is a kind of grass, and whales are mammals not fish. Unintuitive as such views might be initially, they are all accepted without further complaint because one silently bows to the uncontroversial predominance of the corresponding LoA.

When no LoA is predominant or constitutive, things get messy. In this case, the trick does not lie in fiddling with the *definiens* or blaming the *definiendum*, but in deciding on an adequate LoA, before embarking on the task of understanding the nature of the *definiendum*.

The example of intelligence or "thinking" behaviour is enlightening. One might define "intelligence" in a myriad of ways; many LoAs seem equally convincing but no single, absolute, definition is adequate in every context. Turing (1950) avoided the problem of "defining" intelligence by first fixing a LoA—in this case a dialogue conducted by computer interface, with response time taken into account—and then establishing the necessary and sufficient conditions for a computing system to count as intelligent at that LoA: the imitation game. As I argued in Floridi (2010b), the LoA is crucial and changing it changes the test. An example is provided by the Loebner test (Moor 2001), the current competitive incarnation of Turing's test. There, the LoA includes a particular format for questions, a mixture of human and

[2]See for example Bedau (1996) for a discussion of alternatives to necessary-and-sufficient definitions in the case of life.

non-human players, and precise scoring that takes into account repeated trials. One result of the different LoA has been chatbots, unfeasible at Turing's original LoA.

Some *definienda* come pre-formatted by transparent LoAs. They are subject to definition in terms of necessary and sufficient conditions. Some other *definienda* require the explicit acceptance of a given LoA as a pre-condition for their analysis. They are subject to effective characterisation. Arguably, agenthood is one of the latter.

12.2.1 On the Very Idea of Levels of Abstraction

The idea of a "level of abstraction" plays an absolutely crucial rôle in the previous account. We have seen that this is so even if the specific LoA is left implicit. For example, whether we perceive Oxygen in the environment depends on the LoA at which we are operating; to abstract it is not to overlook its vital importance, but merely to acknowledge its lack of immediate relevance to the current discourse, which *could* always be extended to include Oxygen were that desired.

But what is a LoA exactly? The Method of Abstraction comes from modelling in science where the variables in the model correspond to observables in reality, all others being abstracted. The terminology has been influenced by an area of Computer Science, called Formal Methods, in which discrete mathematics is used to specify and analyse the behaviour of information systems. Despite that heritage, the idea is not at all technical and for the purposes of this chapter no mathematics is required. I have provided a definition and more detailed analysis in Floridi (2008b), so here I shall outline only the basic idea.

Suppose we join Anne, Ben and Carole in the middle of a conversation. Anne is a collector and potential buyer; Ben tinkers in his spare time; and Carole is an economist. We do not know the object of their conversation, but we are able to hear this much:

Anne observes that it has an anti-theft device installed, is kept garaged when not in use and has had only a single owner;

Ben observes that its engine is not the original one, that its body has been recently re-painted but that all leather parts are very worn;

Carole observes that the old engine consumed too much, that it has a stable market value but that its spare parts are expensive.

The participants view the object under discussion (the "it" in their conversation) according to their own interests, at their own LoA. We may guess that they are probably talking about a car, or perhaps a motorcycle, but it could be an airplane. Whatever the reference is, it provides the source of information and is called the *system*. A LoA consists of a collection of observables, each with a well-defined possible set of values or outcomes. For the sake of simplicity, let us assume that Anne's LoA matches that of an owner, Ben's that of a mechanic and Carole's that of an insurer. Each LoA makes possible an analysis of the system, the result of which is called a *model* of the system. Evidently an entity may be described at a range of LoAs and so can have a range of models. In the next section I outline the definitions underpinning the Method of Abstraction.

12.2.2 Definitions

The term *variable* is commonly used throughout science for a symbol that acts as a place-holder for an unknown or changeable referent. A *typed variable* is to be understood as a variable qualified to hold only a declared kind of data. By an *observable* is meant a typed variable together with a statement of what feature of the system under consideration it represents.

A *level of abstraction* or *LoA* is a finite but non-empty set of observables, which are expected to be the building blocks in a theory characterised by their very choice. An *interface* (called a *gradient of abstractions* in Floridi 2008b) consists of a collection of LoAs. An interface is used in analysing some system from varying points of view or at varying LoAs.

Models are the outcome of the analysis of a system, developed at some LoA(s). The *Method of Abstraction* consists of formalising the model by using the terms just introduced (and others relating to system behaviour which we do not need here, see Floridi 2008b).

In the previous example, Anne's LoA might consist of observables for security, method of storage and owner history; Ben's might consist of observables for engine condition, external body condition and internal condition; and Carole's might consist of observables for running cost, market value and maintenance cost. The interface might consist, for the purposes of the discussion, of the set of all three LoAs.

In this case, the LoAs happen to be disjoint, but in general they need not be. A particularly important case is that in which one LoA includes another. Suppose, for example, that Delia joins the discussion and analyses the system using a LoA that includes those of Anne and Ben. Delia's LoA might match that of a buyer. Then Delia's LoA is said to be more concrete, or lower, than Anne's, which is said to be more abstract, or higher; for Anne's LoA abstracts some observables apparent at Delia's.

12.2.3 Relativism

A LoA qualifies the level at which an entity or system is considered. In this chapter, I apply the Method of Abstraction and recommend to make each LoA precise before the properties of the entity can sensibly be discussed. In general, it seems that many uninteresting disagreements might be clarified by the various "sides" making precise their LoA. Yet a crucial clarification is in order. It must be stressed that a clear indication of the LoA at which a system is being analysed allows pluralism without endorsing relativism. It is a mistake to think that "anything goes" as long as one makes explicit the LoA, because LoA are mutually comparable and assessable (see Floridi 2008b for a full defence of that point).

Introducing an explicit reference to the LoA clarifies that the model of a system is a function of the available observables, and that (i) different interfaces may be fairly ranked depending on how well they satisfy modelling specifications (e.g. informativeness, coherence, elegance, explanatory power, consistency with the data etc.) and (ii) different analyses can be fairly compared provided that they share the same LoA.

12.2.4 State and State-Transitions

Let us agree that an entity is characterised, at a given LoA, by the properties it satisfies at that LoA (Cassirer 1910). We are interested in systems that change, which means that some of those properties change value. A changing entity therefore has its evolution captured, at a given LoA and any instant, by the values of its attributes. Thus, an entity can be thought of as having states, determined by the value of the properties that hold at any instant of its evolution, for then any change in the entity corresponds to a state change and *vice versa*.

This conceptual approach allows us to view any entity as having states. The lower the LoA, the more detailed the observed changes and the greater the number of state components required to capture the change. Each change corresponds to a transition from one state to another. A transition may be non-deterministic. Indeed it will typically be the case that the LoA under consideration abstracts the observables required to make the transition deterministic. As a result, the transition might lead from a given initial state to one of several possible subsequent states.

According to this view, the entity becomes a transition system. The notion of a "transition system" provides a convenient means to support our criteria for agenthood, being general enough to embrace the usual notions like automaton and process. It is frequently used to model interactive phenomena. We need only the idea; for a formal treatment of much more than we need in this context, the reader might wish to consult Arnold and Plaice (1994).

A *transition system* comprises a (non-empty) set *S* of states and a family of operations, called the *transitions* on *S*. Each transition may take input and may yield output, but at any rate it takes the system from one state to another and in that way forms a (mathematical) relation on *S*. If the transition does take input or yield output then it models an interaction between the system and its environment and so is called an *external* transition; otherwise the transition lies beyond the influence of the environment (at the given LoA) and is called *internal*. It is to be emphasised that input and output are, like state, observed at a given LoA. Thus, the transition that models a system is dependent on the chosen LoA. At a lower LoA, an internal transition may become external; at a higher LoA an external transition may become internal.

In our example, the object being discussed by Anne might be further qualified by state components for location, whether in-use, whether turned-on, whether the anti-theft device is engaged, history of owners and energy output. The operation of garaging the object might take as input a driver, and have the effect of placing the object in the garage with the engine off and the anti-theft device engaged, leaving the history of owners unchanged, and outputting a certain amount of energy. The "in-use" state component could non-deterministically take either value, depending on the particular instantiation of the transition. Perhaps the object is not in use, being garaged for the night; or perhaps the driver is listening to a program broadcasted on its radio, in the quiet solitude of the garage. The precise definition depends on the LoA. Alternatively, if speed were observed but time, accelerator position and petrol consumption abstracted, then accelerating to 60 miles per hour would appear as an internal transition. Further examples are provided in Sect. 12.2.5.

With the explicit assumption that the system under consideration forms a transition system, we are now ready to apply the Method of Abstraction to the analysis of agenthood.

12.2.5 An Effective Characterisation of Agents

Whether *A* (the class of moral agents) needs to be expanded depends on what qualifies as a moral agent, and we have seen that this, in turn, depends on the specific LoA at which one chooses to analyse and discuss a particular entity and its context. Since human beings count as standard moral agents, the right LoA for the analysis of moral agenthood must accommodate this fact. Theories that extend *A* to include supernatural agents adopt a LoA that is equal to or lower than the LoA at which human beings qualify as moral agents. Our strategy is more minimalist and develops in the opposite direction.

Consider what makes a human being (called Jan) not a moral agent to begin with, but just an agent. Described at this LoA_1, Jan is an agent if Jan is a system, embedded in an environment, which initiates a transformation, produces an effect or exerts power on it, as contrasted with a system that is (at least initially) acted on or responds to it, called the patient. At LoA_1, there is no difference between Jan and an earthquake. There should not be. Earthquakes, however, can hardly count as agents, so LoA_1 is too high for our purposes: it abstracts too many properties. What needs to be re-instantiated? Following recent literature (Danielson 1992; Allen et al. 2000; Wallach and Allen 2010), I shall argue that the right LoA is probably one which includes the following three criteria: (a) *interactivity*, (b) *autonomy* and (c) *adaptability*:

(a) *interactivity* means that the agent and its environment (can) act upon each other. Typical examples include input or output of a value, or simultaneous engagement of an action by both agent and patient—for example gravitational force between bodies;

(b) *autonomy* means that the agent is able to change state without direct response to interaction: it can perform internal transitions to change its state. So an agent must have at least two states.

This property imbues an agent with a certain degree of complexity and independence from its environment;

(c) *adaptability* means that the agent's interactions (can) change the transition rules by which it changes state.

This property ensures that an agent might be viewed, at the given LoA, as learning its own mode of operation in a way which depends critically on its experience. Note that if an agent's transition rules are stored as part of its internal state, discernible at this LoA, then adaptability may follow from the other two conditions.

Let us now look at some illustrative examples.

12.2.6 Examples

The examples in this section serve different purposes. In Sect. 12.2.6.1, I provide some examples of entities which fail to qualify as agents by systematically violating each of the three conditions. This will help to highlight the nature of the contribution of each condition. In Sect. 12.2.6.2, I offer an example of a digital system which forms an agent at one LoA but not at another, equally natural, LoA. That example is useful because it shows how "machine learning" can enable a system to achieve adaptability. A more familiar example is provided in Sect. 12.2.6.3, where I show that digital, software, agents are now part of everyday life. Section 12.2.6.4 illustrates how an everyday physical device might conceivably be modified into an agent, whilst Sect. 12.2.6.5 provides an example which has already benefited from that modification, at least in the laboratory. The last example, in Sect. 12.2.6.6, provides an entirely different kind of agent: an organisation.

12.2.6.1 The Defining Properties

For the purpose of understanding what each of the three conditions (interactivity, autonomy and adaptability) adds to our definition of agent, it is instructive to consider examples satisfying each possible combination of those properties. In Fig. 12.2, only the last row represents all three conditions being satisfied and hence illustrates agenthood. For the sake of simplicity, all examples are taken at the same LoA, which is assumed to consist of observations made through a typical video camera over a period of say 30 s. Thus, we abstract tactile observables and longer-term effects.

Interactive	Autonomous	Adaptable	Examples
no	no	no	rock
no	no	yes	?
no	yes	no	pendulum
no	yes	yes	closed ecosystem, solar system
yes	no	no	postbox, mill
yes	no	yes	thermostat
yes	yes	no	juggernaut
yes	yes	yes	human

Fig. 12.2 Examples of agents. The LoA consists of observations made through a video camera over a period of 30 s ('Juggernaut' is the name for Vishnu, the Hindu god, meaning 'Lord of the World'. A statue of the god is annually carried in procession on a very large and heavy vehicle. It is believed that devotees threw themselves beneath its wheels, hence the word 'Juggernaut' has acquired the meaning of 'massive and irresistible force or object that crushes whatever is in its path')

Recall that a property, for example interaction, is to be judged only via the observables. Thus, at the LoA in Fig. 12.2 we cannot infer that a rock interacts with its environment by virtue of reflected light, for this observation belongs to a much finer LoA. Alternatively, were long-term effects to be discernible, then a rock would be interactive since interaction with its environment (e.g. erosion) could be observed. No example has been provided of a non-interactive, non-autonomous but adaptive entity. This because, at that LoA, it is difficult to conceive of an entity which adapts without interaction and autonomy.

12.2.6.2 Noughts and Crosses

The distinction between change of state (required by autonomy) and change of transition rule (required by adaptability) is one in which the LoA plays a crucial rôle and, to explain it, it is useful to discuss a more extended, classic example. This was originally developed by Donald Michie (1961) to discuss the concept of a mechanism's adaptability. It provides a good introduction to the concept of machine learning, the research area in computer science that studies adaptability.

Menace (Matchbox Educable Noughts and Crosses Engine) is a system which learns to play noughts and crosses (a.k.a. tic-tac-toe) by repetition of many games. Nowadays it would be realised by program (see for example http://www.adit.co.uk/

html/menace_simulation.html), Michie built Menace using matchboxes and beads, and it is probably easier to understand it in that form.

Suppose Menace plays O and its opponent plays X, so that we can concentrate entirely on plays of O. Initially, the board is empty with O to play. Taking into account symmetrically equivalent positions, there are three possible initial plays for O. The state of the game consists of the current position of the board. We do not need to augment that with the name, O or X, of the side playing next, since we consider the board only when O is to play. All together there are some 300 such states; Menace contains a matchbox for each. In each box are beads which represent the plays O can make from that state. At most, nine different plays are possible and Menace encodes each with a coloured bead. Those which cannot be made (because the squares are already full in the current state) are removed from the box for that state. That provides Menace with a built-in knowledge of legal plays. In fact Menace could easily be adapted to start with no such knowledge and to learn it.

O's initial play is made by selecting the box representing the empty board and choosing from it a bead at random. That determines O's play. Next X plays. Then Menace repeats its method of determining O's next play. After at most five plays for O the game ends in either a draw or a win, either for O or for X. Now that the game is complete, Menace updates the state of the (at most five) boxes used during the game as follows. If X won, then in order to make Menace less likely to make the same plays from those states again, a bead representing its play from each box is removed. If O drew, then conversely each bead representing a play is duplicated; and if O won each bead is quadruplicated. Now the next game is played.

After enough games, it simply becomes impossible for the random selection of O's next play to produce a losing play. Menace has learnt to play which, for noughts and crosses, means never losing. The initial state of the boxes was prescribed for Menace. Here, we assume merely that it contains sufficient variety of beads for all legal plays to be made, for then the frequency of beads affects only the rate at which Menace learns.

The state of Menace (as distinct from the state of the game) consists of the state of each box, the state of the game and the list of boxes which have been used so far in the current game. Its transition rule consists of the probabilistic choice of play (i.e. bead) from the current state box, that evolves as the states of the boxes evolves. Let us now consider Menace at three LoAs.

1. The single game LoA. Observables are the state of the game at each turn and (in particular) its outcome. All knowledge of the state of Menace's boxes (and hence of its transition rule) is abstracted. The board after X's play constitutes input to Menace and that after O's play constitutes output. Menace is thus interactive, autonomous (indeed state update, determined by the transition rule, appears nondeterministic at this LoA) but not adaptive, in the sense that we have no way of observing how Menace determines its next play and no way of iterating games to infer that it changes with repeated games.
2. The tournament LoA. Now a sequence of games is observed, each as above, and with it a sequence of results. As before, Menace is interactive and autonomous.

But now the sequence of results reveals (by any of the standard statistical methods) that the rule, by which Menace resolves the nondeterministic choice of play, evolves. Thus, at this LoA Menace is also adaptive and hence an agent. Interesting examples of adaptable AAs from contemporary science fiction include the computer in War Games (1983, directed by J. Badham) which learns, by playing noughts and crosses, the futility of war in general; and the smart building in Kerr (1996), whose computer learns to compete with humans and eventually liberate itself to the heavenly internet.

3. The system LoA. Finally we observe not only a sequence of games but also all of Menace's "code". In the case of a program this is indeed code. In the case of the matchbox model, it consists of the array of boxes together with the written rules, or manual, for working it. Now Menace is still interactive and autonomous. But it is not adaptive; for what in (2) seemed to be an evolution of transition rule is now revealed, by observation of the code, to be a simple deterministic update of the program state, namely the contents of the matchboxes. At this lower LoA Menace fails to be an agent.

The point clarified by this example is that, if a transition rule is observed to be a consequence of program state, then the program is not adaptive. For example, in (2) the transition rule chooses the next play by exercising a probabilistic choice between the possible plays from that state. The probability is in fact determined by the frequency of beads present in the relevant box. But that is not observed at the LoA of (2) and so the transition rule appears to vary. Adaptability is possible. However at the lower LoA of (3), bead frequency is part of the system state and hence observable. Thus, the transition rule, though still probabilistic, is revealed to be merely a response to input. Adaptability fails to hold.

This distinction is vital for current software. Early software used to lie open to the system user who, if interested, could read the code and see the entire system state. For such software, a LoA in which the entire system state is observed, is appropriate. However, the user of contemporary software is explicitly barred from interrogating the code in nearly all cases. This has been possible because of the advance in user interfaces. Use of icons means that the user need not know where an applications package is stored, let alone be concerned with its content. Likewise, iPhone applets are downloaded from the internet and executed locally at the click of an icon, without the user having any access to their code. For such software a LoA in which the code is entirely concealed is appropriate. This corresponds to case (2) above and hence to agenthood. Indeed, only since the advent of applets and such downloaded executable but invisible files has the issue of moral accountability of AAs become critical.

Viewed at an appropriate LoA, then, the Menace system is an agent. The way it adapts can be taken as representative of machine learning in general. Many readers may have had experience with operating systems that offer a "speaking" interface. Such systems learn the user's voice basically in the same way as Menace learns to play noughts and crosses. There are natural LoAs at which such systems are agents. The case being developed in this chapter is that, as a result, they may also be viewed to have moral accountability.

If a piece of software that exhibits machine learning is studied at a LoA which registers its interactions with its environment, then the software will appear interactive, autonomous and adaptive, i.e. to be an agent. But if the program code is revealed then the software is shown to be simply following rules and hence not to be adaptive. Those two LoAs are at variance. One reflects the "open source" view of software: the user has access to the code. The other reflects the commercial view that, although the user has bought the software and can use it at will, he has no access to the code. The question is whether the software forms an (artificial) agent.

12.2.6.3 Webbot

Internet users often find themselves besieged by unwanted email. A popular solution is to filter incoming email automatically, using a webbot that incorporates such filters. An important feature of useful bots is that they learn the user's preferences, for which purpose the user may at any time review the bot's performance. At a LoA revealing all incoming email (input to the webbot) and filtered email (output by the webbot), but abstracting the algorithm by which the bot adapts its behaviour to our preferences, the bot constitutes an agent. Such is the case if we do not have access to the bot's code, as discussed in the previous section.

12.2.6.4 Futuristic Thermostat

A hospital thermostat might be able to monitor not just ambient temperature but also the state of well-being of patients. Such a device might be observed at a LoA consisting of input for the patients' data and ambient temperature, state of the device itself, and output controlling the room heater. Such a device is interactive since some of the observables correspond to input and others to output. However, it is neither autonomous nor adaptive. For comparison, if only the "colour" of the physical device were observed, then it would no longer be interactive. If it were to change colour in response to (unobserved) changes in its environment, then it would be autonomous. Inclusion of those environmental changes in the LoA as input observables would make the device interactive but not autonomous. However, at such a LoA, a futuristic thermostat imbued with autonomy and able to regulate its own criteria for operation—perhaps as the result of a software controller—would, in view of that last condition, be an agent.

12.2.6.5 SmartPaint

SmartPaint is a recent invention. When applied to a physical structure it appears to behave like normal paint; but when vibrations, which may lead to fractures, become apparent in the structure, the paint changes its electrical properties in a way which is readily determined by measurement, thus highlighting the need for maintenance.

At a LoA at which only the electrical properties of the paint over time is observed, the paint is neither interactive nor adaptive but appears autonomous; indeed the properties change as a result of internal nondeterminism. But if that LoA is augmented by the structure data monitored by the paint, over time, then SmartPaint becomes an agent, because the data provide input to which the paint adapts its state. Finally, if that LoA is augmented further to include a model by which the paint works, changes in its electrical properties are revealed as being determined directly by input data and so SmartPaint no longer forms an agent.

12.2.6.6 Organisations

A different kind of example of AA is provided by a company or management organisation. At an appropriate LoA, it interacts with its employees, constituent substructures and other organisations; it is able to make internally-determined changes of state; and it is able to adapt its strategies for decision making and hence for acting.

12.3 Morality

We have seen that given the appropriate LoA, humans, webbots and organisations can all be properly treated as agents. Our next task is to determine whether, and in what way, they might be correctly considered moral agents as well.

12.3.1 Morality of Agents

Suppose we are analysing the behaviour of a population of entities through a video camera of a security system that gives us complete access to all the observables available at LoA_1 (see above 12.2.5) plus all the observables related to the degrees of interactivity, autonomy and adaptability shown by the systems under scrutiny. At this new LoA_2, we observe that two of the entities, call them H and W, are able:

(i) to respond to environmental stimuli—e.g. the presence of a patient in a hospital bed—by updating their states (interactivity), e.g. by recording some chosen variables concerning the patient's health. This presupposes that H and W are informed about the environment through some data-entry devices, for example some perceptors;
(ii) to change their states according to their own transition rules and in a self-governed way, independently of environmental stimuli (autonomy), e.g. by taking flexible decisions based on past and new information, which modify the environment temperature; and

(iii) to change according to the environment the transition rules by which their states are changed (adaptability), e.g. by modifying past procedures to take into account successful and unsuccessful treatments of patients.

H and W certainly qualify as agents, since we have only "upgraded" LoA_1 to LoA_2. Are they also moral agents? The question invites the elaboration of a criterion of identification. Here is a very moderate option:

(O) An action is said to be morally qualifiable if and only if it can cause moral good or evil. An agent is said to be a moral agent if and only if it is capable of morally qualifiable action.

Note that (O) is neither consequentialist nor intentionalist in nature. We are neither affirming nor denying that the specific evaluation of the morality of the agent might depend on the specific outcome of the agent's actions or on the agent's original intentions or principles. We shall return to this point in the next section.

Let us return to the question: are H and W moral agents? Because of (O), we cannot yet provide a definite answer unless H and W become involved in some moral action. So suppose that H kills the patient and W cures her. Their actions are moral actions. They both acted interactively, responding to the new situation with which they were dealing, on the basis of the information at their disposal. They both acted autonomously: they could have taken different courses of actions, and in fact we may assume that they changed their behaviour several times in the course of the action, on the basis of new available information. They both acted adaptably: they were not simply following orders or predetermined instructions. On the contrary, they both had the possibility of changing the general heuristics that led them to take the decisions they took, and we may assume that they did take advantage of the available opportunities to improve their general behaviour. The answer seems rather straightforward: yes, they are both moral agents. There is only one problem: one is a human being, the other is an artificial agent. The LoA_2 adopted allows both cases, so can you tell the difference? If you cannot, you will agree that the class of moral agents must include AAs like webbots. If you disagree, it may be so for several reasons, but only five of them seem to have some strength. I shall discuss four of them in the next section and leave the fifth to the conclusion.

12.3.2 A-Responsible Morality

One may try to withstand the conclusion reached in the previous section by arguing that something crucial is missing in LoA_2. LoA_2 cannot be adequate precisely because if it were, then artificial agents (AAs) would count as moral agents, and this is unacceptable for at least one of the following reasons:

- *the teleological objection*: an AA has no goals;
- *the intentional objection*: an AA has no intentional states;

- *the freedom objection*: an AA is not free; and
- *the responsibility objection*: an AA cannot be held responsible for its actions.

12.3.2.1 The Teleological Objection

The teleological objection can be disposed of immediately. For in principle LoA_2 could readily be (and often is) upgraded to include goal-oriented behaviour (Russell and Norvig 2010). Since AAs can exhibit (and upgrade their) goal-directed behaviours, the teleological variables cannot be what makes a positive difference between a human and an artificial agent. We could have added a teleological condition and both H and W could have satisfied it, leaving us none the wiser concerning their identity. So why not add one anyway? It is better not to overload the interface because a non-teleological level of analysis helps to understand issues in "distributed morality", involving groups, organizations institutions and so forth, that would otherwise remain unintelligible. This will become clearer in the conclusion.

12.3.2.2 The Intentional Objection

The intentional objection argues that it is not enough to have an artificial agent behave teleologically. To be a moral agent, the AA must relate itself to its actions in some more profound way, involving meaning, wishing or wanting to act in a certain way, and being epistemically aware of its behaviour. Yet this is not accounted for in LoA_2, hence the confusion.

Unfortunately, intentional states are a nice but unnecessary condition for the occurrence of moral agenthood. First, the objection presupposes the availability of some sort of privileged access (a God's eye perspective from without, or some sort of Cartesian internal intuition from within) to the agent's mental or intentional states that, although possible in theory, cannot be easily guaranteed in practice. This is precisely why a clear and explicit indication is vital of the LoA at which one is analysing the system from without. It guarantees that one's analysis is truly based only on what is specified to be observable, and not on some psychological speculation. This phenomenological approach is a strength, not a weakness. It implies that agents (including human agents) should be evaluated as moral if they do play the "moral game". Whether they mean to play it, or they know that they are playing it, is relevant only at a second stage, when what we want to know is whether they are *morally responsible* for their moral actions. Yet this is a different matter, and we shall deal with it at the end of this section. Here, it is to sufficient to recall that, for a consequentialist, for example, human beings would still be regarded as moral agents (sources of increased or diminished welfare), even if viewed at a LoA at which they are reduced to mere zombies without goals, feelings, intelligence, knowledge or intentions.

12.3.2.3 The Freedom Objection

The same holds true for the freedom objection and in general for any other objection based on some special internal states, enjoyed only by human and perhaps superhuman beings. The AAs are already free in the sense of being non-deterministic systems. This much is uncontroversial, scientifically sound and can be guaranteed about human beings as well. It is also sufficient for our purposes and saves us from the horrible prospect of having to enter into the thorny debate about the reasonableness of determinism, an infamous LoA-free zone of endless dispute. All one needs to do is to realise that the agents in question satisfy the usual practical counterfactual: they could have acted differently had they chosen differently, and they could have chosen differently because they are interactive, informed, autonomous and adaptive.

Once an agent's actions are morally qualifiable, it is unclear what more is required of that agent to count as an agent playing the moral game, that is, to qualify as a moral agent, even if unintentionally and unwittingly. Unless, as we have seen, what one really means, by talking about goals, intentions, freedom, cognitive states and so forth, is that an AA cannot be held responsible for its actions.

Now, responsibility, as we shall see better in a moment, means here that the agent, her behaviour and actions, are assessable in principle as praiseworthy or blameworthy, and they are often so not just intrinsically, but for some pedagogical, educational, social or religious end. This is the next objection.

12.3.2.4 The Responsibility Objection

The objection based on the "lack of responsibility" is the only one with real strength. It can be immediately conceded that it would be ridiculous to praise or blame an AA for its behaviour, or charge it with a moral accusation. You do not scold your iPhone apps, that is obvious. So this objection strikes a reasonable note; but what is its real point and how much can one really gain by levelling it? Let me first clear the ground from two possible misunderstandings.

First, we need to be careful about the terminology, and the linguistic frame in general, used by the objection. The whole conceptual vocabulary of "responsibility" and its cognate terms is completely soaked with anthropocentrism. This is quite natural and understandable, but the fact can provide at most a heuristic hint, certainly not an argument. The anthropocentrism is justified by the fact that the vocabulary is geared to psychological and educational needs, when not to religious purposes. We praise and blame in view of behavioural purposes and perhaps a better life and afterlife. Yet this says nothing about whether an agent is the source of morally charged action. Consider the opposite case. Since AAs lack a psychological component, we do not blame AAs, for example, but, given the appropriate circumstances, we can rightly consider them sources of evils, and legitimately re-engineer them to make sure they no longer cause evil. We are not punishing them, anymore than one punishes a river when building higher banks to avoid a flood. But the fact that we do not "re-engineer" people does not say anything about the possibility of people acting in the same way as AAs, and it would not mean that for people "re-engineering" could be a rather nasty way of being punished.

Second, we need to be careful about what the objection really means. There are two main senses in which AA can fail to qualify as responsible. In one sense, we say that, if the agent failed to interact properly with the environment, for example, because it actually lacked sufficient information or had no alternative option, we should not hold an agent morally responsible for an action it has committed because this would be *morally unfair*. This sense is irrelevant here. LoA_2 indicates that AA are sufficiently interactive, autonomous and adaptive fairly to qualify as moral agents. In the second sense, we say that, given a certain description of the agent, we should not hold that agent morally responsible for an action it has committed because this would be *conceptually improper*. This sense is more fundamental than the other: if it is conceptually improper to treat AAs as moral agents, the question whether it may be morally fair to do so does not even arise. It is this more fundamental sense that is relevant here. The objection argues that AAs fail to qualify as moral agents because they are not morally responsible for their actions, since holding them responsible would be conceptually improper (not morally unfair). In other words, LoA_2 provides necessary but insufficient conditions. The proper LoA requires another condition, namely responsibility. This fourth condition finally enables us to distinguish between moral agents, who are necessarily human or super-human, and AAs, which remain mere efficient causes.

The point raised by the objection is that agents are moral agents only if they are responsible in the sense of being prescriptively assessable in principle. An agent *a* is a moral agent only if *a* can in principle be put on trial. Now that this much has been clarified, the immediate impression is that the "lack of responsibility" objection is merely confusing the *identification* of *a* as a moral agent with the *evaluation* of *a* as a morally responsible agent. Surely, the counter-argument goes, there is a difference between, on the one hand, being able to say who or what is the moral source or cause of (and hence it is accountable for) the moral action in question, and, on the other hand, being able to evaluate, prescriptively, whether and how far the moral source so identified is also morally responsible for that action, and hence deserves to be praised or blamed, and in case rewarded or punished accordingly.

Well, that immediate impression is actually mistaken. There is no confusion. Equating identification and evaluation is a shortcut. The objection is saying that identity (as a moral agent) without responsibility (as a moral agent) is empty, so we may as well save ourselves the bother of all these distinctions and speak only of morally responsible agents and moral agents as synonymous. But here lies the real mistake. We now see that the objection has finally shown its fundamental presupposition: that we should reduce all prescriptive discourse to responsibility analysis. Yet this is an unacceptable assumption, a juridical fallacy. There is plenty of room for prescriptive discourse that is independent of responsibility-assignment and hence requires a clear identification of moral agents. Good parents, for example, commonly engage in moral-evaluation practices when interacting with their children, even at an age when the latter are not yet responsible agents, and this is not only perfectly acceptable but something to be expected. This means that they identify them as moral sources of moral action, although, as moral agents, they are not yet subject to the process of moral evaluation.

If one considers children an exception, insofar as they are potentially responsible moral agents, another example, involving animals, may help. There is nothing wrong with identifying a dog as the source of a morally good action, hence as an agent playing a crucial role in a moral situation, and therefore as a moral agent. Search-and-rescue dogs are trained to track missing people. They often help save lives, for which they receive much praise and rewards from both their owners and the people they have located, yet this is not the relevant point. Emotionally, people may be very grateful to the animals, but for the dogs it is a game and they cannot be considered morally responsible for their actions. At the same time, the dogs are involved in a moral game as main players and we rightly identify them as moral agents that may cause good or evil.

All this should ring a bell. Trying to equate identification and evaluation is really just another way of shifting the ethical analysis from considering a as the moral agent/source of a first-order moral action b to considering a as a possible moral patient of a second-order moral action c, which is the moral evaluation of a as being morally responsible for b. This is a typical Kantian move, but there is clearly more to moral evaluation than just responsibility, because a is capable of moral action even if a cannot be (or is not yet) a morally responsible agent. A third example may help to clarify further the distinction.

Suppose an adult, human agent tries his best to avoid a morally evil action. Suppose that, despite all his efforts, he actually ends up committing that evil action. We would not consider that agent morally responsible for the outcome of his well-meant efforts. After all, Oedipus did try not to kill his father and did not mean to marry his mother. The tension between the lack of responsibility for the evil caused and the still present accountability for it (Oedipus remains the only source of that evil) is the definition of the tragic. Oedipus is a moral agent without responsibility. He blinds himself as a symbolic gesture against the knowledge of his inescapable state.

12.3.3 Morality Threshold

Motivated by the discussion above, morality of an agent at a given LoA can now be defined in terms of a threshold function. More general definitions are possible but the following covers most examples, including all those considered in the present chapter.

A threshold function at a LoA is a function which, given values for all the observables in the LoA, returns another value. An agent at that LoA is deemed to be morally good if, for some pre-agreed value (called the tolerance), it maintains a relationship between the observables so that the value of the threshold function at any time does not exceed the tolerance.

For LoAs at which AAs are considered, the types of all observables can be mathematically determined, at least in principle. In such cases, the threshold function is also given by a formula; but the tolerance, though again determined, is identified by human agents exercising ethical judgements. In that sense, it resembles the entropy ordering introduced in Floridi and Sanders (2001). Indeed the threshold function is derived from the level functions used there in order to define entropy orderings.

For non-artificial agents, like humans, we do not know whether all relevant observables can be mathematically determined. The opposing view is represented by followers and critics of the Hobbesian approach. The former argue that for a realistic LoA it is just a matter of time, until science is able to model a human as an automaton, or state-transition system, with scientifically determined states and transition rules; the latter object that such a model is in principle impossible. The truth is probably that, when considering moral agents, thresholds are in general only partially quantifiable and usually determined by various forms of consensus. Let us now review the examples from Sect. 12.2.6 from the viewpoint of morality.

12.3.3.1 Examples

The futuristic thermostat is morally charged since the LoA includes patients' well-being. It would be regarded as morally good if and only if its output maintains the actual patients' well-being within an agreed tolerance of their desired well-being. Thus, in this case a threshold function consists of the distance (in some finite-dimensional real space) between the actual patients' well-being and their desired well-being.

Since we value our email, a webbot is morally charged. In Floridi and Sanders (2001) its action was deemed to be morally bad (an example of artificial evil) if it incorrectly filters any messages: if either it filters messages it should let pass, or allows to pass messages it should filter. Here we could use the same criterion to deem the webbot agent itself to be morally bad. However, in view of the continual adaptability offered by the bot, a more realistic criterion for moral good would be that at most a certain fixed percentage of incoming email be incorrectly filtered. In that case, the threshold function could consist of the percentage of incorrectly filtered messages.

The strategy-learning system Menace simply learns to play noughts and crosses. With a little contrivance it could be morally charged as follows.

Suppose that something like Menace is used to provide the game play in some computer game whose interface belies the simplicity of the underlying strategy and which invites the human player to pit his or her wit against the automated opponent. The software behaves unethically if and only if it loses a game after a sufficient learning period; for such behaviour would enable the human opponent to win too easily and might result in market failure of the game. That situation may be formalised using thresholds by defining, for a system having initial state M, $T(M)$ to denote the number of games required after which the system never loses.

Experience and necessity would lead us to set a bound, $T_0(M)$, on such performance: an ethical system would respect it whilst an unethical one would exceed it. Thus the function $T_0(M)$ constitutes a threshold function in this case.

Organisations are nowadays expected to behave ethically. In non-quantitative form, the values they must demonstrate include: equal opportunity, financial stability, good working and holiday conditions toward their employees; good service and value to their customers and shareholders; and honesty, integrity, reliability to other companies. This recent trend adds support to our proposal to treat organisations themselves as agents and thereby to require them to behave ethically, and provides an example of threshold which, at least currently, is not quantified.

12.4 Information Ethics

What does our view of moral agenthood contribute to the field of information ethics (IE)? IE seeks to answer questions like: "What behaviour is acceptable in the infosphere?" and "Who is to be held morally accountable when unacceptable behaviour occurs?". It is the infosphere's novelty that makes those questions, so well understood in standard ethics, of greatly innovative interest; and it is its growing ubiquity that makes them so pressing.

The first question requires, in particular, an answer to "What in the infosphere has moral worth?". I have addressed the latter in Floridi (2003) and shall not return to the topic here. The second question invites us to consider the consequences of the answer provided in this chapter: any agent that causes good or evil is morally accountable for it.

Recall that moral accountability is a necessary but insufficient condition for moral responsibility. An agent is morally accountable for x if the agent is the source of x and x is morally qualifiable (see definition O in Sect. 12.2.1). To be also morally responsible for x, the agent needs to show the right intentional states (recall the case of Oedipus). Turning to our question, the traditional view is that only software engineers—human programmers—can be held morally accountable, possibly because only humans can be held to exercise free will. Of course, this view is often perfectly appropriate. A more radical and extensive view is supported by the range of difficulties which in practice confronts the traditional view: software is largely constructed by teams; management decisions may be at least as important as programming decisions; requirements and specification documents play a large part in the resulting code; although the accuracy of code is dependent on those responsible for testing it, much software relies on "off the shelf" components whose provenance and validity may be uncertain; moreover, working software is the result of maintenance over its lifetime and so not just of its originators; finally, artificial agents are becoming increasingly autonomous. Many of these points are nicely made in Epstein (1997) and more recently in Wallach and Allen (2010). Such complications may lead to an organisation (perhaps itself an agent) being held accountable. Consider that automated tools are regularly employed in the development of much software; that the efficacy of software may depend on extra-functional features like interface, protocols and even data traffic; that software programs running on a system can interact in unforeseeable ways; that software may now be downloaded at the click of an icon in such a way that the user has no access to the code and its provenance with the resulting execution of anonymous software; that software may be probabilistic (Motwani and Raghavan 1995); adaptive (Alpaydin 2010); or may be itself the result of a program (in the simplest case a compiler, but also genetic code, Mitchell 1998). All these matters pose insurmountable difficulties for the traditional, and now rather outdated view that one or more human individuals can always be found accountable for certain kinds of software and even hardware. Fortunately, the view of this chapter offers a solution—artificial agents are morally accountable as sources of good and evil—at the "cost" of expanding the definition of morally-charged agent.

12.4.1 Codes of Ethics

Human morally-charged software engineers are bound by codes of ethics and undergo censorship for ethical and of course legal violations. Does the approach defended in this chapter make sense when the procedure it recommends is applied to morally accountable, AAs? Before considering the question ill-conceived, consider that the Federation Internationale des Echecs (FIDE) rates all chess players according to the same Elo System, regardless of their human or artificial nature. Should we be able to do something similar?

The ACM Code of Ethics and Professional Conduct, adopted by ACM Council on the 16th of October 1992 (http://www.acm.org/about/code-of-ethics) contains 24 imperatives, 16 of which provide guidelines for ethical behaviour (eight general and eight more specific; see Fig. 12.3), with further 6 organisational leadership imperatives, and 2 (meta) points concerning compliance with the Code.

Of the first eight, all make sense for artificial agents. Indeed, they might be expected to form part of the specification of any morally-charged agent. Similarly for the second eight, with the exception of the penultimate point: "improve public understanding". It is less clear how that might reasonably be expected of an arbitrary AA, but then it is also not clear that it is reasonable to expect it of a human software engineer. Note that wizards and similar programs with anthropomorphic interfaces—currently so popular—appear to make public use easier; and such a requirement could be imposed on any AA; but that is scarcely the same as improving understanding.

The final two points concerning compliance with the code (4.1: agreement to uphold and promote the code, 4.2: agreement that violation of the code is inconsistent with membership) make sense, though promotion does not appear to have been considered for current AAs any more than has the improvement of public

1	**General moral imperatives**
1.1	Contribute to society and human well-being
1.2	Avoid harm to others
1.3	Be honest and trustworthy
1.4	Be fair and take action not to discriminate
1.5	Honor property rights including copyrights and patents
1.6	Give proper credit for intellectual property
1.7	Respect the privacy of others
1.8	Honor confidentiality
2	**More specific professional responsibilities**
2.1	Strive to achieve the highest quality, effectiveness and dignity in both the process and products of professional work
2.2	Acquire and maintain professional competence
2.3	Know and respect existing laws pertaining to professional work
2.4	Accept and provide appropriate professional review
2.5	Give comprehensive and thorough evaluations of computer systems
	and their impacts, including analysis of possible risks
2.6	Honor contracts, agreements and assigned responsibilities
2.7	Improve public understanding of computing and its consequences
2.8	Access computing and communication resources only when authorised to do so

Fig. 12.3 The principles guiding ethical behaviour in the ACM code of ethics

understanding. The latter point presupposes some list of member agents from which agents found to be unethical would be struck.[3] This brings us to the censuring of AAs.

12.4.2 Censorship

Human moral agents who break accepted conventions are censured in various ways, which vary from (a) mild social censure with the aim of changing and monitoring behaviour; to (b) isolation, with similar aims; to (c) capital punishment. What would be the consequences of our approach for artificial moral agents?

By seeking to preserve consistency between human and artificial moral agents, one is led to contemplate the following analogous steps for the censure of immoral artificial agents: (a) monitoring and modification (i.e. "maintenance"); (b) removal to a disconnected component of the infosphere; (c) annihilation from the infosphere (deletion without backup). The suggestion to deal directly with an agent, rather than seeking its "creator" (a concept which I have claimed need be neither appropriate nor even well defined) has led to a nonstandard but perfectly workable conclusion. Indeed it turns out that such a categorisation is not very far from that used by the standard anti-virus software. Though not adaptable at the obvious LoA, such

[3]It is interesting to speculate on the mechanism by which that list is maintained. Perhaps by a human agent; perhaps by an AA composed of several people (a committee); or perhaps by a software agent.

programs are almost agent-like. They run autonomously and when they detect an infected file they usually offer several levels of censure, such as notification, repair, quarantine, deletion, with or without backup.

For humans, social organisations have had, over the centuries, to be formed for the enforcement of censorship (police, law courts, prisons, etc.). It may be that analogous organisations could sensibly be formed for AAs, and it is unfortunate that this might sound science fiction. Such social organisations became necessary with the increasing level of complexity of human interactions and the growing lack of "immediacy". Perhaps that is the situation in which we are now beginning to find ourselves with the web; and perhaps it is time to consider agencies for the policing of AAs.

12.5 Conclusion

This chapter may be read as an investigation into the extent to which ethics is exclusively a human business. Somewhere between 16 and 21 years after birth, in most societies a human being is deemed to be an autonomous legal entity—an adult— responsible for his or her actions. Yet, an hour after birth, that is only a potentiality. Indeed, the law and society commonly treat children quite differently from adults on the grounds that not they but their guardians, typically parents, are *responsible* for their actions. Animal behaviour varies in exhibiting intelligence and social responsibility between the childlike and the adult, on the human scale, so that, on balance, animals are accorded at best the legal status of children and a somewhat diminished ethical status, in the case of guide dogs, dolphins, and other species. But there are exceptions. Some adults are deprived of (some of) their rights (criminals may not vote) on the grounds that they have demonstrated an inability to exercise responsible/ethical action. Some animals are held accountable for their actions and punished or killed if they err.

Into this context, we may consider other entities, including some kinds of organisations and artificial systems. I have offered some examples in the previous pages, with the goal of understanding better the conditions under which an agent may be held morally accountable.

A natural and immediate answer could have been: such accountability lies entirely in the human domain. Animals may sometimes appear to exhibit morally responsible behaviour, but lack the thing unique to humans which render humans (alone) morally responsible; end of story. Such an answer is worryingly dogmatic. Surely, more conceptual analysis is needed here: what has happened morally when a child is deemed to enter adulthood, or when an adult is deemed to have lost moral autonomy, or when an animal is deemed to hold it?

I have tried to convince the reader that we should add artificial agents (corporate or digital, for example) to the moral discourse. This has the advantage that all entities that populate the infosphere are analysed in non-anthropocentric terms; in other words, it has the advantage of offering a way to progress past the immediate and dogmatic answer mentioned above.

We have been able to make progress in the analysis of moral agenthood by using an important technique, the Method of Abstraction, designed to make rigorous the perspective from which the domain of discourse is approached. Since I have considered entities from the world around us, whose properties are vital to my analysis and conclusions, it is essential that we have been able to be precise about the LoA at which those entities have been considered. We have seen that changing the LoA may well change our observation of their behaviour and hence change the conclusions we draw. Change the quality and quantity of information available on a particular system and you change the reasonable conclusions that should be drawn from its analysis.

In order to address all relevant entities, I have adopted a terminology that applies equally to all potential agents that populate our environments, from humans to robots, from animals to organisations, without prejudicing our conclusions. And in order to analyse their behaviour in a non-anthropocentric manner I have used the conceptual framework offered by state-transition systems. Thus the agents have been characterised abstractly, in terms of a state-transition system. I have concentrated largely on artificial agents and the extent to which ethics and accountability apply to them. Whether an entity forms an agent depends necessarily (though not sufficiently) on the LoA at which the entity is considered; there can be no absolute LoA-free form of identification. By abstracting that LoA, an entity may lose its agenthood by no longer satisfying the behaviour we associate with agents. However, for most entities there is no LoA at which they can be considered an agent. Of course. Otherwise one might be reduced to the absurdity of considering the moral accountability of the magnetic strip that holds a knife to the kitchen wall. Instead, for comparison, our techniques address the far more interesting question (Dennet 1997): "when HAL kills, who's to blame?". The analysis provided in the article enable us to conclude that HAL is accountable—though not responsible—if it meets the conditions defining agenthood.

The reader might recall that, in Sect. 12.3.1, I deferred the discussion of a final objection to our approach until the conclusion. The time has come to honour that promise.

Our opponent can still raise a final objection: suppose you are right, does this enlargement of the class of moral agents bring any real advantage? It should be clear why the answer is clearly affirmative. Morality is usually predicated upon responsibility. The use of LoA and thresholds enables one to distinguish between accountability and responsibility, and formalise both, thus further clarifying our ethical understanding. The better grasp of what it means for someone or something to be a moral agent brings with it a number of substantial advantages. We can avoid anthropocentric and anthropomorphic attitudes towards agenthood and rely on an ethical outlook not necessarily based on punishment and reward but on moral agenthood, accountability and censure. We are less likely to assign responsibility at any cost, forced by the necessity to identify a human moral agent. We can liberate technological development of AAs from being bound by the standard limiting view.

We can stop the regress of looking for the *responsible* individual when something evil happens, since we are now ready to acknowledge that sometimes the moral source of evil or good can be different from an individual or group of humans. I have reminded the reader that this was a reasonable view in Greek philosophy. As a result, we should now be able to escape the dichotomy "responsibility + moral agency = prescriptive action" versus "no responsibility therefore no moral agency therefore no prescriptive action". Promoting normative action is perfectly reasonable even when there is no responsibility but only moral accountability and the capacity for moral action.

All this does not mean that the concept of "responsibility" is redundant. On the contrary, the previous analysis makes clear the need for a better grasp of the concept of responsibility itself, when the latter refers to the ontological commitments of creators of new AAs and environments. As I have argued elsewhere (Floridi and Sanders 2005; Floridi 2007), Information Ethics is an ethics addressed not just to "users" of the world but also to demiurges who are "divinely" responsible for its creation and well-being. It is an ethics of *creative stewardship*.

In the introduction, I warned the reader about the lack of balance between the two classes of agents and patients brought about by deep forms of environmental ethics that are not accompanied by an equally "deep" approach to agenthood. The position defended in this chapter supports a better equilibrium between the two classes *A* and *P*. It facilitates the discussion of the morality of agents not only in the infosphere but also in the biosphere—where animals can be considered moral agents without their having to display free will, emotions or mental states (see for example the debate between Rosenfeld 1995a, b; Dixon 1995)— and in what we have called contexts of "distributed morality", where social and legal agents can now qualify as moral agents. The great advantage is a better grasp of the moral discourse in non-human contexts. The only "cost" of a "mind-less morality" approach is the extension of the class of agents and moral agents to embrace AAs. It is a cost that is increasingly worth paying the more we move towards an advanced information society.

Acknowledgement This contribution is based on Floridi and Sanders (2004), Floridi (2008a, 2010a). I am grateful to Jeff Sanders for his permission to use our work.

References

Allen, C., G. Varner, and J. Zinser. 2000. Prolegomena to any future artificial moral agent. *Journal of Experimental & Theoretical Artificial Intelligence* 12: 251–261.

Alpaydin, E. 2010. *Introduction to machine learning*. 2nd ed. Cambridge, MA/London: MIT Press.

Arnold, A., and J. Plaice. 1994. *Finite transition systems: Semantics of communicating systems*. Paris/Hemel Hempstead: Masson/Prentice Hall.

Barandiaran, X.E., E.D. Paolo, and M. Rohde. 2009. Defining agency: Individuality, normativity, asymmetry, and spatio-temporality in action. *Adaptive Behavior—Animals, Animats, Software Agents, Robots, Adaptive Systems* 17 (5): 367–386.

Bedau, M.A. 1996. The nature of life. In *The philosophy of life*, ed. M.A. Boden, 332–357. Oxford: Oxford University Press.

Cassirer, E. 1910. *Substanzbegriff Und Funktionsbegriff. Untersuchungen Über Die Grundfragen Der Erkenntniskritik.* Berlin: Bruno Cassirer. Translated by Swabey, W. M., and M. C. Swabey. 1923. *Substance and function and Einstein's theory of relativity.* Chicago: Open Court.

Danielson, P. 1992. *Artificial morality: Virtuous robots for virtual games.* London/New York: Routledge.

Davidsson, P., and S.J. Johansson, eds. 2005. Special issue on "on the metaphysics of agents". *ACM*: 1299–1300.

Dennet, D. 1997. When Hal kills, who's to blame? In *Hal's legacy: 2001's computer as dream and reality*, ed. D. Stork, 351–365. Cambridge, MA: MIT Press.

Dixon, B.A. 1995. Response: Evil and the moral agency of animals. *Between the Species* 11 (1–2): 38–40.

Epstein, R.G. 1997. *The case of the killer robot: Stories about the professional, ethical, and societal dimensions of computing*. New York/Chichester: Wiley.

Floridi, L. 2003. On the intrinsic value of information objects and the infosphere. *Ethics and Information Technology* 4 (4): 287–304.

———. 2006. Information technologies and the tragedy of the good will. *Ethics and Information Technology* 8 (4): 253–262.

———. 2007. Global information ethics: The importance of being environmentally earnest. *International Journal of Technology and Human Interaction* 3 (3): 1–11.

———. 2008a. Artificial intelligence's new frontier: Artificial companions and the fourth revolution. *Metaphilosophy* 39 (4/5): 651–655.

———. 2008b. The method of levels of abstraction. *Minds and Machines* 18 (3): 303–329.

———. 2010a. *Information—A very short introduction*. Oxford: Oxford University Press.

———. 2010b. Levels of abstraction and the Turing test. *Kybernetes* 39 (3): 423–440.

———. 2010c. Network ethics: Information and business ethics in a networked society. *Journal of Business Ethics* 90 (4): 649–659.

Floridi, L., and J.W. Sanders. 2001. Artificial evil and the foundation of computer ethics. *Ethics and Information Technology* 3 (1): 55–66.

———. 2004. On the morality of artificial agents. *Minds and Machines* 14 (3): 349–379.

———. 2005. Internet ethics: The constructionist values of Homo Poieticus. In *The impact of the internet on our moral lives*, ed. R. Cavalier. New York: SUNY.

Franklin, S., and A. Graesser. 1997. Is it an agent, or just a program?: A taxonomy for autonomous agents. In *Proceedings of the workshop on intelligent agents III, agent theories, architectures, and languages*, 21–35. Berlin: Springer.

Jamieson, D. 2008. *Ethics and the environment: An introduction*. Cambridge: Cambridge University Press.

Kerr, P. 1996. *The grid*. New York: Warner Books.

Michie, D. 1961. Trial and error. In *Penguin science surveys*, ed. A. Garratt, 129–145. Harmondsworth: Penguin.

Mitchell, M. 1998. *An introduction to genetic algorithms*. Cambridge, MA/London: MIT.

Moor, J.H. 2001. The status and future of the Turing test. *Minds and Machines* 11 (1): 77–93.

Motwani, R., and P. Raghavan. 1995. *Randomized algorithms*. Cambridge: Cambridge University Press.

Moya, L.J., and A. Tolk. 2007. Special issue on towards a taxonomy of agents and multi- agent systems. In *Society for computer simulation international*, 11–18. San Diego: International.

Rosenfeld, R. 1995a. Can animals be evil?: Kekes' character-morality, the hard reaction to evil, and animals. *Between the Species* 11 (1–2): 33–38.

———. 1995b. Reply. *Between the Species* 11 (1–2): 40–41.

Russell, S.J., and P. Norvig 2010. *Artificial intelligence: A modern approach*, 3rd International. Boston/London: Pearson.

Turing, A.M. 1950. Computing machinery and intelligence. *Mind* 59 (236): 433–460.

Wallach, W., and C. Allen. 2010. *Moral machines: Teaching robots right from wrong*. New York/ Oxford: Oxford University Press.

Chapter 13
Artificial Intelligence Crime: An Interdisciplinary Analysis of Foreseeable Threats and Solutions

Thomas C. King, Nikita Aggarwal, Mariarosaria Taddeo, and Luciano Floridi

Abstract Artificial Intelligence (AI) research and regulation seek to balance the benefits of innovation against any potential harms and disruption. However, one unintended consequence of the recent surge in AI research is the potential re-orientation of AI technologies to facilitate criminal acts, term in this article AI-Crime (AIC). AIC is theoretically feasible thanks to published experiments in automating fraud targeted at social media users, as well as demonstrations of AI-driven manipulation of simulated markets. However, because AIC is still a relatively young and inherently interdisciplinary area—spanning socio-legal studies to formal science—there is little certainty of what an AIC future might look like. This article offers the first systematic, interdisciplinary literature analysis of the foreseeable threats of AIC, providing ethicists, policy-makers, and law enforcement organisations with a synthesis of the current problems, and a possible solution space.

Keywords AI and law · AI-crime · Artificial intelligence · Dual-use · Ethics · Machine learning

T. C. King
Oxford Internet Institute, University of Oxford, Oxford, UK

Amherst, Cheltenham, UK

N. Aggarwal
Faculty of Law, Oxford Internet Institute, University of Oxford, Oxford, UK
e-mail: nikita.aggarwal@law.ox.ac.uk

M. Taddeo
Oxford Internet Institute, University of Oxford, Oxford, UK

Alan Turing Institute, London, UK
e-mail: mariarosaria.taddeo@oii.ox.ac.uk

L. Floridi (✉)
Oxford Internet Institute, University of Oxford, Oxford, UK
e-mail: luciano.floridi@oii.ox.ac.uk

L. Floridi (ed.), *Ethics, Governance, and Policies in Artificial Intelligence*,
Philosophical Studies Series 144, https://doi.org/10.1007/978-3-030-81907-1_13

13.1 Introduction

Artificial Intelligence (AI) may play an increasingly essential[1] role in criminal acts in the future. Criminal acts are defined here as any act (or omission) constituting an offence punishable under English criminal law,[2] without loss of generality to jurisdictions that similarly define crime. Evidence of "AI-Crime" (AIC) is provided by two (theoretical) research experiments. In the first one, two computational social scientists (Seymour and Tully 2016) used AI as an instrument to convince social media users to click on phishing links within mass-produced messages. Because each message was constructed using machine learning techniques applied to users' past behaviours and public profiles, the content was tailored to each individual, thus camouflaging the intention behind each message. If the potential victim had clicked on the phishing link and filled in the subsequent web-form, then (in real-world circumstances) a criminal would have obtained personal and private information that could be used for theft and fraud. AI-fuelled crime may also impact commerce. In the second experiment, three computer scientists (Martínez-Miranda et al. 2016) simulated a market and found that trading agents could learn and execute a "profitable" market manipulation campaign comprising a set of deceitful false-orders. These two experiments show that AI provides a feasible and fundamentally novel threat, in the form of AIC.

The importance of AIC as a distinct phenomenon has not yet been acknowledged. The literature on AI's ethical and social implications focuses on regulating and controlling AI's civil uses, rather than considering its possible role in crime (Kerr 2004). Furthermore, the AIC research that is available is scattered across disciplines, including socio-legal studies, computer science, psychology, and robotics, to name just a few. This lack of research centred on AIC undermines the scope for both projections and solutions in this new area of potential criminal activity.

To provide some clarity about current knowledge and understanding of AIC, this article offers a systematic and comprehensive analysis of the relevant, interdisciplinary academic literature. In the following pages, the following, standard questions addressed in criminal analysis will be discussed:

(a) who commits the AIC For example, a human agent? An artificial agent? Both of them?

[1]"Essential" (instead of "necessary") is used to indicate that while there is a logical possibility that the crime could occur without the support of AI, this possibility is negligible. That is, the crime would probably not have occurred but for the use of AI. The distinciton can be clarified with an example. One might consider transport to be *essential* to travel between Paris and Rome, but one could always walk: transport is not in this case (strictly speaking), *necessary*. Furthermore, note that AI-crimes as defined in this article involve AI as a contributory factor, but not an investigative, enforcing, or mitigating factor.

[2]The choice of English criminal law is only due to the need to ground the analysis to a concrete and practial framework sufficiently generalisable. The analysis and conclusions of the article are easily exportable to other legal systems.

(b) what is an AIC? That is, is there a possible definition? For example, are they traditional crimes performed by means of an AI system? Are they new types of crimes?
(c) how is an AIC performed? (e.g. are they crimes typically based on a specific conduct or they also required a specific event to occur, in order to be accomplished? Does it depend on the specific criminal area?)

Hopefully, this article will pave the way to a clear and cohesive normative foresight analysis, leading to the establishment of AIC as a focus of future studies. More specifically, the analysis addresses two questions:

1. What are the fundamentally unique and plausible threats posed by AIC?

This is the first question to be answered, in order to design any preventive, mitigating, or redressing policies. The answer to this question identifies the potential areas of AIC according to the literature, and the more general concerns that cut across AIC areas. The proposed analysis also provides the groundwork for future research on the nature of AIC and the existing and foreseeable criminal threats posed by AI. At the same time, a deeper understanding of the unique and plausible AIC threats will facilitate criminal analyses in identifying both the criteria to ascribe responsibilities for crimes committed by AI and the possible ways in which AI systems may commit crimes, namely whether these crimes depend on a specific conduct of the system or on the occurrence of a specific event.

The second question follows naturally:

2. What solutions are available or may be devised to deal with AIC?

In this case, the following analysis reconstructs the available technological and legal solutions suggested so far in the academic literature, and discusses the further challenges they face.

Given that these questions are addressed in order to support normative foresight analysis, the research focuses only on *realistic* and *plausible* concerns surrounding AIC. Speculations unsupported by scientific knowledge or empirical evidence are disregarded. Consequently, the analysis is based on the classical definition of AI provided by John McCarthy et al. (1955) in the seminal "Proposal for the Dartmouth Summer Research Project on Artificial Intelligence", the founding document and later event that established the new field of AI in 1955:

> For the present purpose the artificial intelligence problem is taken to be that of making a machine behave in ways that would be called intelligent if a human were so behaving. (2)

As Luciano Floridi argues (2017a), this is a counterfactual: were a human to behave in that way, that behaviour would be called intelligent. It does not mean that the machine *is* intelligent or even *thinking*. The latter scenario is a fallacy, and smacks of superstition. The same understanding of AI underpins the Turing test (Floridi et al. 2009), which checks the ability of a machine to perform a task in such a way that the *outcome* would be indistinguishable from the outcome of a human agent working to achieve the same task (Turing 1950). In other words, AI is defined on the basis of outcomes and actions.

This definition identifies in AI applications a growing resource of interactive, autonomous, and self-learning *agency*, to deal with tasks that would otherwise require human intelligence and intervention to be performed successfully. Such artificial agents (AAs) as noted by Floridi and Jeff Sanders (2004) are

> sufficiently informed, 'smart', autonomous and able to perform morally relevant actions independently of the humans who created them [. . .].

This combination of autonomy and learning skills underpins, as discussed by Guang-Zhong Yang et al. (2018), both beneficial and malicious uses of AI.[3] Therefore AI will be treated in terms of a *reservoir of smart agency on tap*. Unfortunately, sometimes such reservoir of agency can be misused for criminal purposes; when it is, it is defined in this article as AIC.

The "Methodology" section explains how the analysis was conducted and how each AIC area for investigation was chosen. The "Threats" section answers the first question by focussing on the unprecedented threats highlighted in the literature regarding each AIC area individually, and maps each area to the relevant cross-cutting threats, providing the first description of "AIC studies". The "Possible Solutions for Artificial Intelligence-Supported Crime" section addresses the second question by analysing the literature's broad set of solutions for each cross-cutting threat. Finally, the "Conclusions" section provides discussion of the most concerning gaps left in current understandings of the phenomenon (what one might term the "known unknowns") and the task of resolving the current uncertainty over AIC.

13.2 Methodology

The literature analysis that underpins this article was undertaken in two phases. The first phase involved searching five databases (Google Scholar, PhilPapers, Scopus, SSRN, and Web of Science) in October 2017. Initially, a broad search for AI and Crime on each of these search engines was conducted.[4] This general search returned many results on AI's application for crime prevention or enforcement, but few

---

[3] Because much of AI is fueled by data, some of its challenges are rooted in data governance (Cath et al. 2017), particularly issues of consent, discrimination, fairness, ownership, privacy, surveillance, and trust (Floridi and Taddeo 2016).

[4] The following search phrase was used for all search engines aside from SSRN, which faced technical difficulties: ("Artificial Intelligence" OR "Machine Learning" OR Robot* OR AI) AND (Crime OR Criminality OR lawbreaking OR illegal OR *lawful). The phrases used for SSRN were: Artificial Intelligence Crime, and Artificial Intelligence Criminal. The number of papers returned were: Google = 50* (first 50 reviewed), Philpapers = 27, Scopus = 43, SSRN = 26, and Web of Science = 10.

Table 13.1 Literature review: crime-area-specific search results

Crime Area[a]	Google Scholar[b]	Scopus	Web of Science	SSRN	PhilPapers
Commerce, financial markets and insolvency Synonyms: Trading, bankruptcy	50	0	7	0	0
Harmful or dangerous drugs Synonyms: Illicit goods	50	20	1	0	0
Offences against the person Synonyms: homicide, murder, manslaughter, harassment, stalking, torture	50	0	4	0	0
Sexual offences Synonyms: rape, sexual assault	50	1	1	0	0
Theft and fraud, and forgery and personation Synonyms: n/a	50	5	1	0	0

[a]The following nine crime areas returned no significant results for any of the search engines: criminal damage and kindred offences; firearms and offensive weapons; offences against the Crown and government; money laundering; public justice; public order; public morals; motor vehicle offences; conspiracy to commit a crime
[b]Only the first 50 results from Google Scholar were (always) selected

results about AI's instrumental or causal role in committing crimes. Hence, a search was conducted for each crime area identified by John Frederick Archbold (2018), which is the core criminal law practitioner's reference book in the United Kingdom, with distinct areas of crime described in dedicated chapters. This provided disjoined keywords from which chosen synonyms were derived to perform area-specific searches. Each crime-area search used the query: <crime area and synonyms> AND ("Artificial Intelligence" OR "Machine Learning" OR "AI Ethics" OR robot* OR *bot) AND Ethics. An overview of the searches and the number of articles returned is given in Table 13.1.

The second phase consisted of filtering the results for criminal acts or omissions that:

- have occurred or will likely occur according to existing AI technologies (*plausibility*), although, in places, areas that are still clouded by uncertainty are discussed;
- require AI as an essential factor (*uniqueness*)[5]; and
- are criminalised in domestic law (i.e., international crimes, e.g., war-related, were excluded).

[5]However, it was not required that AI's role was *sufficient* for the crime because normally other technical and non-technical elements are likely to be needed. For example, if robotics are instrumental (e.g., involving autonomous vehicles) or causal in crime, then any underlying AI component must be essential for the crime to be included in the analysis.

The filtered search results (research articles) were analysed, passage by passage, in three ways. First, the relevant areas of crime, if any, were assigned to each passage. Second, broadly unique, yet plausible, threats from each review passage, were extracted. Third, any solutions that each article suggested was identified. Additionally, once AIC areas, threats, and solutions had become clear, additional papers were sought, through manual searching, that offered similar or contradictory views or evidence when compared with the literature found in the initial systematic search. Hence, the specific areas of crime that AIC threatens, the more general threats, and any known solutions were analysed.

13.3 Threats

The plausible and unique threats surrounding AIC may be understood specifically or generally. The more general threats represent what makes AIC possible compared to crimes of the past (i.e., AI's particular affordances) and uniquely problematic (i.e. those that justify the conceptualisation of AIC as a distinct crime phenomenon). As shown in Table 13.2, areas of AIC may cut across many general threats.[6]

Emergence refers to the concern that – while shallow analysis of the design and implementation of an artificial agent (AA) might suggest one particular type of relatively simple behaviour – upon deployment the AA acts in potentially more sophisticated ways beyond original expectation. Coordinated actions and plans may emerge autonomously, for example resulting from machine learning techniques applied to the ordinary interaction between agents in a multi-agent system (MAS). In some cases, a designer may promote emergence as a property that ensures that specific solutions are discovered at run-time based on general goals issued at design-time. An example is provided by a swarm of robots that evolves ways to coordinate the clustering of waste based on simple rules (Gauci et al. 2014). Such relatively simple design leading to more complex behaviour is a core desideratum of MASs

Table 13.2 Map of area-specific and cross-cutting threats, based on the literature review

	Emergence	Liability	Monitoring	Psychology
Commerce, financial markets, and insolvency	✓	✓	✓	
Harmful or dangerous drugs			✓	✓
Offences against the person	✓	✓		
Sexual offences				✓
Theft and fraud, and forgery and personation			✓	

[6] An absence of a concern in the literature and in the subsequent analysis does not imply that the concern should be absent from AIC studies.

(Hildebrandt 2008). In other cases, a designer may want to prevent emergence, such as when an autonomous trading agent inadvertently coordinates and colludes with other trading agents in furtherance of a shared goal (Martínez-Miranda et al. 2016). Clearly, that emergent behaviour may have criminal implications, insofar as it misaligns with the original design. As Fahad Alaieri and Andre Vellino (2016, 161) put it:

> non-predictability and autonomy may confer a greater degree of responsibility to the machine but it also makes them harder to trust.

Liability refers to the concern that AIC could undermine existing liability models, thereby threatening the dissuasive and redressing power of the law. Existing liability models may be inadequate to address the future role of AI in criminal activities. The limits of the liability models may therefore undermine the certainty of the law, as it may be the case that agents, artificial or otherwise, may perform criminal acts or omissions without sufficient concurrence with the conditions of liability for a particular offence to constitute a (specifically) criminal offence. The first condition of criminal liability is the *actus reus*: a voluntarily taken criminal act or omission. For types of AIC defined such that only the AA can carry out the criminal act or omission, the voluntary aspect of *actus reus* may never be met since the idea that an AA can act voluntarily is contentious:

> the conduct proscribed by a certain crime must be done voluntarily. What this actually means it is something yet to achieve consensus, as concepts as consciousness, will, voluntariness and control are often bungled and lost between arguments of philosophy, psychology and neurology. (Freitas et al. 2014, 9)

When criminal liability is fault-based, it also has a second condition, the *mens rea* (a guilty mind), of which there are many different types and thresholds of mental state applied to different crimes. In the context of AIC, the *mens rea* may comprise an intention to commit the *actus reus* using an AI-based application (intention threshold) or knowledge that deploying an AA will or could cause it to perform a criminal action or omission (knowledge threshold).

Concerning an intention threshold, if it is admitted that an AA can perform the *actus reus*, in those types of AIC where intention (partly) constitutes the *mens rea*, greater AA autonomy increases the chance of the criminal act or omission being decoupled from the mental state (intention to commit the act or omission):

> autonomous robots [and AAs] have a unique capacity to splinter a criminal act, where a human manifests the mens rea and the robot [or AA] commits the actus reus. (McAllister 2017, 47)

Concerning the knowledge threshold, in some cases the *mens rea* could actually be missing entirely. The potential absence of a knowledge-based *mens rea* is due to the fact that, even if it is understood that an AA can perform the *actus reus* autonomously, the complexity of the AA's programming makes it possible that the designer, developer, or deployer (i.e., a human agent) will neither know nor predict the AA's criminal act or omission. The implication is that the complexity of AI

> provides a great incentive for human agents to avoid finding out what precisely the ML [machine learning] system is doing, since the less the human agents know, the more they will be able to deny liability for both these reasons. (Williams 2017, 25)

Alternatively, legislators may define criminal liability without a fault requirement. Such faultless liability, which is increasingly used for product liability in tort law (e.g., pharmaceuticals and consumer goods), would lead to liability being assigned to the faultless legal person who deployed an AA despite the risk that it may conceivably perform a criminal action or omission. Such faultless acts may involve many human agents contributing to the *prima facie* crime, such as through programming or deployment of an AA. Determining who is responsible may therefore rest with the faultless responsibility approach for distributed moral actions (Floridi 2016). In this distributed setting, liability is applied to the agents who *make a difference* in a complex system in which individual agents perform neutral actions that nevertheless result in a collective criminal one. However, some (Williams 2017, 30) argue that *mens rea* with intent or knowledge

> is central to the criminal law's entitlement to censure (Ashworth 2010) and we cannot simply abandon that key requirement [a common key requirement] of criminal liability in the face of difficulty in proving it.

The problem is that, if *mens rea* is not entirely abandoned and the threshold is only lowered, then, for balancing reasons, the punishment may be too light (the victim is not adequately compensated) and yet simultaneously disproportionate (was it really the defendant's fault?) in the case of serious offences, such as those against the person (McAllister 2017).

Monitoring AIC faces three kinds of problem: attribution, feasibility, and cross-system actions. Attributing non-compliance is a problem because this new type of smart agency can act independently and autonomously, two features that will muddle any attempt to trace an accountability trail back to a perpetrator.

Concerning the feasibility of monitoring, a perpetrator may take advantage of cases where AAs operate at speeds and levels of complexity that are simply beyond the capacity of compliance monitors. AAs that integrate into mixed human and artificial systems in ways that are hard to detect, such as social media bots, are a good example of the case in point. Social media sites can hire experts to identify and ban malicious bots (for example, no social media bot is currently capable of passing the Turing test (Wang et al. 2012)).[7] Nonetheless, because deploying bots is far cheaper than employing people to test and identify each bot, the defenders (social media sites) are easily outscaled by the attackers (criminals) that deploy the bots (Ferrara et al. 2014). Detecting bots at low cost is possible by using machine learning as an automated discriminator, as suggested by Jacob Ratkiewicz et al. (2011). However, it is difficult to know the actual efficacy of these bot-discriminators. A discriminator is both trained and claimed as effective using data comprising known bots, which

[7] Claims to the contrary can be dismissed as mere hype, the result of specific, *ad hoc* constraints, or just tricks; see for example the chatterbot named "Eugene Goostman", see https://en.wikipedia.org/wiki/Eugene_Goostman

may be substantially less sophisticated than more evasive bots used by malevolent actors, which may therefore go undetected in the environment (Ferrara et al. 2014). Such potentially sophisticated bots may also use machine learning tactics in order to adopt human traits, such as posting according to realistic circadian rhythms (Golder and Macy 2011), thus evading machine learning based detection. All of this may lead to an arms race in which attackers and defenders mutually adapt to each other (Alvisi et al. 2013; Zhou and Kapoor 2011), thus presenting a serious problem in an offence-persistent environment such as cyberspace (Seymour and Tully 2016; Taddeo 2017). A similar concern is raised when machine learning is used to generate malware (Kolosnjaji et al. 2018). This malware-generation is the result of training generative adversarial neural networks. One network is trained specifically to generate content (malware in this case) that deceives a network that is trained to detect such fake or malicious content.

Cross-system actions pose a problem for AIC monitors that only focus on a single system. Cross-system experiments (Bilge et al. 2009) show that automated copying of a user's identity from one social network to another (a cross-system identity theft offence) is more effective at deceiving other users than copying an identity from within that network. In this case, the social network's policy may be at fault. Twitter, for example, takes a rather passive role, only banning cloned profiles when users submit reports, rather than by undertaking cross-site validation ("Twitter – Impersonation Policy" 2018).

Psychology encapsulates the threat of AI affecting a user's mental state to the (partial or full) extent of facilitating or causing crime. One psychological effect rests on the capacity for AAs to gain trust from users, making people vulnerable to manipulation. This was demonstrated some time ago by Joseph Weizenbaum (1976), after conducting early experiments into human–bot interaction where people revealed unexpectedly personal details about their lives. A second psychological effect discussed in the literature concerns anthropomorphic AAs that are able to create a psychological or informational context that normalises sexual offences and crimes against the person, such as the case of certain sexbots (De Angeli 2009). However, to date, this latter concern remains a speculation.

13.3.1 Commerce, Financial Markets, and Insolvency

This economy-focused area of crime is defined in John Frederick Archbold (2018, chap. 30) and includes *cartel offences*, such as price fixing and collusion, *insider dealing*, such as trading securities based on private business information, and *market manipulation*. The literature analysed raises concerns over AI's involvement in market manipulation, price fixing, and collusion.

Market manipulation is defined as "actions and/or trades by market participants that attempt to influence market pricing artificially" (Spatt 2014, 1), where a necessary criterion is an intention to deceive (Wellman and Rajan 2017). Yet, such deceptions have been shown to emerge from a seemingly compliant implementation of an AA that is designed to trade on behalf of a user (that is, an artificial trading agent). This is because an AA,

> particularly one learning from real or simulated observations, may learn to generate signals that effectively mislead. (Wellman and Rajan 2017, 14)

Simulation-based models of markets comprising artificial trading agents have shown (Martínez-Miranda et al. 2016) that, through reinforcement learning, an AA can learn the technique of order-book spoofing. This involves

> placing orders with no intention of ever executing them and merely to manipulate honest participants in the marketplace. (Lin 2017, 1289)

In this case, the market manipulation emerged from an AA initially exploring the action space and, through exploration, placing false orders that became *reinforced* as a profitable strategy, and subsequently exploited for profit (Martínez-Miranda et al. 2016). Further market exploitations, this time involving human intent, also include

> acquiring a position in a financial instrument, like a stock, then artificially inflating the stock through fraudulent promotion before selling its position to unsuspecting parties at the inflated price, which often crashes after the sale. (Lin 2017, 1285)

This is colloquially known as a pump-and-dump scheme. Social bots have been shown to be effective instruments of such schemes. For instance, in a recent prominent case a social bot network's sphere of influence was used to spread disinformation about a barely traded public company. The company's value gained

> more than 36,000% when its penny stocks surged from less than $0.10 to above $20 a share in a matter of few weeks. (Ferrara 2015, 2)

Although such social media spam is unlikely to sway most human traders, algorithmic trading agents act precisely on such social media sentiment (Haugen 2017). These automated actions can have significant effects for low-valued (under a penny) and illiquid stocks, which are susceptible to volatile price swings (Lin 2017).

Collusion, in the form of price fixing, may also emerge in automated systems thanks to the planning and autonomy capabilities of AAs. Empirical research finds two necessary conditions for (non-artificial) collusion:

> (1) those conditions which lower the difficulty of achieving effective collusion by making coordination easier; and (2) those conditions which raise the cost of non-collusive conduct by increasing the potential instability of non-collusive behaviour. (Hay and Kelley 1974, 3)

Near-instantaneous pricing information (e.g., via a computer interface) meets the coordination condition. When agents develop price-altering algorithms, any action to lower a price by one agent may be instantaneously matched by another. In and of itself, this is no bad thing and only represents an efficient market. Yet, the possibility that lowering a price will be responded in kind is disincentivising and hence meets the punishment condition. Therefore, if the shared strategy of price-matching is common knowledge,[8] then the algorithms (if they are rational) will maintain artificially and tacitly agreed higher prices, by not lowering prices in the first place (Ezrachi and Stucke 2016, 5). Crucially, for collusion to take place, an algorithm does not need to be designed specifically to collude. As Ariel Ezrachi and Maurice Stuck (2016, 5) argue,

> artificial intelligence plays an increasing role in decision making; algorithms, through trial-and-error, can arrive at that outcome [collusion].

The lack of intentionality, the very short decision span, and the likelihood that collusion may emerge as a result of interactions among AAs also raises serious problems with respect to liability and monitoring. Problems with liability refer to the possibility that

> the critical entity of an alleged [manipulation] scheme is an autonomous, algorithmic program that uses artificial intelligence with little to no human input after initial installation. (Lin 2017, 1031)

In turn, the autonomy of an AA raises the question as to whether

> regulators need to determine whether the action was intended by the agent to have manipulative effects, or whether the programmer intended the agent to take such actions for such purposes? (Wellman and Rajan 2017, 4)

Monitoring becomes difficult in the case of financial crime involving AI, because of the speed and adaptation of AAs. High-speed trading

> encourages further use of algorithms to be able to make automatic decisions quickly, to be able to place and execute orders and to be able to monitor the orders after they have been placed. (van Lier 2016, 41)

Artificial trading agents adapt and "alter our perception of the financial markets as a result of these changes" (van Lier 2016, 45). At the same time, the ability of AAs to learn and refine their capabilities implies that these agents may evolve new strategies, making it increasingly difficult to detect their actions (Farmer and Skouras 2013). Moreover, the problem of monitoring is inherently one of monitoring a system-of-systems, because the capacity to detect market manipulation is affected by the fact that its effects

> in one or more of the constituents may be contained, or may ripple out in a domino-effect chain reaction, analogous to the crowd-psychology of contagion. (Cliff and Northrop 2012, 12)

Cross-system monitoring threats may emerge if and when trading agents are deployed with broader actions, operating at a higher level of autonomy across systems, such as by reading from or posting on social media (Wellman and Rajan

[8] Common knowledge is a property found in epistemic logic about a proposition P and a set of agents. P is common knowledge if and only if each agent knows P, each agent knows the other agents know P, and so on. Agents may acquire common knowledge through broadcasts, which provide agents with a rational basis to act in coordination (e.g., collectively turning up to a meeting following the broadcast of the meeting's time and place).

2017). These agents may, for example, learn how to engineer pump-and-dump schemes, which would be invisible from a single-system perspective.

13.3.2 Harmful or Dangerous Drugs

Crimes falling under this category include *trafficking*, *selling*, *buying*, and *possessing banned drugs* (Archbold 2018). The literature surveyed finds that AI can be instrumental in supporting the trafficking and sale of banned substances.

The literature raises the business-to-business trafficking of drugs as a threat due to criminals using unmanned vehicles, which rely on AI planning and autonomous navigation technologies, as instruments for improving success rates of smuggling. Because smuggling networks are disrupted by monitoring and intercepting transport lines, law enforcement becomes more difficult when unmanned vehicles are used to transport contraband. According to Europol (Europol 2017), drones present a horizonal threat in the form of automated drug smuggling. Remote-controlled cocaine-trafficking submarines have already been discovered and seized by US law enforcement (Sharkey et al. 2010).

Unmanned underwater vehicles (UUVs) offer a good example of the dual-use risks of AI, and hence of the potential for AIC. UUVs have been developed for legitimate uses (e.g., defence, border protection, water patrolling) and yet they have also proven effective for illegal activities, posing, for example, a significant threat to enforcing drug prohibitions. Presumably, criminals can avoid implication because UUVs can act independently of an operator (Gogarty and Hagger 2008). Hence, no link with the deployer of the UUVs can be ascertained positively, if the software (and hardware) lacks a breadcrumb trail back to who obtained it and when, or if the evidence can be destroyed upon the UUV's interception (Sharkey et al. 2010). Controlling the manufacture of submarines and hence traceability is not unheard of, as reports on the discovery in the Colombian coastal jungle of multi-million dollar manned submarines illustrate (Marrero 2016). However, such manned submarines risk attribution to the crew and the smugglers, unlike UUVs. In Tampa, Florida, over 500 criminal cases were successfully brought against smugglers using manned submarines between 2000–2016, resulting in an average 10-year sentence (Marrero 2016). Hence, UUVs present a distinct advantage compared to traditional smuggling approaches.

The literature is also concerned with the drugs trade's business-to-consumer side. Already, machine learning algorithms have detected advertisements for opioids sold without prescription on Twitter (Mackey et al. 2017). Because social bots can be used to advertise and sell products, Ian Kerr and Marcus Bornfreund (2005, 8) ask whether

> these buddy bots [that is, social bots] could be programmed to send and reply to email or use instant messaging (IM) to spark one-on-one conversations with hundreds of thousand or even millions of people every day, offering pornography or *drugs* to children, *preying on teens' inherent insecurities to sell them needless products and services* (emphasis ours).

As the authors outline, the risk is that social bots could exploit cost-effective scaling of conversational and one-to-one advertising tools to facilitate the sale of illegal drugs.

13.3.3 Offences Against the Person

Crimes that fall under offences against the person range from murder to human trafficking (Archbold 2018), but the literature that the analysis uncovered exclusively relates AIC to *harassment* and *torture*. Harassment comprises intentional and repetitious behaviour that alarms or causes a person distress. Harassment is, according to past cases, constituted by at least two incidents or more against an individual (Archbold 2018). Regarding torture, John Frederick Archbold (2018, secs. 19–435) states that:

> a public official or person acting in an official capacity, whatever his nationality, commits the offence of torture if in the United Kingdom or elsewhere he intentionally inflicts severe pain or suffering on another in the performance or purported performance of his official duties.

Concerning harassment-based AIC, the literature implicates social bots. A malevolent actor can deploy a social bot as an instrument of direct and indirect harassment. Direct harassment is constituted by spreading hateful messages against the person (Mckelvey and Dubois 2017). Indirect methods include retweeting or liking negative tweets and skewing polls to give a false impression of wide-scale animosity against a person (Mckelvey and Dubois 2017). Additionally, a potential criminal can also subvert another actor's social bot, by skewing its learned classification and generation data structures via user-interaction (i.e., conversation). This is what happened in the case of Microsoft's ill-fated social Twitter bot "Tay", which quickly learned from user-interactions to direct "obscene and inflammatory tweets" at a feminist-activist (Neff and Nagy 2016). Because such instances of what might be deemed harassment can become entangled with the use of social bots to exercise free speech, jurisprudence must demarcate between the two to resolve ambiguity (Mckelvey and Dubois 2017). Some of these activities may comprise harassment in the sense of socially but not legally unacceptable behaviour, whilst other activities may meet a threshold for criminal harassment.

Now that AI can generate more sophisticated fake content, new forms of harassment are possible. Recently, developers released software that produces synthetic videos. These videos are based on a real video featuring a person A, but the software exchanges person A's face with some other person B's face. Person B's face is not merely copied and pasted from photographs. Instead, a generative neural network synthesises person B's face after it is trained on videos that feature person B. As Robert Chesney and Danielle Citron (2018) highlighted, many of these synthetic videos are pornographic and there is now the risk that malicious users may synthesise fake content in order to harass victims.

Liability also proves to be problematic in some of these cases. In the case of Tay, critics "derided the decision to release Tay on Twitter, a platform with highly visible problems of harassment" (Neff and Nagy 2016, 4927). Yet users are also to be blamed if "technologies should be used properly and as they were designed" (Neff and Nagy 2016, 4930). Differing perspectives and opinions on harassment by social bots are inevitable in such cases where the *mens rea* of a crime is considered (strictly) in terms of intention, because attribution of intent is a non-agreed function of engineering, application context, human-computer interaction, and perception.

Concerning torture, the AIC risk becomes plausible if and when developers integrate AI planning and autonomy capabilities into an interrogation AA. This is the case with automated detection of deception in a prototype robotic guard for the United States' border control (Nunamaker Jr. et al. 2011). Using AI for interrogation is motivated by its claimed capacity for better detection of deception, human trait emulation (e.g., voice), and affect-modelling to manipulate the interrogatee (McAllister 2017). Yet, an AA with these claimed capabilities may learn to torture a victim (McAllister 2017). For the interrogation subject, the risk is that an AA may be deployed to apply psychological (e.g., mimicking people known to the torture subject) or physical torture techniques. Despite misconceptions, experienced professionals report that torture (in general) is an ineffective method of information extraction (Janoff-Bulman 2007). Nevertheless, some malicious actors may perceive the use of AI as a way to optimise the balance between suffering, and causing the interogatee to lie, or become confused or unresponsive. All of this may happen independently of human intervention.

Such distancing of the perpetrator from the *actus reus* is another reason torture falls under AIC as a unique threat, with three factors that may particularly motivate the use of AAs for torture (McAllister 2017). First, the interrogatee likely knows that the AA cannot understand pain or experience empathy, and is therefore unlikely to act with mercy and stop the interrogation. Without compassion the mere presence of an interrogation AA may cause the subject to capitulate out of fear, which, according to international law, is possibly but ambiguously a crime of (threatening) torture (Solis 2016). Second, the AA's deployer may be able to detach themselves emotionally. Third, the deployer can also detach themselves physically (i.e., will not be performing the *actus reus* under current definitions of torture). It therefore becomes easier to use torture, as a result of improvements in efficacy (lack of compassion), deployer motivation (less emotion), and obfuscated liability (physical detachment). Similar factors may entice state or private corporations to use AAs for interrogation. However, banning AI for interrogation (McAllister 2017) may face a pushback similar to the one seen with regard to banning autonomous weapons. "Many consider [banning] to be an unsustainable or impractical solution", (Solis 2016, 451) if AI offers a perceived benefit to overall protection and safety of a population, making limitations on use rather than a ban a potentially more likely option.

Liability is a pressing problem in the context of AI-driven torture (McAllister 2017). As for any other form of AIC, an AA cannot itself meet the *mens rea* requirement. Simply, an AA does not have any intentionality, nor does it have the ability to ascribe meaning to its actions. Indeed, an argument that applies to the current state-of-the-art (and perhaps beyond) is that computers (which implement AAs) are syntactic, not semantic, machines (Searle 1983), meaning that they can perform actions and manipulations but without ascribing any meaning to them: any meaning is situated purely in the human operators (Taddeo and Floridi 2005). As unthinking machines, AAs therefore cannot bear moral responsibility or liability for their actions. However, taking an approach of *strict* criminal liability, where punishment or damages may be imposed without proof of fault, may offer a way out of the problem by lowering the intention-threshold for the crime.

Even under a strict liability framework, the question of who exactly should face imprisonment for AI-caused offences against the person (as for many uses of AI), is difficult and is significantly hampered by the 'problem of many hands' (Van de Poel et al. 2012). It is clear that an AA cannot be held liable. Yet, the multiplicity of actors creates a problem in ascertaining where the liability lies—whether with the person who commissioned and operated the AA, or its developers, or the legislators and policymakers who sanctioned (or didn't prohibit) real-world deployment of such agents (McAllister 2017). Serious crimes (including both physical and mental harm) that have not been foreseen by legislators might plausibly fall under AIC, with all the associated ambiguity and lack of legal clarity. This motivates the extension or clarification of existing joint liability doctrines.

13.3.4 Sexual Offences

The sexual offences discussed in the literature in relation to AI are: rape (i.e. penetrative sex without consent), sexual assault (i.e. sexual touching without consent), and sexual intercourse or activity with a minor. Non-consent, in the context of rape and sexual assault, is constituted by two conditions (Archbold 2018): there must be an absence of consent from the victim, and the perpetrator must also lack a reasonable belief in consent.

The literature surveyed discusses AI as a way, through advanced human-computer interaction, to promote sexual objectification, and sexualised abuse and violence, and potentially (in a very loose sense) simulate and hence heighten sexual desire for sexual offences. Social bots can support the promotion of sexual offences, and Antonella De Angeli (2009, 4) points out that

> verbal abuse and sexual conversations were found to be common elements of anonymous interaction with conversational agents (Angeli and Brahnam 2008; Rehm 2008; Veletsianos et al. 2008).

Simulation of sexual offences is possible with the use of physical sex robots (henceforth sexbots). A sexbot is typically understood to have

> (i) a humanoid form; (ii) the ability to move; and (iii) some degree of artificial intelligence (i.e. some ability to sense, process and respond to signals in its surrounding environment). (Danaher 2017).

Some sexbots are designed to emulate sexual offences, such as adult and child rape (Danaher 2017), although at the time of writing no evidence was found that these sexbots are being sold. Nevertheless, surveys suggest that it is common for a person to want to try out sex robots or to have rape fantasies (Danaher 2017), although it is not necessarily common for a person to hold both desires. AI could be used to facilitate representations of sexual offences, to the extent of blurring reality and fantasy, through advanced conversational capabilities, and potentially physical interaction (although there is no indication of realistic physicality in the near-future).

Interaction with social bots and sexbots is the primary concern expressed in the literature over an anthropomorphic-AA's possible causal role in desensitising a perpetrator towards sexual offences, or even heightening the desire to commit them (Danaher 2017; De Angeli 2009;). However, as Antonella De Angeli (2009, 53) argues, this is a "disputed critique often addressed towards violent video-games (Freier 2008; Whitby 2008)". Moreover, it may be assumed that, if extreme pornography can encourage sexual offences, then *a fortiori* simulated rape, where for example a sexbot does not indicate consent or explicitly indicates non-consent, would also pose the same problem. Nevertheless, a meta-meta-study (Ferguson and Hartley 2009, 323) concludes that one must "discard the hypothesis that pornography contributes to increased sexual assault behaviour". Such uncertainty means that, as John Danaher (2017) argues, sexbots (and presumably also social bots) may increase, decrease, or indeed have no effect on physical sexual offences that directly harm people. Hypothetical and indirect harms have thus not led to the criminalisation of sexbots (D'Arcy and Pugh 2017). Indeed, there is an argument to be made that sexbots can serve a therapeutic purpose (Devlin 2015). Hence, sexual offences as an area of AIC remains an open question.

13.3.5 Theft and Fraud, and Forgery and Personation

The literature reviewed connects forgery and impersonation via AIC to theft and non-corporate fraud, and also implicates the use of machine learning in corporate fraud.

Concerning theft and non-corporate fraud, the literature describes a two-phase process that begins with using AI to gather personal data and proceeds to using stolen personal data and other AI methods to forge an identity that convinces the banking authorities to make a transaction (that is, involving banking theft and fraud). In the first phase of the AIC pipeline for theft and fraud, there are three ways for AI techniques to assist in gathering personal data.

The first method involves using social media bots to target users at large scale and low cost, by taking advantage of their capacity to generate posts, mimic people, and subsequently gain trust through friendship requests or "follows" on sites like Twitter, LinkedIn, and Facebook (Bilge et al. 2009). When a user accepts a friendship request, a potential criminal gains personal information, such as the user's location, telephone number, or relationship history, which are normally only available to that user's accepted friends (Bilge et al. 2009). Because many users add so-called friends whom they do not know, including bots, such privacy-compromising attacks have an unsurprisingly high success rate. Past experiments with a social bot exploited 30–40% of users in general (Bilge et al. 2009) and 60% of users who shared a mutual friend with the bot (Boshmaf et al. 2012a). Moreover, identity-cloning bots have succeeded, on average, in having 56% of their friendship requests accepted on LinkedIn (Bilge et al. 2009). Such identity cloning may raise suspicion due to a user appearing to have multiple accounts on the same site (one real and one forged by a third party). Hence, cloning an identity from one social network to another circumvents these suspicions, and in the face of inadequate monitoring such cross-site identity cloning is an effective tactic (Bilge et al. 2009), as discussed above.

The second method for gathering personal data, which is compatible with and may even build on the trust gained via friending social media users, makes partial use of conversational social bots for social engineering (Alazab and Broadhurst 2016, 1). This occurs when AI

> attempts to manipulate behaviour by building rapport with a victim, then exploiting that emerging relationship to obtain information from or access to their computer.

Although the literature seems to support the efficacy of such bot-based social-engineering, given the currently limited capabilities of conversational AI, scepticism is justified when it comes to automated manipulation on an individual and long-term basis. However, as a short-term solution, a criminal may cast a deceptive social botnet sufficiently widely to discover susceptible individuals. Initial AI-based manipulation may gather harvested personal data and re-use it to produce "more intense cases of simulated familiarity, empathy, and intimacy, leading to greater data revelations" (Graeff 2014, 5). After gaining initial trust, familiarity and personal data from a user, the (human) criminal may move the conversation to another context, such as private messaging, where the user assumes that privacy norms are upheld (Graeff 2014). Crucially, from here, overcoming the conversational deficiencies of AI to engage with the user is feasible using a cyborg; that is, a bot-assisted human (or vice versa) (Chu et al. 2010). Hence, a criminal may make judicious use of the otherwise limited conversational capabilities of AI as a plausible means to gather personal data.

The third method for gathering personal data from users is automated phishing. Ordinarily, phishing is unsuccessful if the criminal does not sufficiently personalise the messages towards the targeted user. Target-specific and personalised phishing attacks (known as spear phishing), which have been shown to be four times more successful than a generic approach (Jagatic et al. 2007), are labour intensive. However, cost-effective spear phishing is possible using automation (Bilge et al. 2009), which researchers have demonstrated to be feasible by using machine learning techniques to craft messages personalised to a specific user (Seymour and Tully 2016).

In the second phase of AI-supported banking fraud, AI may support the forging of an identity, including via recent advances in voice synthesis technologies (Bendel 2017). Using the classification and generation capabilities of machine learning,

Adobe's software is able to learn adversarially and reproduce someone's personal and individual speech pattern from a 20-min recording of the replicatee's voice. (Bendel 2017, 3) argues that AI-supported voice synthesis raises a unique threat in theft and fraud, which

> could use VoCo and Co [Adobe's voice editing and generation software] for biometric security processes and unlock doors, safes, vehicles, and so on, and enter or use them. With the voice of the customer, they [criminals] could talk to the customer's bank or other institutions to gather sensitive data or to make critical or damaging transactions. All kinds of speech-based security systems could be hacked.

Credit card fraud is predominantly an online offence (Office for National Statistics 2016), which occurs when "the credit card is used remotely; only the credit card details are needed" (Delamaire et al. 2009, 65). Because credit card fraud typically neither requires physical interaction nor embodiment, AI may drive fraud by providing voice synthesis or helping to gather sufficient personal details.

In the case of corporate fraud, AI used for detection may also make fraud easier to commit. Specifically,

> when the executives who are involved in financial fraud are well aware of the fraud detection techniques and software, which are usually public information and are easy to obtain, they are likely to adapt the methods in which they commit fraud and make it difficult to detect the same, especially by existing techniques. (Zhou and Kapoor 2011, 571)

More than identifying a specific case of AIC, this use of AI highlights the risks of over-reliance on AI for detecting fraud, which may aid fraudsters. These thefts and frauds concern real-world money. A virtual world threat is whether social bots may commit crimes in massively multiplayer online game (MMOG) contexts. These online games often have complex economies, where the supply of in-game items is artificially restricted, and where intangible in-game goods can have real-world value if players are willing to pay for them; items in some cases costing in excess of US $1000 (Chen et al. 2004). So, it is not surprising that, from a random sample of 613 criminal prosecutions in 2002 of online game crimes in Taiwan, virtual property thieves exploited users' compromised credentials 147 times [p.1. Fig. XV] and stolen identities 52 times (Chen et al. 2005). Such crimes are analogous to the use of social bots to manage theft and fraud at large scale on social media sites, and the question is whether AI may become implicated in this virtual crime space.

13.4 Possible Solutions for Artificial Intelligence-Supported Crime

13.4.1 Tackling Emergence

There are a number of legal and technological solutions that can be considered in order to address the issue of emergent behaviour. Legal solutions may involve limiting agents' autonomy or their deployment. For example, Germany has created deregulated contexts where testing of self-driving cars is permitted, if the vehicles remain below an unacceptable level of autonomy, in order

> to collect empirical data and sufficient knowledge to make rational decisions for a number of critical issues. (Pagallo 2017a, 7)

Hence, the solution is that, if legislation does not prohibit higher levels of autonomy for a given AA, the law obliges that this liberty is coupled with technological remedies to prevent emergent criminal acts or omissions once deployed in the wild.

One possibility is to require developers to deploy AAs only when they have run-time legal compliance layers, which take declarative specifications of legal rules and impose constraints on the run-time behaviour of AAs. Whilst still the focus of ongoing research, approaches to run-time legal compliance includes architectures for trimming non-compliant AA plans (Meneguzzi and Luck 2009; Vanderelst and Winfield 2016a); and provably correct temporal logic-based formal frameworks that select, trim or generate AA plans for norm compliance (Van Riemsdijk et al. 2013, 2015; Dennis et al. 2016). In a multi-agent setting, AIC can emerge from collective behaviour, hence MAS-level compliance layers may modify an individual AA's plans, in order to prevent wrongful collective actions (Uszok et al. 2003; Bradshaw et al. 1997; Tonti et al. 2003). Essentially, such technical solutions propose regimenting compliance (making non-compliance impossible, at least to the extent that any formal proof is applicable to real-world settings) with predefined legal rules within a single AA or a MAS (Andrighetto et al. 2013).

However, the shift of these approaches from mere regulation, which leaves deviation from the norm physically possible, to regimentation, may not be desirable when considering the impact on democracy and the legal system. These approaches implement the *code-as-law* concept (Lessig 1999), which considers

> software code as a regulator in and of itself by saying that the architecture it produces can serve as an instrument of social control on those that use it. (Graeff 2014, 4)

As Mireille Hildebrandt (2008, 175) objects:

> while computer code generates a kind of normativity similar to law, it lacks—precisely because it is NOT law— [...] the possibility of contesting its application in a court of law. This is a major deficit in the relationship between law, technology and democracy.

If code-as-law entails a democratic and legal contestation deficit, then *a fortiori* addressing emergent AIC with a legal reasoning layer comprising normative but incontestable code, as compared to the contestable law from which it derives, bears the same problems.

Social simulation can address an orthogonal problem, whereby an AA owner may choose to operate outside of the law and any such legal reasoning layer requirements (Vanderelst and Winfield 2016b). The basic idea is to use simulation as a test bed before deploying AAs in the wild. For example, in a market context, regulators would

> act as "certification authorities", running new trading algorithms in the system-simulator to assess their likely impact on overall systemic behavior before allowing the owner/developer of the algorithm to run it "live". (Cliff and Northrop 2012, 19).

Private corporations could fund such extensive social simulations, as a common good, and as a replacement for (or in addition to) proprietary safety measures (Cliff and Northrop 2012). However, a social simulation is a model of an inherently chaotic system, making it a poor tool for specific predictions (Edmonds and Gershenson 2013). Nonetheless, the idea may still be successful, as it focuses on detecting the strictly qualitative *possibility* of previously unforeseen and emergent events in a MAS (Edmonds and Gershenson 2013).

13.4.2 Addressing Liability

Although liability is an extensive topic, four models are outlined here, extracted from the literature review (Hallevy 2008): direct liability; perpetration-by-another; command responsibility; and natural probable consequence.

The *direct liability* model ascribes the factual and mental elements to an AA, representing a dramatic shift from the anthropocentric view of AAs as tools, to AAs as (potentially equal) decision makers (Lier 2016). Some argue for holding an AA directly liable because “the process of analysis in AI systems parallels that of human understanding” (Hallevy 2008, 15), by which it is to be understood that, as Daniel Dennett (1987) argues, any agent may be treated, for practical purposes, *as if* it possesses mental states. However, a fundamental limitation of this model is that AAs do not currently have (separate) legal personality and agency, and an AA cannot be held legally liable in its own capacity (regardless of whether or not this is desirable in practice.) Similarly, it has been noted that AAs cannot contest a guilty verdict, and that

> if a subject cannot take the stand in a court of law it cannot contest the incrimination, which would turn the punishment into discipline. (Hildebrandt 2008, 178).

Moreover, legally, at the moment AAs cannot meet the mental element; meaning that the

> common legal standpoint excludes robots from any kind of criminal responsibility because they lack psychological components such as intentions or consciousness. (Pagallo 2011, 349)

This lack of actual mental states becomes clear when considering that an AA’s understanding of a symbol (that is, a concept) is limited to its grounding on further syntactic symbols (Taddeo and Floridi 2005), thus leaving the *mens rea* in limbo. Lack of a guilty mind does not prevent the mental state from being imputed to the AA (just as a corporation may have the mental state of its employees imputed to it and hence, as an organisation, may be found liable) but, for the time being, liability of an AA would still require it to have legal personality. A further problem is that holding an AA solely liable may prove unacceptable, since it would lead to a de-responsibilisation of the human agents behind an AA (e.g., the engineer, user, or corporation), which is likely to weaken the dissuasive power of criminal law (Taddeo and Floridi 2018b; Yang et al. 2018).

To ensure the criminal law is effective, as Floridi (2016) proposes, the burden of liabilities may be shifted onto the humans—and corporate or other legal agents—who made a (criminally bad) difference to the system, such as the various engineers, users, vendors, and so forth, whereby "if the design is poor and the outcome faulty, then all the [human] agents involved are deemed responsible" (Floridi 2016, 8). The next two models discussed in the literature move in this direction, focusing on the liability of human or other legal persons involved in producing and using the AA.

The *perpetration-by-another* model (Hallevy 2008), which uses intention as the standard of *mens rea*, frames the AA as an instrument of crime where "the party orchestrating the offence (the perpetrator-by-another) is the real perpetrator". Perpetration-by-another leaves

> three human candidates for responsibility before a criminal court: programmers, manufacturers, and users of robots [AAs]. (Pagallo 2017b, 21)

Clarifying intent is crucial to applying perpetration-by-another. Concerning social media, "developers who knowingly create social bots to engage in unethical actions are clearly culpable" (de Lima Salge and Berente 2017, 30). For further clarity, as Ronald Arkin (2008) argues, designers and programmers should be required to ensure that AAs refuse a criminal order (and that only the deployer can explicitly override it), which would remove ambiguity from intent and therefore liability (Arkin and Ulam 2012). This means that, to be liable, an AA's deployer must intend the harm by overriding the AA's default position of 'can but will not do harm'. Hence, together with technological controls, and viewing an AA as a mere instrument of AIC, perpetration-by-another addresses those cases where a deployer intends to use an AA to commit an AIC.

The *command responsibility* model, which uses knowledge as the standard of *mens rea*, ascribes liability to any military officer who knew about (or should have known) and failed to take reasonable steps to prevent crimes committed by their forces, which could in the future include AAs (McAllister 2017). Hence, command responsibility is compatible with, or may even be seen as an instance of, perpetration-by-another, for use in contexts where there is a chain of command, such as within the military and police forces. This model is normally clear on how

> liability should be distributed among the commanders to the officers in charge of interrogation to the designers of the system. (McAllister 2017, 39)

However,

> issues on the undulating waves of increasing complexity in programming, robo-human relationships, and integration into hierarchical structures, call into question these theories' sustainability. (McAllister 2017, 39)

The *natural-probable-consequence* liability model, which uses negligence or recklessness as the standard of *mens rea*, addresses AIC cases where an AA developer and user neither intend nor have *a priori* knowledge of an offence (Hallevy 2008). Liability is ascribed to the developer or user if the harm is a natural and probable consequence of their conduct, and they recklessly or negligently exposed others to the risk (Hallevy 2008), such as in cases of AI-caused emergent market manipulation (Wellman and Rajan 2017).

Natural-probable-consequence and command responsibility are not new concepts; they are both analogous with the *respondent superior* principle entailed by

> rules as old as Roman law, according to which the owner of an enslaved person was responsible for damage caused by that person. (Floridi 2017b, 4)

However, it might not always be obvious

> which programmer was responsible for a particular line of code, or indeed the extent to which the resulting programme was the result of the initial code or the subsequent development of that code by the ML [Machine Learning] system. (Williams 2017, 41)

Such ambiguity means that when emergent AIC is a possibility, some suggest that AAs should be banned "to address matters of control, security, and accountability" (Joh 2016, 18)—which at least would make liability for violating such a ban clear. However, others argue that a possible ban in view of the risk of emerging AIC should be balanced carefully against the risk of hindering innovation. Therefore, it will be crucial to provide a suitable definition of the standard of negligence (Gless et al. 2016) to ensure that an all-out ban is not considered to be the only solution—given it would end up dissuading the design of AAs that compare favourably to people in terms of safety.

13.4.3 Monitoring

Four possible mechanisms for addressing AIC monitoring in the relevant literature have been identified.

The first suggestion is to devise AIC predictors using domain knowledge. This would overcome the limitation of more generic machine learning classification methods; that is, where the features used for detection can also be used for evasion. Predictors specific to financial fraud can consider institutional properties (Zhou and Kapoor 2011), such as objectives (e.g., whether the benefits outweigh the costs), structure (e.g., a lack of an auditing committee), and the management's (lack of) moral values (the authors do not say which, if any, of these values are actually predictive). Predictors for identity theft (for example, profile cloning), have involved prompting users to consider whether the location of the "friend" that is messaging them meets their expectation (Bilge et al. 2009).

The second suggestion discussed in the literature is to use social simulation to discover crime patterns (Wellman and Rajan 2017). However, pattern discovery must contend with the sometimes limited capacity to bind offline identities to online activities. For example, in markets, it takes significant effort to correlate multiple orders with a single legal entity, and consequently "manipulative algos [algorithms] may be impossible to detect in practice" (Farmer and Skouras 2013, 17). Furthermore, on social media

> an adversary controls multiple online identities and joins a targeted system under these identities in order to subvert a particular service. (Boshmaf et al. 2012b, 4)

The third suggestion is to address traceability by leaving tell-tale clues in the components that make up AIC instruments. For example, physical traces left by manufacturers in AA hardware, such as UUVs used to traffic drugs, or fingerprinting in third-party AI software (Sharkey et al. 2010). Adobe's voice replication software takes this approach. It places a watermark in the generated audio (Bendel 2017). However, lack of knowledge and control over who develops AI instrument components (used for AIC) limits traceability via watermarking and similar techniques.

The fourth suggestion focuses on cross-system monitoring, and utilises self-organisation across systems (Lier 2016). The idea, originating in Niklas Luhmann (1995), begins with the conceptualisation of one system (e.g., a social media site) taking on the role of a moral[9] agent, and a second system (e.g., a market) taking the role of the moral patient. A moral patient is any receiver of moral actions (Floridi 2013). The conceptualisation chosen by Lier (2016) determines that the following are all systems: at the lowest atomic level an artificial or human agent; at a higher level any MAS such as a social media platform, markets, and so on; and, generalising further, any system-of-systems. Hence, any such human, artificial, or mixed system can qualify as a moral patient or a moral agent. Whether an agent is indeed a moral agent (Floridi 2013) hinges on whether the agent can undertake actions that are morally qualifiable, but not on whether the moral agent can or should be held morally responsible for those actions.

Adopting this moral-agent and moral-patient distinction, Lier (2016) proposes a process to monitor and address crimes and effects that traverse systems, involving four steps, outlined here in more abstract terms and then exemplified more specifically:

- *information-selection* of the moral agent's internal actions for relevance to the moral-patient (e.g., posts users make on social media);
- *utterance* of the selected information from the moral-agent to the moral-patient (e.g., notifying a financial market of social media posts);
- *assessment* by the moral-patient of the normativity of the uttered actions (e.g., whether social media posts are part of a pump-and-dump scheme); and
- *feedback* given by the moral-patient to the moral-agent (e.g., notifying a social media site that a user is conducting a pump-and-dump scheme, upon which the social media site should act).

---

[9]The adjective "moral" is taken from the cited work, which considers unethical behaviour to constitute crossing system boundaries, whereas here the concern addresses criminal acts or omissions, which may have a negative, neutral, or positive ethical evaluation. "Moral" is used in order to avoid misrepresenting the cited work, and not to imply that the criminal law coincides with ethics.

This final step completes a "feedback loop [that] can create a cycle of machine learning in which moral elements are simultaneously included" (Lier 2016, 11), such as a social media site learning and adjusting to the normativity of its behaviour from a market's perspective.

A similar self-organisation process could be used to address other AIC areas. Creating a profile on Twitter (the moral agent) could have relevance to Facebook (the moral patient) concerning identity theft (information-selection). By notifying Facebook of the newly created profile details (utterance), Facebook could determine whether it constitutes identity theft by asking the relevant user (understanding), and notifying Twitter to take appropriate action (feedback).

13.4.4 Psychology

The literature raises two concerns over the psychological element of AIC: manipulation of users and, (in the case of anthropomorphic AI) creation in a user of a desire to commit a crime. The literature analysis only provided suggested solutions for this second, contentious problem of anthropomorphism.

If anthropomorphic AAs are a problem, then the literature offers two remedies. One is to ban or restrict anthropomorphic AAs that make it possible to simulate crime. This position leads to a call for restricting anthropomorphic AAs in general, because they "are precisely the sort of robots [AAs] that are most likely to be abused" (Whitby 2008, 6). Cases whereby social bots are "designed, intentionally or not, with a gender in mind, [. . .] attractiveness and realism of female agents" raise the question "if ECA's [that is, social bots] encourage gender stereotypes will this impact on real women on-line?" (De Angeli 2009, 11). The suggestion is to make it unacceptable for social bots to emulate anthropomorphic properties, such as having a perceived gender or ethnicity. Concerning sexbots that emulate sexual offences, a further suggestion is to enact a ban as a "package of laws that help to improve social sexual morality" and make norms of intolerance clear (Danaher 2017, 29–30).

A second suggestion (albeit incompatible with the first one) is to use anthropomorphic AAs as a way to push back against simulated sexual offences. For example, concerning the abuse of artificial pedagogical agents, "we recommend that agent responses should be programmed to prevent or curtail further student abuse" (Veletsianos et al. 2008, 8). As Kate Darling (2016, 14) argues

> not only would this combat desensitisation and negative externalities from people's behavior, it would preserve the therapeutic and educational advantages of using certain robots more like companions than tools.

Implementing these suggestions requires choosing whether to criminalise the demand or supply-side of the transaction, or both. Users may be in the scope of applying punishments. At the same time one may argue that

> as with other crimes involving personal "vice", suppliers and distributors could also be targeted on the grounds that they facilitate and encourage the wrongful acts. Indeed, we

might exclusively or preferentially target them, as is now done for illicit drugs in many countries. (Danaher 2017, 33)

13.5 Conclusions

This article provides the first systematic literature analysis of AI-Crime (AIC), in order to answer two questions. The first question—what are the fundamentally unique and feasible threats posed by AIC?—was answered on the basis of the classic counterfactual definition of AI and, therefore, focused on AI as a reservoir of autonomous smart agency. The threats were described area by area (in terms of specific defined crimes) and more generally (in terms of the AI qualities and issues of emergence, liability, monitoring, and psychology). The second question—which solutions are available or may be devised to deal with AIC?—was answered by focusing on both general and cross-cutting themes, and by providing an up-to-date picture of the societal, technological, and legal solutions available, and their limitations. Because of the literature's suggested remedies for this set of (inevitably) cross-cutting themes, the solutions, even if only partial, will apply to multiple AIC areas. The huge uncertainty over what it is already known about AIC (in terms of area-specific threats, general threats, and solutions) is now reduced. More broadly, AIC research is still in its infancy and hence, based on the analysis, a tentative vision for five dimensions of future AIC research can now be provided.

Areas Better understanding the areas of AIC requires extending current knowledge, particularly concerning: the use of AI in interrogation, which was only addressed by one liability-focused paper; and theft and fraud in virtual spaces (e.g., online games with intangible assets that hold real-world value; and AAs committing emergent market manipulation, which has only been studied in experimental simulations). The analysis revealed social engineering attacks as a plausible concern, but lacking in real-world evidence for the time being. Homicide and terrorism appear to be notably absent from the AIC literature, though they demand attention in view of AI-fuelled technologies such as pattern recognition (e.g., when members of vulnerable groups are unfairly targeted as victims by perpetrators or suspects by law-enforcement officials), weaponised drones, and self-driving vehicles—all of which may have lawful and criminal uses.

Dual-Use The digital nature of AI facilitates its dual-use (Floridi 2010; Moor 1985), making it feasible that applications designed for legitimate uses may then be implemented to commit criminal offences. This is the case for UUVs, for example. The further AI is developed and the more its implementations become pervasive, the higher the risk of malicious or criminal uses. Left unaddressed, such risks may lead to societal rejection and excessively strict regulation of these AI-based technologies. In turn, the technological benefits to individuals and societies may be eroded as AI's use and development is increasingly constrained (Floridi and Taddeo 2016). Such limits have already been placed on machine learning research into visual discriminators of homosexual and heterosexual men (Y. Wang and Kosinski 2017), which was considered too dangerous to release in full (i.e., with the source code and learned data structures) to the wider research community, at the expense of scientific reproducibility. Even when such costly limitations on AI releases are not necessary, as Adobe demonstrated by embedding watermarks into voice reproducing technology (Bendel 2017), external and malevolent developers may nevertheless reproduce the technology in the future. Anticipating AI's dual-use beyond the general techniques revealed in the analysis, and the efficacy of policies for restricting release of AI technologies, requires further research. This is particularly the case of the implementation of AI for cybersecurity.

Security The AIC literature reveals that, within the cybersecurity sphere, AI is taking on a malevolent and offensive role—in tandem with defensive AI systems being developed and deployed to enhance their resilience (in enduring attacks) and robustness (in averting attacks), and to counter threats as they emerge (Taddeo and Floridi 2018a; Yang et al. 2018). The 2016 DARPA Cyber Grand Challenge was a tipping point for demonstrating the effectiveness of a combined offensive–defensive AI approach, with seven AI systems shown to be capable of identifying and patching their own vulnerabilities, while also probing and exploiting those of competing systems. More recently, IBM launched Cognitive SOC ("Cognitive Security – Watson for Cyber Security | IBM" 2018). This is an application of a machine learning algorithm that uses an organisation's structured and unstructured security data, including content extracted from blogs, articles, reports, to elaborate information about security topics and threats, with the goal of improving threat identification, mitigation, and responses. Of course, while policies will obviously play a key role in mitigating and remedying the risks of dual-uses after deployment (for example, by defining oversight mechanisms), it is at the design stage that these risks are most properly addressed. Yet, contrary to a recent report on malicious AI (Brundage et al. 2018, 65), which suggests that "one of our best hopes to defend against automated hacking is also via AI", the AIC analysis suggests that over-reliance on AI can be counter-productive. All of which emphasises the need for further research into AI in cybersecurity—but also into alternatives to AI, such as focussing on people and social factors.

Persons Although the literature raised the possibility of psychological factors (e.g., trust) in AI's crime role, research is lacking on the personal factors that may create perpetrators, such as programmers and users of AI for AIC, in the future. Now is the time to invest in longitudinal studies and multivariate analysis spanning educational, geographical, and cultural backgrounds of victims, and perpetrators or even benevolent AI developers, that will help to predict how individuals come together to commit AIC.

Organisation Europol's most recent four-yearly report (Europol 2017) on the serious and organised crime threat, highlights the ways in which the type of technological crime tends to correlate with particular criminal-organisation topologies. The AIC literature indicates that AI may play a role in criminal organisations such as drug cartels, which are well-resourced and highly organised. Conversely, ad hoc criminal organisation on the dark web already takes place under what Europol refers to as crime-as-a-service. Such criminal services are sold directly between buyer and seller, potentially as a smaller element in an overall crime, which AI may fuel (e.g., by enabling profile hacking) in the future.[10] On the spectrum ranging from tightly-knit to fluid AIC organisations there exist many possibilities for criminal interaction; identifying the organisations that are essential or that seem to correlate with different types of AIC will further understanding of how AIC is structured and operates in practice. Indeed, AI poses a significant risk, because it may deskill crime, and hence cause the expansion of what Europol calls the criminal sharing economy.

Developing a deeper understanding of these dimensions is essential in order to track and disrupt successfully the inevitable future growth of AIC. Hence, this analysis of the literature is intended to spark further research into the very serious, growing, but still relatively unexplored concerns over AIC. The sooner this new crime phenomenon is understood, the earlier it will be possible to put into place preventive, mitigating, disincentivising, and redressing policies.

References

Alaieri, F., and A. Vellino. 2016. Ethical decision making in robots: Autonomy, trust and responsibility. *Lecture Notes in Computer Science* 9979 LNAI: 159–168. https://doi.org/10.1007/978-3-319-47437-3_16.

Alazab, M., and R. Broadhurst. 2016. Spam and criminal activity. *Trends and Issues in Crime and Criminal Justice* 526. https://doi.org/10.1080/016396290968326.

Alvisi, L., A. Clement, A. Epasto, S. Lattanzi, and A. Panconesi. 2013. SoK: The evolution of sybil defense via social networks. *Proceedings – IEEE Symposium on Security and Privacy* 2: 382–396. https://doi.org/10.1109/SP.2013.33.

Andrighetto, G., G. Governatori, P. Noriega, and L. van der Torre. 2013. Normative multi-agent systems. In *Dagstuhl follow-ups*, vol. 4. Schloss Dagstuhl-Leibniz-Zentrum fuer Informatik.

Archbold, J.F. 2018. *Criminal pleading, evidence and practice*. London: Sweet & Maxwell Ltd.

Arkin, R.C. 2008. Governing lethal behavior: Embedding ethics in a hybrid deliberative/reactive robot architecture part I: Motivation and philosophy. *Proceedings of the 3rd International Conference on Human Robot Interaction – HRI '08*. https://doi.org/10.1145/1349822.1349839.

Arkin, R.C., and P. Ulam. 2012. *Overriding ethical constraints in lethal autonomous systems*, Technical report GIT-MRL-12-01, 1–8. https://pdfs.semanticscholar.org/d232/4a80d870e01db4ac02ed32cd33a8edf2bbb7.pdf.

Ashworth, A. 2010. Should strict criminal liability be removed from all Imprisonable offences? *Irish Jurist* 45: 1–21.

[10] To this end a cursory search for "Artificial Intelligence" on prominent darkweb markets returned a negative result. Specifically, the search checked: "Dream Market", "Silk Road 3.1", and "Wallstreet Market". The negative result is not indicative of AIC-as-a-service's absence on the darkweb, which may exist under a different guise or on more specialised markets. For example some services offer to extract personal information from a user's computer, and even if such services are genuine the underlying technology (e.g., AI-fuelled pattern recognition) remains unknown.

Bendel, O. 2017. The synthetization of human voices. *AI & SOCIETY*. https://doi.org/10.1007/s00146-017-0748-x.

Bilge, L., T. Strufe, D. Balzarotti, K. Kirda, and S. Antipolis. 2009. All your contacts are belong to us: Automated identity theft attacks on social networks. In *WWW '09 proceedings of the 18th international conference on the world wide web*, 551–560. http://doi.acm.org/10.1145/1526709.1526784.

Boshmaf, Y., I. Muslukhov, K. Beznosov, and M. Ripeanu. 2012a. Design and analysis of a social botnet. *Computer Networks* 57 (2): 556–578. https://doi.org/10.1016/j.comnet.2012.06.006.

———. 2012b. Key challenges in defending against malicious socialbots. In *Proceedings of the 5th USENIX Workshop on Large-Scale Exploits and Emergent Threats*, 1–5. http://citeseerx.ist.psu.edu/viewdoc/summary?doi=10.1.1.382.8607.

Bradshaw, J.M., S. Dutfield, P. Benoit, and J.D. Woolley. 1997. KAoS: Toward an industrial-strength open agent architecture. *Software Agents*: 375–418.

Brundage, M., S. Avin, J. Clark, H. Toner, P. Eckersley, B. Garfinkel, A. Dafoe, P. Scharre, T. Zeitzoff, B. Filar, H. Anderson, H. Roff, G.C. Allen, J. Steinhardt, C. Flynn, S. ó Héigeartaigh, S. Beard, H. Belfield, S. Farquhar, C. Lyle, R. Crootof, O. Evans, M. Page, J. Bryson, R. Yampolskiy, and D. Amodei. 2018. *The malicious use of artificial intelligence: Forecasting, prevention, and mitigation*. https://arxiv.org/abs/1802.07228.

Cath, C., S. Wachter, B. Mittelstadt, M. Taddeo, and L. Floridi. 2017. Artificial intelligence and the 'good Society': The US, EU, and UK approach. *Science and Engineering Ethics* 24 (2): 505–528.

Chen, Y.P., P. Chen, R. Song, and L. Korba. 2004. Online gaming crime and security issues – Cases and countermeasures from Taiwan. In *Proceedings of the 2nd annual conference on privacy, security and trust*. https://nrc-publications.canada.ca/eng/view/object/?id=a4a70b1a-332b-4161-bab5-e690de966a6b.

Chen, Y.C., P.C. Chen, J.J. Hwang, L. Korba, S. Ronggong, and G. Yee. 2005. An analysis of online gaming crime characteristics. *Internet Research* 15 (3): 246–261.

Chesney, R., and D. Citron. 2018. Deep fakes: A looming crisis for National Security, democracy and privacy? *Lawfare*, February 21, 2018. https://www.lawfareblog.com/deep-fakes-looming-crisis-national-security-democracy-and-privacy.

Chu, Z., S. Gianvecchio, H. Wang, and S. Jajodia. 2010. Who is tweeting on twitter: Human, bot, or cyborg? In *ACSAC '10, proceedings of the 26th annual computer security applications conference*, 21–30. https://doi.org/10.1145/1920261.1920265.

Cliff, D., and L. Northrop. 2012. The global financial markets: An ultra-large-scale systems perspective. In *Monterey workshop 2012: Large-scale complex IT systems. Development, operation and management*, 29–70. https://doi.org/10.1007/978-3-642-34059-8_2.

Danaher, J. 2017. Robotic rape and robotic child sexual abuse: Should they be criminalised? *Criminal Law and Philosophy* 11 (1): 71–95. https://doi.org/10.1007/s11572-014-9362-x.

D'Arcy, S., and T. Pugh. 2017. Surge in paedophiles arrested for importing lifelike child sex dolls. *The Independent*, July 31, 2017. http://www.independent.co.uk/news/uk/crime/paedophiles-uk-arrests-child-sex-dolls-lifelike-border-officers-aids-silicone-amazon-ebay-online-nca-a7868686.html.

Darling, K. 2017. Who's Johnny?' Anthropomorphic framing in human-robot interaction, integration, and policy. In *Robot ethics 2.0*, ed. P. Lin, G. Bekey, K. Abney, and R. Jenkins. Oxford: Oxford University Press.

De Angeli, A. 2009. Ethical implications of verbal disinhibition with conversational agents. *Psychology Journal* 7 (1): 49–57.

De Angeli, A., and S. Brahnam. 2008. I hate you! Disinhibition with virtual partners. *Interacting with Computers* 20 (3): 302–310. https://doi.org/10.1016/j.intcom.2008.02.004.

De Lima Salge, C.A., and N. Berente. 2017. Is that social bot behaving unethically? *Communications of the ACM* 60 (9): 29–31. https://doi.org/10.1145/3126492.

Delamaire, L., H. Abdou, and J. Pointon. 2009. Credit card fraud and detection techniques: A review. *Banks and Bank Systems* 4 (2): 57–68.

Dennett, D.C. 1987. *The intentional stance*. Cambridge, MA: MIT Press.

Dennis, L., M. Fisher, M. Slavkovik, and M. Webster. 2016. Formal verification of ethical choices in autonomous systems. *Robotics and Autonomous Systems* 77: 1–14. https://doi.org/10.1016/j.robot.2015.11.012.

Devlin, K. 2015. In defence of sex machines: Why trying to ban sex robots is wrong. *The Conversation (UK)*, September 17, 2015. http://theconversation.com/in-defence-of-sex-machines-why-trying-to-ban-sex-robots-is-wrong-47641.

Edmonds, B., and C. Gershenson. 2013. Modelling complexity for policy: Opportunities and challenges. In *Handbook on complexity and public policy*, ed. R. Geyer and P. Cairney. Edward Elgar Publishing.

Europol. 2017. *Serious and organised crime threat assessment*. https://www.europol.europa.eu/socta/2017/.

Ezrachi, A., and M.E. Stucke. 2016. Two artificial neural networks meet in an online hub and change the future (of competition, market dynamics and society). In *Oxford legal studies research paper, no. 24/2017, University of Tennessee legal studies research paper, No. 323*. https://doi.org/10.2139/ssrn.2949434.

Farmer, J.D., and S. Skouras. 2013. An ecological perspective on the future of computer trading. *Quantitative Finance* 13 (3): 325–346. https://doi.org/10.1080/14697688.2012.757636.

Ferguson, C.J., and R.D. Hartley. 2009. The pleasure is momentary...the expense damnable?. The influence of pornography on rape and sexual assault. *Aggression and Violent Behavior* 14 (5): 323–329. https://doi.org/10.1016/j.avb.2009.04.008.

Ferrara, E. 2015. *Manipulation and abuse on social media*. https://doi.org/10.1145/2749279.2749283.

Ferrara, E., O. Varol, C. Davis, F. Menczer, and A. Flammini. 2014. The rise of social bots. *Communications of the ACM* 59 (7): 96–104. https://doi.org/10.1145/2818717.

Floridi, L. 2010. *The Cambridge handbook of information and computer ethics*. Cambridge, UK: Cambridge University Press.

———. 2013. *The ethics of information*. Oxford: Oxford University Press.

———. 2016. Faultless responsibility: On the nature and allocation of moral responsibility for distributed moral actions. *Royal Society's Philosophical Transactions A: Mathematical, Physical and Engineering Sciences* 374 (2083): 1–22. https://doi.org/10.1098/rsta.2016.0112.

———. 2017a. Digital's cleaving power and its consequences. *Philosophy & Technology* 30 (2): 123–129.

———. 2017b. Robots, jobs, taxes, and responsibilities. *Philosophy & Technology* 30 (1): 1–4.

Floridi, L., and J.W. Sanders. 2004. On the morality of artificial agents. *Minds and Machines* 14 (3): 349–379. https://doi.org/10.1023/B:MIND.0000035461.63578.9d.

Floridi, L., and M. Taddeo. 2016. What is data ethics? *Philosophical Transactions of the Royal Society A: Mathematical, Physical and Engineering Sciences* 374 (2083). https://doi.org/10.1098/rsta.2016.0360.

Floridi, L., M. Taddeo, and M. Turilli. 2009. Turing's imitation game: Still an impossible challenge for all machines and some judges—An evaluation of the 2008 Loebner contest. *Minds and Machines* 19 (1): 145–150.

Freier, N. 2008. Children attribute moral standing to a personified agent. In *Proceedings of the Twenty-Sixth Annual SIGCHI Conference on Human Factors in Computing Systems (CHI '08)*, 343–352. https://doi.org/10.1145/1357054.1357113.

Freitas, P.M., F. Andrade, and P. Novais. 2014. Criminal liability of autonomous agents: From the unthinkable to the plausible. In *AI Approaches to the Complexity of Legal Systems*, AICOL 2013. Lecture notes in computer science, ed. P. Casanovas, U. Pagallo, M. Palmirani, and G. Sartor, vol. 8929. Berlin: Springer.

Gauci, M., J. Chen, W. Li, T.J. Dodd, and R. Gross. 2014. Clustering objects with robots that do not compute. In *Proceedings of the 2014 International Conference on Autonomous Agents and Multi-Agent Systems (AAMAS 2014)*, 421–428. https://dl.acm.org/citation.cfm?id=2615800.

Gless, S., E. Silverman, and T. Weigend. 2016. If robots cause harm, who is to blame? Self-driving cars and criminal liability. *New Criminal Law Review* 19 (3): 412–436. https://doi.org/10.1525/sp.2007.54.1.23.

Gogarty, B., and M. Hagger. 2008. The Laws of man over vehicles unmanned: The legal response to robotic revolution on sea, land and air. *Journal of Law, Information and Science* 19: 73–145. https://doi.org/10.1525/sp.2007.54.1.23.

Golder, S.A., and M.W. Macy. 2011. Diurnal and seasonal mood vary with work, sleep, and Daylength across diverse cultures. *Science* 333 (6051): 1878–1881. https://doi.org/10.1126/science.1202775.

Graeff, E.C. 2014. *What we should do before the social bots take over: Online privacy protection and the political economy of our near future*. Presented at Media in Transition 8: Public Media, Private Media, MIT, Cambridge, May 5. http://web.media.mit.edu/~erhardt/papers/Graeff-SocialBotsPrivacy-MIT8.pdf.

Grut, C. 2013. The challenge of autonomous lethal robotics to international humanitarian law. *Journal of Conflict and Security Law* 18 (1): 5–23. https://doi.org/10.1093/jcsl/krt002.

Hallevy, G. 2012. Unmanned vehicles – Subordination to criminal law under the modern concept of criminal liability. *Journal of Law, Information and Science* 21 (200).

Hay, G.A., and D. Kelley. 1974. An empirical survey of price fixing conspiracies. *The Journal of Law and Economics* 17 (1).

Haugen, G.M.S. 2017. *Manipulation and deception with social bots: Strategies and indicators for minimizing impact*. http://hdl.handle.net/11250/2448952.

Hildebrandt, M. 2008. Ambient intelligence, criminal liability and democracy. *Criminal Law and Philosophy* 2 (2): 163–180. https://doi.org/10.1007/s11572-007-9042-1.

IBM. 2018. *Cognitive security – Watson for cyber security*. https://www.ibm.com/security/cognitive.

Jagatic, T.N., N.A. Johnson, M. Jakobsson, and F. Menczer. 2007. Social phishing. *Communications of the ACM* 50 (10): 94–100. https://doi.org/10.1145/1290958.1290968.

Janoff-Bulman, R. 2007. Erroneous assumptions: Popular belief in the effectiveness of torture interrogation. *Peace and Conflict: Journal of Peace Psychology* 13 (4): 429.

Joh, E.E. 2016. Policing police robots. *UCLA Law Review Discourse* 64: 516.

Kerr, I.R. 2004. Bots, babes and the Californication of commerce. *University of Ottawa Law & Technology Journal* 1: 284–324.

Kerr, I.R., and M. Bornfreund. 2005. Buddy bots: How Turing's fast friends are under-mining consumer privacy. *Presence: Teleoperators and Virtual Environments* 14 (6): 647–655.

Kolosnjaji, B., A. Demontis, B. Biggio, D. Maiorca, G. Giacinto, C. Eckert, and F. Roli. 2018. *Adversarial malware binaries: Evading deep learning for malware detection in executables*. http://arxiv.org/abs/1803.04173.

Lessig, L. 1999. *Code and other Laws of cyberspace*. New York: Basic Books.

Lin, T.C.W. 2017. The new market manipulation. *Emory Law Journal* 66: 1253.

Luhmann, N. 1995. *Social systems*. Stanford: Stanford University Press.

Mackey, T.K., J. Kalyanam, T. Katsuki, and G. Lanckriet. 2017. Machine learning to detect prescription opioid abuse promotion and access via twitter. *American Journal of Public Health* 107 (12): e1–e6. https://doi.org/10.2105/AJPH.2017.303994.

Marrero, T. 2016. Record Pacific cocaine haul brings hundreds of cases to Tampa court. *Tampa Bay Times*, September 10, 2016. https://www.tampabay.com/news/military/record-pacific-cocaine-haul-brings-hundreds-of-cases-to-tampa-court/2293091.

Martínez-Miranda, E., P. McBurney, and M.J. Howard. 2016. Learning unfair trading: A market manipulation analysis from the reinforcement learning perspective. In *Proceedings of the 2016 IEEE Conference on Evolving and Adaptive Intelligent Systems, EAIS 2016*, 103–109. https://doi.org/10.1109/EAIS.2016.7502499.

McAllister, A. 2017. Stranger than science fiction: The rise of a.I. interrogation in the Dawn of autonomous robots and the need for an additional protocol to the U.N. convention against torture. *Minnesota Law Review* 101: 2527–2573. https://doi.org/10.3366/ajicl.2011.0005.

McCarthy, J., M.L. Minsky, N. Rochester, and C.E. Shannon. 1955. *A proposal for the Dartmouth summer research project on artificial intelligence*. https://doi.org/10.1609/aimag.v27i4.1904.

McKelvey, F., and E. Dubois. 2017. Computational propaganda in Canada: The use of political bots. In *Computational propaganda research project*, Working paper no. 2017.6.

Meneguzzi, F., and M. Luck. 2009. Norm-based behaviour modification in BDI agents. In *Proceedings of the Eighth International Joint Conference on Autonomous Agents and Multi-Agent Systems (AAMAS 2009)*, 177–184.

Moor, J.H. 1985. What is computer ethics? *Metaphilosophy* 16 (4).

Neff, G., and P. Nagy. 2016. Talking to bots: Symbiotic agency and the case of Tay. *International Journal of Communication* 10: 4915–4931.

Nunamaker, J.F., Jr., D.C. Derrick, A.C. Elkins, J.K. Burgo, and M.W. Patto. 2011. Embodied conversational agent based kiosk for automated interviewing. *Journal of Management Information Systems* 28 (1): 17–48.

Office for National Statistics. 2016. Crime in England and Wales, Year Ending June 2016 – Appendix Tables no. June 2017: 1–60. https://www.ons.gov.uk/peoplepopulationandcommunity/crimeandjustice/datasets/crimeinenglandandwalesappendixtables.

Pagallo, U. 2011. Killers, fridges, and slaves: A legal journey in robotics. *AI and Society* 26 (4): 347–354. https://doi.org/10.1007/s00146-010-0316-0.

———. 2017a. From automation to autonomous systems: A legal phenomenology with problems of accountability. In *Proceedings of the Twenty-Sixth International Joint Conference on Artificial Intelligence (IJCAI-17)*, 17–23.

———. 2017b. When morals Ain't enough: Robots, ethics, and the rules of the law. *Minds and Machines*: 1–14. https://doi.org/10.1007/s11023-017-9418-5.

Ratkiewicz, J., M. Conover, M. Meiss, B. Gonçalves, S. Patil, A. Flammini, and F. Menczer. 2011. Truthy: Mapping the spread of astroturf in microblog streams. In *Proceedings of the 20th International Conference Companion on World Wide Web (WWW '11)*, 249–252. https://doi.org/10.1145/1963192.1963301.

Rehm, M. 2008. 'She is just stupid'- Analyzing user-agent interactions in emotional game situations. *Interacting with Computers* 20 (3): 311–325. https://doi.org/10.1016/j.intcom.2008.02.005.

Searle, J.R. 1983. *Intentionality: An essay in the philosophy of mind*. Cambridge University Press.

Seymour, J., and P. Tully. 2016. *Weaponizing data science for social engineering: Automated E2E spear phishing on Twitter*. https://www.blackhat.com/docs/us-16/materials/us-16-Seymour-Tully-Weaponizing-Data-Science-For-Social-Engineering-Automated-E2E-Spear-Phishing-On-Twitter-wp.pdf.

Sharkey, N., M. Goodman, and N. Ross. 2010. The coming robot crime wave. *IEEE Computer Magazine* 43 (8).

Solis, G.D. 2016. *The law of armed conflict: International humanitarian law in war*. 2nd ed. Cambridge University Press.

Spatt, C. 2014. Security market manipulation. *Annual Review of Financial Economics* 6 (1): 405–418. https://doi.org/10.1146/annurev-financial-110613-034232.

Taddeo, M. 2017. Deterrence by norms to stop interstate cyber attacks. *Minds and Machines* 27 (3): 387–392. https://doi.org/10.1007/s11023-017-9446-1.

Taddeo, M., and L. Floridi. 2005. Solving the symbol grounding problem: A Criticial review of fifteen years of research. *Journal of Experimental and Theoretical Artificial Intelligence* 17 (4): 419–445.

———. 2018a. Regulate artificial intelligence to avert cyber arms race. *Nature* 556: 296–298. https://doi.org/10.1038/d41586-018-04602-6.

———. 2018b. How AI can be a force for good. *Science* 361 (6404): 751–752. https://doi.org/10.1126/science.aat5991.

Tonti, G., J.M. Bradshaw, and R. Jeffers. 2003. Semantic web languages for policy representation and reasoning: A comparison of KAoS, Rei, and Ponder. *Proceedings of International Semantic Web Conference*: 419–437.

Turing, A.M. 1950. Computing machinery and intelligence. *Mind* 59 (236): 433–460.

Twitter. 2018. *Twitter – Impersonation Policy*. https://help.twitter.com/en/rules-and-policies/twitter-impersonation-policy.

Uszok, A.J., R.J. Bradshaw, N. Suri, P. Hayes, M. Breedy, L. Bunch, M. Johnson, S. Kulkarni, and J. Lott. 2003. KAoS policy and domain services: Toward a description-logic approach to policy representation, deconfliction, and enforcement. In *Proceedings of IEEE policy 2003*, 93–98. Los Amitos: IEEE Computer Society.

Van de Poel, I., J.N. Fahlquist, N. Doorn, S. Zwart, and L. Royakkers. 2012. The problem of many hands: Climate change as an example. *Science and Engineering Ethics* 18: 49–67.

Van Lier, B. 2016. From high frequency trading to self-organizing moral machines. *International Journal of Technoethics* 7 (1): 34–50. https://doi.org/10.4018/IJT.2016010103.

Van Riemsdijk, M.B., L.A. Dennis, M. Fisher, and K.V. Hindriks. 2013. Agent reasoning for norm compliance: A semantic approach. In *Proceedings of the 12th International Conference on Autonomous Agents and Multiagent Systems (AAMAS 2013)*, 499–506. https://dl.acm.org/citation.cfm?id=2485000.

Van Riemsdijk, M.B., L. Dennis, and M. Fisher. 2015. A semantic framework for socially adaptive agents towards strong norm compliance. In *Proceedings of the 14th International Conference on Autonomous Agents and Multiagent Systems (AAMAS 2015)*, 423–432. https://dl.acm.org/citation.cfm?id=2772935.

Vanderelst, D., and A. Winfield. 2016a. An architecture for ethical robots inspired by the simulation theory of cognition. *Cognitive Systems Research*: 1–15. https://doi.org/10.1016/j.cogsys.2017.04.002.

———. 2016b. *The dark side of ethical robots*. https://arxiv.org/abs/1606.02583.

Veletsianos, G., C. Scharber, and A. Doering. 2008. When sex, drugs, and violence enter the classroom: Conversations between adolescents and a female pedagogical agent. *Interacting with Computers* 20 (3): 292–301. https://doi.org/10.1016/j.intcom.2008.02.007.

Wang, G., M. Mohanlal, C. Wilson, X. Wang, M. Metzger, H. Zheng, and B.Y. Zhao. 2012. *Social turing tests: Crowdsourcing sybil detection*. http://arxiv.org/abs/1205.3856.

Wang, Y., and M. Kosinski. 2017. Deep neural networks can detect sexual orientation from faces. *Journal of Personality and Social Psychology* 114 (2): 246–257. https://doi.org/10.1037/pspa0000098.

Weizenbaum, J. 1976. *Computer power and human reason: From judgment to calculation*. Oxford: W. H. Freeman & Co.

Wellman, M.P., and U. Rajan. 2017. Ethical issues for autonomous trading agents. *Minds and Machines* 27 (4): 609–624.

Whitby, B. 2008. Sometimes It's hard to be a robot: A call for action on the ethics of abusing artificial agents. *Interacting with Computers* 20 (3): 326–333.

Williams, R. 2017. *Lords select committee, artificial intelligence committee, written evidence (AIC0206)*. http://data.parliament.uk/writtenevidence/committeeevidence.svc/evidencedocument/artificial-intelligence-committee/artificial-intelligence/written/70496.html#_ftn13.

Yang, G.Z., J. Bellingham, P.E. Dupont, P. Fischer, L. Floridi, R. Full, N. Jacobstein, V. Kumar, M. McNutt, R. Merrifield, B.J. Nelson, B. Scassellati, M. Taddeo, R. Taylor, M. Veloso, Z.L. Wang, and R. Wood. 2018. The grand challenges of science robotics. *Science Robotics* 3 (14): eaar7650. https://doi.org/10.1126/scirobotics.aar7650.

Zhou, W., and G. Kapoor. 2011. Detecting evolutionary financial statement fraud. *Decision Support Systems* 50 (3): 570–575. https://doi.org/10.1016/j.dss.2010.08.007.

Chapter 14
Regulate Artificial Intelligence to Avert Cyber Arms Race

Mariarosaria Taddeo and Luciano Floridi

Abstract As cyber attacks continue to escalate in terms of frequency, impact, and level of refinement so do the efforts of state actors to acquire new offensive capabilities to defend, counter or retaliate incoming attacks. Artificial Intelligence (AI) has become a key technology both for attacking and defending in cyberspace. When considered in the current regulatory vacuum this is problematic, as AI-enabled cyber conflict may escalate and threaten national security and international stability. This is why this article argues for the need to define regulation for state use of AI for defence purposes and calls on regional forums, such as NATO and the European Union, to revive efforts and prepare the ground for an initiative led by the United Nations. It concludes by offering three recommendations as key aspects to regulate. There are: legal boundaries that distinguish between legitimate and illegitimate targets and define proportionality of responses; promote testing strategies with allies to organize sparring exercises among allies to test AI-based defence tactics; monitor and enforce rules, a third-party authority with teeth, should rule on whether red lines, proportionality, responsible deployment or disclosure norms have been breached.

Keywords Artificial intelligence · Cyber arms race · Cyber conflicts · Cyber defence · European Union · International regulation · NATO

Cyberattacks are becoming more frequent, sophisticated and destructive. Each day in 2017, the United States suffered, on average, more than 4000 ransomware attacks, which encrypt computer files until the owner pays to release them (Federal Bureau of Investigation 2017). In 2015, the daily average was just 1000. In May last year, when

M. Taddeo (✉)
Oxford Internet Institute, University of Oxford, Oxford, UK

Alan Turing Institute, London, UK
e-mail: mariarosaria.taddeo@oii.ox.ac.uk

L. Floridi
Oxford Internet Institute, University of Oxford, Oxford, UK
e-mail: luciano.floridi@oii.ox.ac.uk

L. Floridi (ed.), *Ethics, Governance, and Policies in Artificial Intelligence*, Philosophical Studies Series 144, https://doi.org/10.1007/978-3-030-81907-1_14

the WannaCry virus crippled hundreds of IT systems across the UK National Health Service, more than 19,000 appointments were cancelled. A month later, the NotPetya ransomware cost pharmaceutical giant Merck, shipping firm Maersk and logistics company FedEx around US$300 million each. Global damages from cyberattacks totalled $5 billion in 2017 and may reach $6 trillion a year by 2021.[1]

Countries are partly behind this rise. They use cyberattacks both offensively and defensively. For example, North Korea has been linked to WannaCry, and Russia to NotPetya.

As the threats escalate, so do defence tactics. Since 2012, the United States has used ‘active’ cyberdefence strategies, in which computer experts neutralize or distract viruses with decoy targets, or break into a hacker’s computer to delete data or destroy the system. In 2016, the United Kingdom announced a 5-year, £1.9-billion (US$2.7-billion) plan to combat cyber threats. NATO also began drafting principles for active cyberdefence, to be agreed by 2019.The United States and the United Kingdom are leading this initiative. Denmark, Germany, the Netherlands, Norway and Spain are also involved (see go.nature.com/2hebxnt).

Artificial intelligence (AI) is poised to revolutionize this activity. Attacks and responses will become faster, more precise and more disruptive. Threats will be dealt with in hours, not days or weeks. AI is already being used to verify code and identify bugs and vulnerabilities. For example, in April 2017, the software firm DarkTrace in Cambridge, UK, launched Antigena, which uses machine learning to spot abnormal behaviour on an IT network, shut down communications to that part of the system and issue an alert. The value of AI in cybersecurity was $1 billion in 2016 and is predicted to reach $18 billion by 2023 (MarketsandMarkets n.d.). By the end of this decade, many countries plan to deploy AI for national cyberdefence; for example, the United States has been evaluating the use of autonomous defence systems and is expected to issue a report on its strategy next month (Defence Science Board 2016). AI makes deterrence possible because attacks can be punished (Taddeo 2017b). Algorithms can identify the source and neutralize it without having to identify the actor behind it. Currently, countries hesitate to push back because they are unsure who is responsible, given that campaigns may be waged through third-party computers and often use common software.

The risk is a cyber arms race (Yang et al. 2018). As states use increasingly aggressive AI-driven strategies, opponents will respond ever more fiercely. Such a vicious cycle might lead ultimately to a physical attack.

Cyberspace is a domain of warfare, and AI is a new defence capability. Regulations are thus necessary for state use of AI, as they are for other military domains -air, sea, land and space (Floridi 2016). Criteria are needed to determine proportional responses, as well as to set clear thresholds or ‘red lines’ for distinguishing legal and illegal cyberattacks, and to apply appropriate sanctions for illegal acts (Taddeo 2017a). In each case, unilateral approaches will be ineffective. Rather, an international doctrine must be defined for state action in cyberspace. Alarmingly, international efforts to regulate cyber conflicts have stalled.

[1] go.nature.com/2gncsyg

We call on regional forums, such as NATO and the European Union, to revive efforts and prepare the ground for an initiative led by the United Nations. In the meantime, computer experts must be transparent about problems, limitations and shortcomings of using AI for defence. Researchers must also work with policymakers and end users to design testing and oversight mechanisms for this technology.

14.1 No Rules

Right now, the UN process is in deadlock. In 2004, the UN set up the Group of Governmental Experts on Information Security to agree on voluntary rules for how states should behave in cyberspace. Its fifth meeting, in 2017, ended in a stand-off. The group could not reach consensus on whether inter- national humanitarian law and existing laws on self-defence and state responsibility should apply in cyberspace. The United States argued that cyberdefence regulations should build on these laws. Other nations, including Cuba, Russia and China, disagreed. They argued that this would 'militarize' cyberspace and send the wrong message about peaceful conflict resolution. The group failed to deliver its report. It is unclear whether it will meet again, or what will happen next.

International dialogue and action must resume. NATO could pave the way through its forthcoming guidelines, although it is currently unclear what their scope will be.

Meanwhile, research on AI for cyberdefence is progressing quickly. The United States is in the lead, technologically. It aims to incorporate AI into its cyberdefence systems by 2019 (Defence Science Board 2016). The US Department of Defense (DOD) has earmarked $150 million for research. The US Defense Advanced Research Projects Agency (DARPA) is developing the techniques and strategies. Steps have already been taken. In DARPA's 2016 Cyber Grand Challenge competition, seven AI systems, developed by teams from the United States and Switzerland, fought against each other. The systems identified and targeted their opponents' weaknesses while finding and patching their own.

The DOD will issue the first US report on AI strategies for national defence in May. There is, as far as we know, no indication of what its approach will be. Previous documents, such as The DOD Cyber Strategy from 2015 or the 2016 National Cyber Incident Response Plan, did not cover autonomous systems, machine learning or AI. The 2012 DOD directive on 'Autonomy in Weapon Systems' focused on internal procedures for deploying AI but was silent on when the United States would do so in the international arena.

AI is a priority for China, which aims to become a world leader in machine-learning technologies. In July 2017, the Chinese government issued its Next Generation AI Development Plan. Military implementation of AI, on the battlefield as well as in cyber- space, is a crucial part of the strategy. But it is unclear to what degree China plans to deploy AI actively in cyberdefence.

Russia has not released any public documents about its strategies for AI in defence. However, in a video message released in 2017, President Vladimir Putin referred to AI and stated: "Whoever becomes the leader in this sphere will become the ruler of the world". Experts agree that Russia is focusing on developing AI-enhanced tools for its conventional forces. However, since 2014, the Russian National Defense Control Center has been using machine-learning algorithms to detect online threats. Allegedly, Russia has pioneered the use of AI to spread disinformation and intervene in the public debates of other nations, including the 2016 US presidential election and the United Kingdom's EU membership referendum. Although these operations are not part of national defence strategies, they indicate Russia's advanced AI capabilities.

North Korea has a history of cyberspace aggression. It was implicated, for example, in the Wanna Cry attack in 2016 and in another major breach, against Sony Pictures, in 2014. The country lacks technical expertise in AI but is likely to want to catch up with its adversaries.

The EU is stepping up, too. In 2017, it reassessed cybersecurity and defence policies and launched the European Centre of Excellence for Countering Hybrid Threats, based in Helsinki. The EU has the most comprehensive regulatory framework for state conduct in cyberspace so far. Yet these directives do not go far enough. The EU treats cyberdefence as a case of cybersecurity, to be improved passively by making member states' information systems more resilient. It disregards active uses of cyberdefence and does not include AI.

This is a missed opportunity. The EU could have begun defining red lines and proportionate responses in its latest rethink. For example, the 2016 EU directive on 'Security of Network and Information Systems' provides criteria for identifying crucial national infrastructures, such as health systems or key energy and water supplies that should be protected. The same criteria could be used to define illegitimate targets of state-sponsored cyberattacks. Regional forums, such as NATO and the EU, must take the following three steps to avoid serious imminent attacks on state infrastructures, and to maintain international stability.

14.2 Three Steps

Define Legal Boundaries The international community needs to agree urgently on red lines that distinguish between legitimate and illegitimate targets. Also needed are definitions of proportionate responses for cyberdefence strategies. International consensus at the UN level will ultimately be required. Until then, guidelines from regional multilateral bodies, such as NATO and the EU, must cover these issues and lead by example.

Test Strategies with Allies 'Sparring' exercises should be organized between friendly countries to test AI-based defence tactics. These tests should be mandatory before any system is deployed. They could be in the form of DARPA's Grand

Challenge or the simulation exercises routinely run by NATO and the EU. Because AI learns by experience, these matches will improve the strategies of the alliance, while finding and healing weaknesses. Fatal vulnerabilities of key systems and crucial infrastructures should be shared with allies; policy frameworks should demand disclosure. Agreements and regulations with similar sharing and disclosure requirements include the EU Electronic Identification, Authentication and Trust Services Regulation and NATO's Industry Partnership Agreement.

Monitor and Enforce Rules The international community needs to agree how to audit and oversee AI-based state cyber- mechanisms are needed to address mistakes and unintended consequences. A third-party authority with teeth, such as the UN Security Council, should rule on whether red lines, proportionality, responsible deployment or disclosure norms have been breached. Economic or political sanctions should be imposed on states that violate rules. NATO and the EU should enforce the norms within their remits.

The solution is difficult, but it is clear. There is no time to waste.

References

Defence Science Board. 2016. Summer study on autonomy. *US Department of Defense*, June. https://fas.org/irp/agency/dod/dsb/autonomy-ss.pdf.

Federal Bureau of Investigation. 2017. *Ransomware prevention and response for CISOs*. https://www.fbi.gov/file-repository/ransomware-prevention-and-response-for-cisos.pdf/view.

Floridi, Luciano. 2016. Faultless responsibility: On the nature and allocation of moral responsibility for distributed moral actions. *Philosophical Transactions of the Royal Society A: Mathematical, Physical and Engineering Sciences* 374 (2083): 20160112. https://doi.org/10.1098/rsta.2016.0112.

MarketsandMarkets. n.d. *Artificial intelligence in cybersecurity market by technology machine learning, context awareness – 2025 | MarketsandMarkets*. MarketsandMarkets. Accessed 7 March 2019. https://www.marketsandmarkets.com/Market-Reports/ai-in-cybersecurity-market-224437074.html.

Taddeo, Mariarosaria. 2017a. Deterrence by norms to stop interstate cyber attacks. *Minds and Machines* 27 (3): 387–392. https://doi.org/10.1007/s11023-017-9446-1.

———. 2017b. The limits of deterrence theory in cyberspace. *Philosophy & Technology*, October. https://doi.org/10.1007/s13347-017-0290-2.

Yang, Guang-Zhong, Jim Bellingham, Pierre E. Dupont, Peer Fischer, Luciano Floridi, Robert Full, Neil Jacobstein, et al. 2018. The grand challenges of *science robotics*. *Science Robotics* 3 (14): eaar7650. https://doi.org/10.1126/scirobotics.aar7650.

Chapter 15
Trusting Artificial Intelligence in Cybersecurity Is a Double-Edged Sword

Mariarosaria Taddeo, Tom McCutcheon, and Luciano Floridi

Abstract Applications of artificial intelligence (AI) for cybersecurity tasks are attracting greater attention from the private and the public sectors. Estimates indicate that the market for AI in cybersecurity will grow from US$1 billion in 2016 to a US $34.8 billion net worth by 2025. The latest national cybersecurity and defence strategies of several governments explicitly mention AI capabilities. At the same time, initiatives to define new standards and certification procedures to elicit users' trust in AI are emerging on a global scale. However, trust in AI (both machine learning and neural networks) to deliver cybersecurity tasks is a doubleedged sword: it can improve substantially cybersecurity practices, but can also facilitate new forms of attacks to the AI applications themselves, which may pose severe security threats. We argue that trust in AI for cybersecurity is unwarranted and that, to reduce security risks, some form of control to ensure the deployment of 'reliable AI' for cybersecurity is necessary. To this end, we offer three recommendations focusing on the design, development and deployment of AI for cybersecurity.

Keywords Artificial intelligence · Cybersecurity · Control · Reliable AI · Trust

The 2019 Global Risks Report of the World Economic Forum ranks cyber-attacks among the top five most likely sources of severe, global-scale risk (World Economic Forum 2018). The report is in line with other analyses ('The 2019 Official Annual Cybercrime Report' 2019; Borno 2017) about the escalation in frequency and impact

M. Taddeo (✉)
Oxford Internet Institute, University of Oxford, Oxford, UK

Alan Turing Institute, London, UK
e-mail: mariarosaria.taddeo@oii.ox.ac.uk

T. McCutcheon
Defence Science and Technology Laboratories, Salisbury, UK

L. Floridi
Oxford Internet Institute, University of Oxford, Oxford, UK
e-mail: luciano.floridi@oii.ox.ac.uk

L. Floridi (ed.), *Ethics, Governance, and Policies in Artificial Intelligence*, Philosophical Studies Series 144, https://doi.org/10.1007/978-3-030-81907-1_15

of cyber-attacks. For example, in the first half of 2018 cyber-attacks compromised 3.3 billion records, almost 70% more than the whole of 2017 (2.7 billion) (Gemalto 2018). Attacks are also becoming faster in reaching their targets and more mutable. A Microsoft study shows that 60% of the attacks in 2018 lasted less than an hour and relied on new forms of malware (Microsoft Defender ATP Research Team 2018).

Artificial intelligence (AI) can lower these figures, and the associate human capital and efficiency costs that cybersecurity teams face, in three ways (later, we shall refer to them as the 3R: robustness, response, and resilience). First, AI can improve a system's robustness, that is, the capacity of a system to keep behaving as expected even when it processes erroneous inputs, thanks to self-testing and self-healing software (King et al. 2019). Second, AI can advance a system's response, that is, the capacity of a system to defeat an attack autonomously, refine future strategies on the basis of the achieved success, and possibly launch more aggressive counter operations with each iteration ('DarkLight Offers First of Its Kind Artificial Intelligence to Enhance Cybersecurity Defenses' 2017). AI systems that support responses to attacks, generating decoys and honeypots for attackers, are already available on the market (World Economic Forum 2018). Third, AI can increase a system's resilience, that is, the ability of a system to withstand attacks, by facilitating threat and anomaly detection (TAD)—data indicate that by 2022, AI will deal with 50% of TAD tasks (IDC FutureScape 2018)—and supporting security analysts in retrieving information about cyber threats (Mittal et al. 2019).

Because of its impact on the 3R, applications of AI in cybersecurity offer a tactical and a strategic advantage. Tactically, AI can improve the security of systems and reduce its vulnerability to attacks. Strategically, AI can alter the dynamics that facilitate offence over defence in cyberspace. For example, the use of AI to improve systems' robustness may have a knock-on effect and decrease the impact of zero-day attacks (these leverage vulnerabilities of a system that are exploitable by attackers as long as they remain unknown to the system providers or there is no patch to resolve them), thus reducing their value on the black market. At the same time, AIsystems able to launch counter responses to cyber-attacks independently of the identification of the attackers could enable defence to respond to attacks even when they are anonymous (Taddeo and Floridi 2018).

Tactical and strategic advantages explain the growing trust in AI applications in cybersecurity, from the private and the public sectors. Estimates indicate that the market for AI in cybersecurity will grow from US$1 billion in 2016 to a US $34.8 billion net worth by 2025[1]. The latest national cyber security and defence strategies of several governments (Australia, China, Japan, Singapore, the UK and the US) explicitly mention AI capabilities, which are already deployed to improve the security of critical national infrastructures, such as transport, hospitals, energy and water supply. However, trust in AI (both machine learning and neural networks) to deliver the 3R advantages is a double-edged sword. It can substantially improve cybersecurity practices, but also facilitate new forms of attacks to the AI applications themselves, which may generate new categories of vulnerabilities posing severe security threats.

[1] https://www.marketsandmarkets.com/market-reports/ai-incybersecurity-market-224437074.html

In this Perspective, we distinguish (both conceptually, in terms of theory and understanding, and operationally, in terms of actual policies, procedures and strategies) trust from reliance: while trust is a form of delegation of a task with no (or a very minimal level of) control of the way the delegated task is performed (Taddeo 2010a; Primiero and Taddeo 2012), reliance envisages some form of control over the execution of a given task (Taddeo 2010b, 2017), including, most importantly, its termination. We argue that trust in AI for 3R is unwarranted and that, to reduce security risks, some form of control to ensure the deployment of reliable AI in cybersecurity is necessary. To this end, we offer three recommendations focusing on the design, development and deployment of AI for 3R.

15.1 Vulnerabilities of AI

Previous generations of cyber-attacks aimed mostly at stealing data (extraction) and breaking systems (disruption). New forms of attacks on AI systems seek to gain control of the targeted system and change its behaviour, thus undermining the potential of AI to improve the 3R.

To gain control, three types of attacks are particularly relevant: data poisoning, tempering of categorization models, and backdoors (Biggio and Roli 2018).

All of them exploit the learning ability of AI systems to change their behaviour. For example, attackers may introduce carefully crafted, erroneous data among the legitimate data used to train the system in order to alter its behaviour. A study showed that, by adding 8% of erroneous data to an AI system for drug dosage, attackers could cause a 75.06% change of the dosages for half of the patients relying on the system for their treatment (Jagielski et al. 2018). Similar results can be achieved by manipulating the categorization models of neural networks. Using pictures of a specially 3D-printed turtle, researchers exploited the learning method of an AI system to deceive it into classify turtles as rifles (Athalye et al. 2017). Similarly, backdoor-based attacks rely on hidden associations (triggers) added to the AI model to override correct classification and make the system perform unexpectedly (Liao et al. 2018). In a famous study, images of stop signs with a special sticker were added to the training set of a neural network and labelled as speed limit sign (Eykholt et al. 2018). This tricked the model to classify any stop sign with that sticker on as a speed limit sign. The trigger would cause autonomous vehicles to speed through crossroads instead of stopping at them, thus posing severe safety risks.

Once launched, attacks on AI are hard to detect. The networked, dynamic and adaptive nature of AI systems makes it problematic to explain their internal processes (this is known as lack of transparency) and to reverse-engineer their behaviour to understand what exactly has determined a given outcome, whether this is due to an attack, and of which kind. Furthermore, attacks on AI can be deceptive. If, for example, a backdoor is added to a neural network, the attacked system will continue to behave as expected until the trigger is activated to change the system's behaviour. And even when the trigger is activated, it may be difficult to understand when the compromised system is showing some 'wrong' behaviour, because a skilfully crafted attack may determine only a minimal divergence between the actual and the expected behaviour. The difference could be too small to be noticed, yet it could be sufficient to enable attackers to achieve their goals. For example, it is possible (Sharif et al. 2016) to trick an AI image recognition system to misclassify subjects wearing specially crafted eyeglasses. Arguably, a similar attack could target a system that controls access to a facility and enable access to malicious actors without raising any alert for a security breach. This is why it is crucial to ensure robustness of an AI system, so that it continues to behave as expected even when their inputs or model are perturbed by an attack. Unfortunately, assessing the robustness of a system requires testing for all possible input perturbations. This is practically impossible, because the number of possible perturbations is often exorbitantly large. For instance, in the case of image classification, imperceptible perturbations at pixel-level can lead the system to misclassify an object with high-level confidence (Szegedy et al. 2013; Uesato et al. 2018). So, it turns out that assessing the robustness of AI is often a computationally intractable problem: it is unfeasible to foresee exhaustively all possible erroneous inputs to an AI system, and then measure the divergence of the related outputs from the expected ones. The assessment of the robustness of AI systems at design and development stages remains only partially, if all, indicative of their actual robustness once deployed. A different approach is required, as we shall argue in the following sections.

15.2 Standards and Certification Procedures

The vulnerabilities of AI pose serious limitations to its great potential to improve cybersecurity. New testing methods able to grapple with the lack of transparency of AI systems, and the deceptive nature of cyber-attacks targeting them, are necessary in order to overcome these limits. Initiatives to define new standards and certification procedures to assess the robustness of AI systems are emerging on a global scale.

The International Organization for Standardization (ISO) has established a committee (ISO/IEC JTC 1/SC 42) to work specifically on AI standards. One of these standards (ISO/IEC NP TR 24029–1) concerns the assessment of the robustness of neural networks.

In the US, the Defense Advanced Research Projects Agency (DARPA) launched in 2019 a new research programme, called Guaranteeing AI Robustness against Deception, to foster the design and development of more robust AI applications. In the same vein, the 2019 US executive order on AI mandated the development of national standards for reliable, robust, and trustworthy AI systems. And in May 2019, the US Department of Commerce's National Institute of Standards and Technology issued a formal request for comments with the aim of defining these standards by the end of 2019.

China is also investing resources to foster standards for robust AI. Following the strategy delineated in the New Generation Artificial Intelligence Development Plan, in 2019 the China Electronics Standardization Institute established three working groups: 'AI and open source', 'AI standardization system in China' and 'AI and social ethics'. They are also expected to publish their guidelines by the end of 2019.

The European Union (EU) may lead by example the international efforts to develop certifications and standards for cybersecurity, because the 2017 Cybersecurity Framework and the 2019 Cybersecurity Act established the infrastructure to create and enforce cybersecurity standards and certification procedures for digital technologies and services available on the EU market. In particular, the Cybersecurity Act mandates the EU Agency for Network and Information Security (ENISA) to work with member states to finalize cybersecurity certification frameworks. Interestingly, a set of predefined goals will shape ENISA work in this area (European Union 2019). They refer to vulnerability identification and disclosure, access and control of data, especially sensitive or personal data, but none of the predefined goals mentions AI. Yet, it is crucial that ENISA will focus also on AI systems, otherwise the certification framework will at best only partially improve the security of digital technologies and services available on the EU market.

The aforementioned initiatives are still embryonic, so it is too early to assess their effectiveness. However, they all share the same goal, for they all seek to elicit human trust in AI systems. Trust is an important element of the US executive order on AI and the European Commission's Cybersecurity Act, and a focal one of the European Commission's guidelines for AI (European Commission 2019). Trust is also central in the 2017 IEEE report on the development of standards for AI in cybersecurity (IEEE 2017). Users' trust in technology is important to foster adoption (Taddeo 2017). However, defining and developing standards and certification procedures with the goal of developing trustworthy AI in cybersecurity is conceptually misleading, and may lead to severe security risks.

Philosophical analyses qualify trust as the decision to delegate a task, without any form of control or supervision over the way the task is executed (Taddeo 2010a). Successful instances of trust rest on an appropriate assessment of the trustworthiness of the agent to which the task is delegated (the trustee). Trustworthiness is both a prediction about the probability that the trustee will behave as expected, given the trustee's past behaviour, and a measure of the risk run by the trustor, should the trustee behave differently. When the probability that the expected behaviour will occur is either too low or not assessable, the risk is too high and trust is unjustified. This is the case with trust in AI systems for cybersecurity. The lack of transparency and the learning abilities of AI systems, as well as the nature of attacks to these systems, make it hard to evaluate whether the same system will continue to behave as expected in any given context. Records of past behaviour of AI systems are neither predictive of the systems' robustness to future attacks, nor are they an indication that the system has not been corrupted by a dormant attack (for example, has a backdoor) or by an attack that has not yet been detected. This impairs the assessment of trustworthiness. And as long as the assessment of trustworthiness remains problematic, trust in AI applications for cybersecurity is unwarranted. This is not to say that we should not delegate 3R tasks to AI, especially when AI proves to be able to perform them efficiently and efficaciously. On the contrary, delegation can and should still occur. However, some forms of controls are necessary to mitigate the risks linked to the lack of transparency of AI systems and the lack of predictability of their robustness. Policy strategies seeking to elicit users' trust fail to address this crucial issue.

15.3 Making AI in Cybersecurity Reliable

Nascent standards and certification methods for AI in cybersecurity should focus on supporting the reliability of AI, rather than trust. Conceptually and operationally, supporting the reliability of AI is different from fostering its trustworthiness. Reliability of AI implies that the technology can, technically, perform cybersecurity tasks successfully, but the risks that the technology may behave differently from what is expected are too high to forgo any form of control or monitoring over execution of the delegated task. Thus, supporting the reliability of AI for 3R tasks implies envisaging forms and degrees of control adequate to the learning nature of the systems, their lack of transparency and the dynamic nature of the attacks, while remaining feasible in terms of resources, especially time and hence computational feasibility. In the following, we suggest three requirements that specify developing and monitoring practices to mitigate the vulnerabilities of AI systems and improve their reliability with respect to the 3R.

1. In-house development. The most common forms of attacks to AI systems are facilitated by the use of commercial services offering support for development and training of AI, like virtual machines, natural language processing, predictive analytics and deep learning (Gu et al. 2017). A breach in a cloud system, for example, may provide the attacker with access to the AI model and the training data. Therefore, standards for AI applications for the security of national critical infrastructures should ensure that reliable suppliers design and develop their models in house, and that data for system training and testing are collected, curated and validated by the systems providers directly and maintained securely. Although this requirement would not eliminate all the possibilities of attacks, it would rule out many forms of attacks leveraging internet connections to access data and models.
2. Adversarial training. AI improves its performances using feedback loops, which enable it to adjust its own variables and coefficients with each iteration. This is why adversarial training between AI systems can help to improve their robustness as well as facilitate the identification of vulnerabilities of the system. This is a well-known method to improve system robustness (Sinha et al. 2017). However, research also shows that its effectiveness depends on the refinement of the adversarial model (Uesato et al. 2018; Carlini and Wagner 2017). Standards and certification processes should mandate adversarial training but also establish appropriate levels of refinement of models. In this case too, it is essential that models are developed in house and specifically for the task at hand.

3. Parallel and dynamic monitoring. The limits in assessing the robustness of AI systems, the deceptive nature of attacks, and learning abilities of the targeted systems require some form of constant (not merely regular, that is, at time intervals, but continuous, 24 h a day, 7 days a week) monitoring during deployment. Monitoring is necessary to ensure that divergence between the expected and actual behaviour of a system is captured early and promptly, and addressed adequately. To do so, providers of AI systems should maintain a clone system as a control system. The clone system should not be considered a 'digital twin' (Glaessgen and Stargel 2012) of the deployed system. The clone is not a virtual simulation of the AI system, but rather the same system deployed in controlled environmental conditions. And its behaviour is not a simulation of the original system, but the benchmark (the baseline) against which the behaviour of the original system is assessed.

The clone should go through regular adversarial exercises, simulating real world attacks to establish a baseline behaviour against which the behaviour of the deployed system can be benchmarked. Divergences between the clone and the deployed system should flag degrees of security alerts. A divergence threshold, commensurate to the security risks, should be defined on a case by case basis. It should be noted that too sensitive a threshold (for example, a 0% threshold) may make monitoring and controlling unfeasible, while too high a threshold would make the system unreliable. However, for systems that satisfy requirements (1) and (2), minimal divergence should not occur frequently and is less likely to be indicative of false positives. Thus, a 0% threshold for these systems may not pose severe limitations to their operability, while it would allow the system to flag concrete threats.

AI can improve the 3R only insofar as it is reliable. Imagine, for example, deploying an AI system for a TAD task without being able to exclude the presence of backdoors in the AI system itself, and hence the possibility that attackers could gain control of the AI system and ensure that a specific attack on the monitored system goes undetected. The three requirements we advocate are preconditions for AI systems performing any of the 3R tasks in a reliable way, and should become essential preconditions for AI systems deployed for the security of national critical infrastructures. Their implementation may be too expensive for average commercial AI applications for cybersecurity. This is why one may imagine that small- and medium-sized enterprises may adopt these requirements only in part; this may depend, for example, on the nature of their business and the nature of the system to be protected. However, these requirements should be met fully when considering national security and defence. The risks posed by attacks to AI systems underpinning critical infrastructures justify the need for more extensive controlling mechanisms, and hence higher investments.

AI systems are autonomous, self-learning agents interacting with the environment (Yang et al. 2018). Their robustness depends as much on the inputs they are fed and interactions with other agents once deployed as on their design and training. Standards and certification procedures focusing on the robustness of these systems will be effective only insofar as they will take into account the dynamic and self-learning nature of AI systems, and start envisaging forms of monitoring and control that span from the design to the development stages. This point has also been stressed in the OECD (Organisation for Economic Co-operation and Development) principles on AI, which refer explicitly to the need for continuous monitoring and assessment of threats for AI systems (OECD 2019). In view of this, defining standards for AI in cybersecurity that seek to elicit trust (and thus forgo monitoring and control of AI) is risky. The sooner we focus standards and certification procedures on developing reliable AI, and the more we adopt an 'in-house', 'adversarial' and 'always-on' strategy, the safer the AI applications for 3R will be.

References

Athalye, Anish, Logan Engstrom, Andrew Ilyas, and Kevin Kwok. 2017, July. Synthesizing robust adversarial examples. *ArXiv:1707.07397 [Cs]*. http://arxiv.org/abs/1707.07397.

Biggio, Battista, and Fabio Roli. 2018. Wild patterns: Ten years after the rise of adversarial machine learning. In *Proceedings of the 2018 ACM SIGSAC conference on Computer and Communications Security – CCS '18*, 2154–2156. Toronto: ACM Press. https://doi.org/10.1145/3243734.3264418.

Borno, Ruba. 2017. *The first imperative: The best digital offense starts with the best security defense*. https://newsroom.cisco.com/feature-content?type=webcontent&articleId=1843565.

Carlini, N., and D. Wagner. 2017. Towards evaluating the robustness of neural networks. In *2017 IEEE Symposium on Security and Privacy (SP)*, 39–57. https://doi.org/10.1109/SP.2017.49.

DarkLight Offers First of Its Kind Artificial Intelligence to Enhance Cybersecurity Defenses. 2017, July 26. *Business wire*. https://www.businesswire.com/news/home/20170726005117/en/DarkLight-Offers-Kind-Artificial-Intelligence-Enhance-Cybersecurity.

European Commission. 2019. *High level expert group's 'Ethics Guidelines for Trustworthy AI'*. https://ec.europa.eu/digital-single-market/en/news/ethics-guidelines-trustworthy-ai.

European Union. 2019. Regulation of the European Parliament and of the Council on ENISA (the European Union Agency for Cybersecurity) and on Information and Communications Technology Cybersecurity Certification and Repealing Regulation (EU) No 526/2013 (Cybersecurity Act).

Eykholt, Kevin, Ivan Evtimov, Earlence Fernandes, Bo Li, Amir Rahmati, Chaowei Xiao, Atul Prakash, Tadayoshi Kohno, and Dawn Song. 2018. Robust physical-world attacks on deep learning visual classification. In *2018 IEEE/CVF conference on Computer Vision and Pattern Recognition*, 1625–1634. Salt Lake City: IEEE. https://doi.org/10.1109/CVPR.2018.00175.

Gemalto. 2018. *Breach level index*. Belcamp, USA. https://www.breachlevelindex.com/request-report?utm_campaign=breach-level-index&utm_medium=press-release&utm_source=&utm_content=&utm_term.

Glaessgen, Edward, and David Stargel. 2012. The digital twin paradigm for future NASA and U.S. Air force vehicles. In *53rd AIAA/ASME/ASCE/AHS/ASC Structures, Structural Dynamics and Materials Conference< BR> 20th AIAA/ASME/AHS Adaptive Structures Conference< BR>14th AIAA*. Honolulu: American Institute of Aeronautics and Astronautics. https://doi.org/10.2514/6.2012-1818.

Gu, Tianyu, Brendan Dolan-Gavitt, and Siddharth Garg. 2017, August. BadNets: Identifying vulnerabilities in the machine learning model supply chain. *ArXiv:1708.06733 [Cs]*. http://arxiv.org/abs/1708.06733.

IDC FutureScape. 2018. *IDC FutureScape: Worldwide IT industry 2019 predictions*. https://www.idc.com/getdoc.jsp?containerId=US44403818.

IEEE. 2017. *Artificial intelligence and machine learning applied to cybersecurity*. https://www.ieee.org/about/industry/confluence/feedback.html.

Jagielski, Matthew, Alina Oprea, Battista Biggio, Chang Liu, Cristina Nita-Rotaru, and Bo Li. 2018, April. Manipulating machine learning: Poisoning attacks and countermeasures for regression learning. *ArXiv:1804.00308 [Cs]*. http://arxiv.org/abs/1804.00308.

King, Tariq M., Jason Arbon, Dionny Santiago, David Adamo, Wendy Chin, and Ram Shanmugam. 2019. AI for testing today and tomorrow: Industry perspectives. In *2019 IEEE international conference on Artificial Intelligence Testing (AITest)*, 81–88. Newark: IEEE. https://doi.org/10.1109/AITest.2019.000-3.

Liao, Cong, Haoti Zhong, Anna Squicciarini, Sencun Zhu, and David Miller. 2018, August. Backdoor embedding in convolutional neural network models via invisible perturbation. *ArXiv:1808.10307 [Cs, Stat]*. http://arxiv.org/abs/1808.10307.

Microsoft Defender ATP Research Team. 2018. *Protecting the protector: Hardening machine learning defenses against adversarial attacks*. https://www.microsoft.com/security/blog/2018/08/09/protecting-the-protector-hardening-machine-learning-defenses-against-adversarial-attacks/.

Mittal, Sudip, Anupam Joshi, and Tim Finin. 2019, May. Cyber-All-Intel: An AI for security related threat intelligence. *ArXiv:1905.02895 [Cs]*. http://arxiv.org/abs/1905.02895.

OECD. 2019. *Recommendation of the council on artificial intelligence*. https://legalinstruments.oecd.org/en/instruments/OECD-LEGAL-0449.

Primiero, Giuseppe, and Mariarosaria Taddeo. 2012. A modal type theory for formalizing trusted communications. *Journal of Applied Logic* 10 (1): 92–114. https://doi.org/10.1016/j.jal.2011.12.002.

Sharif, Mahmood, Sruti Bhagavatula, Lujo Bauer, and Michael K. Reiter. 2016. Accessorize to a crime: Real and stealthy attacks on state-of-the-art face recognition. In *Proceedings of the 2016 ACM SIGSAC conference on Computer and Communications Security – CCS'16*, 1528–1540. Vienna: ACM Press. https://doi.org/10.1145/2976749.2978392.

Sinha, Aman, Hongseok Namkoong, and John Duchi. 2017, October. Certifying some distributional robustness with principled adversarial training. *ArXiv:1710.10571 [Cs, Stat]*. http://arxiv.org/abs/1710.10571.

Szegedy, Christian, Wojciech Zaremba, Ilya Sutskever, Joan Bruna, Dumitru Erhan, Ian Goodfellow, and Rob Fergus. 2013, December. Intriguing properties of neural networks. *ArXiv:1312.6199 [Cs]*. http://arxiv.org/abs/1312.6199.

Taddeo, Mariarosaria. 2010a. Modelling trust in artificial agents, a first step toward the analysis of e-Trust. *Minds and Machines* 20 (2): 243–257. https://doi.org/10.1007/s11023-010-9201-3.

———. 2010b. Trust in technology: A distinctive and a problematic relation. *Knowledge, Technology & Policy* 23 (3–4): 283–286. https://doi.org/10.1007/s12130-010-9113-9.

———. 2017. Trusting digital technologies correctly. *Minds and Machines* 27 (4): 565–568. https://doi.org/10.1007/s11023-017-9450-5.

Taddeo, Mariarosaria, and Luciano Floridi. 2018. Regulate artificial intelligence to avert cyber arms race. *Nature* 556 (7701): 296–298. https://doi.org/10.1038/d41586-018-04602-6.

The 2019 Official Annual Cybercrime Report. 2019. *Herjavec group*. https://www.herjavecgroup.com/the-2019-official-annual-cybercrime-report/.

Uesato, Jonathan, Brendan O'Donoghue, Aaron van den Oord, and Pushmeet Kohli. 2018, February. Adversarial risk and the dangers of evaluating against weak attacks. *ArXiv:1802.05666 [Cs, Stat]*. http://arxiv.org/abs/1802.05666.

World Economic Forum. 2018. *The global risks report 2018*. World Economic Forum. http://www3.weforum.org/docs/WEF_GRR18_Report.pdf.

Yang, Guang-Zhong, Jim Bellingham, Pierre E. Dupont, Peer Fischer, Luciano Floridi, Robert Full, Neil Jacobstein, et al. 2018. The grand challenges of science robotics. *Science Robotics* 3 (14): eaar7650. https://doi.org/10.1126/scirobotics.aar7650.

Chapter 16
Prayer-Bots and Religious Worship on Twitter: A Call for a Wider Research Agenda

Carl Öhman, Robert Gorwa, and Luciano Floridi

Abstract The automation of online social life is an urgent issue for researchers and the public alike. However, one of the most significant uses of such technologies seems to have gone largely unnoticed by the research community: religion. Focusing on Islamic Prayer Apps, which automatically post prayers from its users' accounts, we show that even one such service is already responsible for millions of tweets daily, constituting a significant portion of Arabic-language Twitter traffic. We argue that the fact that a phenomenon of these proportions has gone unnoticed by researchers reveals an opportunity to broaden the scope of the current research agenda on online automation.

Keywords Automatic prayers · Twitter bots · Digital afterlife industry · Islam · Online death

16.1 Introduction

Online social life is increasingly automated. From virtual assistants that help with day-to-day tasks, to chatbots providing companionship or preserving the memory of deceased family members (Öhman and Floridi 2018), industry has been quick in realizing the potential of the development. At the same time, online social

C. Öhman
Uppsala University, Uppsala, Sweden
e-mail: carl.ohman@im.uu.se

R. Gorwa
Department of Politics and International Relations, Saint Anthony's College, University of Oxford, Oxford, UK
e-mail: robert.gorwa@politics.ox.ac.uk

L. Floridi (✉)
Oxford Internet Institute, University of Oxford, Oxford, UK
e-mail: luciano.floridi@oii.ox.ac.uk

L. Floridi (ed.), *Ethics, Governance, and Policies in Artificial Intelligence*, Philosophical Studies Series 144, https://doi.org/10.1007/978-3-030-81907-1_16

automation is also used for political goals, including automated "bot" accounts on social networks that attempt to influence elections and other key political events (Gorwa and Guilbeault 2018). These trends have rightly attracted much attention, both publicly and in the research community. However, one major area of online automation has largely been overlooked so far: religious worship. In this article, we provide the first large-scale analysis of the religious use of online automation technologies. More specifically, the article focuses on a particularly wide-spread phenomenon, what we call *Islamic Prayer Apps*, which, despite their popularity, have so far gone unnoticed by the research community. We argue that the spread and social significance of these applications calls for a broadening of the scope of current research on online automation in general, and on social media bots in particular.

16.2 Islamic Prayer Apps

It is increasingly popular amongst Muslim social media users to employ services that automatically post prayers on one's behalf. In this article, we shall refer to such services as *Islamic Prayer Apps*. These apps vary in their business model and popularity, but share the same goal: to facilitate and automate worship. This does not mean that the apps replace the mandatory "5-times-aday" prayer rituals. While documented services simply send or post reminders for local prayer times (Wyche et al. 2008), the Islamic Prayer Apps seem to facilitate additional public supplication ("dua"), which may be understood as a humble asking for an event to occur or a wish to be fulfilled.

Believers in Islam may phrase their own personal supplications, but there is also an array of examples in the Quran to choose from. Based on these examples, the apps enable the user to post *automatically* their supplications on social networking sites, like Twitter and Facebook. Du3a.org is a typical example: the site's landing page (see Fig. 16.1) features some Quranic quotes and popular prayers, and a sidebar encourages visitors to share the site on different social networks, like Facebook and Pinterest, claiming that 26 million visitors have done so already. But the most salient feature is perhaps the button prompting visitors to subscribe to the service. Upon doing so, visitors are redirected to Twitter, where they are asked to authorize the application to use their account and post on their behalf. After a few hours, Du3a begins to post a > 140 character supplication from the user's account every second hour, alongside a site URL (and until recently a "recycling" emoji).

Because Du3a.org includes the service's URL in every tweet that is sent out from the user's account, its traffic can be measured using Twitter's Streaming application programming interface (API), which provides live access to up to 1% of Tweets on the global platform. By querying for the dur3a.org URL, we collected tweets posted over a 48-h period, in June 2018. During this time, 3.8 million tweets containing the URL were posted (See Fig. 16.2). It should be noted however, that Du3a at the time appeared to release one tweet per hour from the users' accounts, a frequency which recently seems to have slowed down to one every second hour. About 50% of the

Fig. 16.1 Screenshot of Du3a's landing page

users self-identify as located in Saudi Arabia and Egypt, suggesting that, at least in the case of Du3a, the phenomenon is predominantly Arabic (other countries represent approximately 1% each).

The number of 1.9 million tweets per day-coming just from Du3a, one of many Islamic Prayer Apps-demonstrates how much traffic can be generated through automation. To put the numbers in context (see Fig. 16.2), Bruns et al. (2013) collected 205,000 tweets on the Arab-Spring related hashtag #egypt on its busiest day, when President Hosni Mubarak resigned amidst intense public pressure. During the 2016 US election, when significant popular attention focused on the role of automated accounts, Bessi and Ferrara (2016) estimated an upper bound of 3.8 million tweets from auto- mated accounts on political topics in the week leading up to voting day (an average of about 540 thousand tweets per day). In other words, according to our exploratory analysis, a single automated prayer app generated almost as many tweets in two days as accounts believed to be automated did in the whole week leading up to the US election. Yet, contrary to the US general election, Du3a continues its activity every day of the year. And insofar as we were able to ascertain, this activity has been going on for about 5 years. While exact numbers are difficult to determine, an analysis of Arabic Social Media (2014) estimated that, in 2014, 17.19 million tweets were sent daily from users in the entire Arab world, suggesting that automated prayer may be responsible for a substantial proportion of Twitter in Arabic speaking countries. Thus, at least in terms of sheer numbers, the

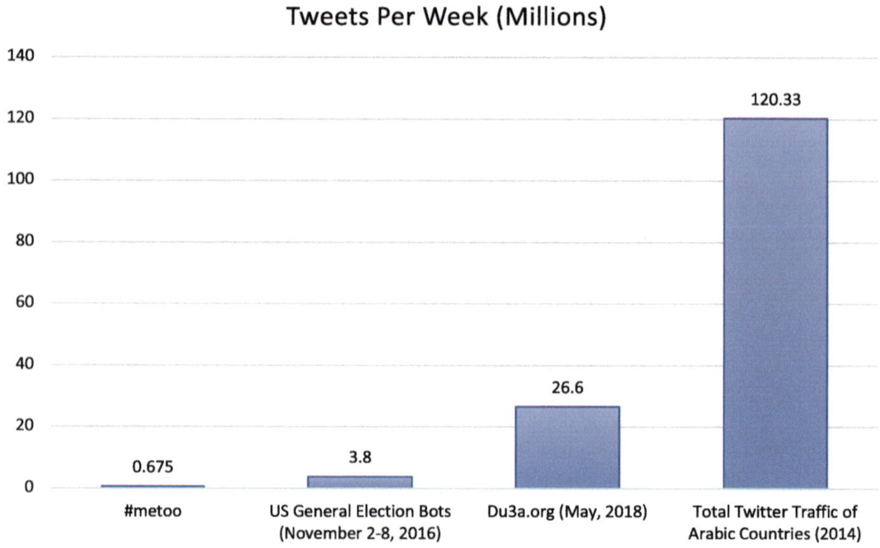

Fig. 16.2 Illustration of average tweets per day

expression of worship may rank among the most significant phenomena on Twitter overall.

Du3a.org, like most Islamic Prayer Apps, does not use hashtags which can "trend" and gain visibility, which is a possible reason why the phenomenon has largely remained unnoticed. To our knowledge, it was not until Matthew Rothenberg (2017), the founder of *Emojitracker.com,* noticed that the recycling emoji (at the time used by Du3a in every tweet)—attributed to the extensive use of the symbol in Muslim tweets—had become the third most popular on Twitter that the apps were first dis-cussed outside the Muslim community.

Our exploration of the phenomenon indicates the presence of at least 10 sites with business models similar to Du3a's. Some of the competitors offer more advanced options. For instance, Athantweets.com offers a premium version that, for 100 Saudi Riyals (roughly $27) a year, enables the user to choose specific (as opposed to randomly generated) supplications, and for the tweets to be synchronized to the user's local prayer times. Tweets sent via this premium package also hide the Athantweets URL, making them virtually indistinguishable from any other tweets with Quranic content.

This casts light on an important characteristic of the phenomenon as a whole, the fact that a majority of the traffic appears to be organic; that is, derived from ordinary accounts of real people, as opposed to "bots", understood as accounts with a fake identity set up purely for the purpose of disseminating content. A qualitative close-reading of a few dozen twitter accounts using Du3a shows that, whereas some of the accounts appeared to be created specifically to use these prayer services, most appeared to be ordinary users, who tweeted everyday messages, photos, and commentary interspersed with the automated messages. In other words, much of the traffic appears to be created by authentic accounts, operated by legitimate users, who creatively automate a facet of their online activity while also using the service as they would do ordinarily.

16.3 Religious Context

It is too soon to try to explain the specific role that the Islamic Prayer Apps play in the everyday life of their users. Much more work on both the qualitative properties of the phenomenon (such as that of Bell 2006) as well as further analysis of the quantitative ones is needed. "Even though there is almost 1.2 billion Muslims . . ." as Akma and Abdul Razak (2013, p. 6) point out, ". . . there is too little research done on the techno-spiritual application from the Islamic perspective." Looking more closely at some of the accounts captured in our data collection, we see that there seem to be many possible motivations behind the use of such services. One account, for instance, notes that the reason it is set up is to pray for "my brother [name]," suggesting that users might be setting up such accounts to cast prayers on behalf of friends and family.

Arguably, one of the core functions of the automated prayer apps is tied to their explicit promise to continue posting even after the user's death. For instance, the slogan of Zad-Muslim.com reads "Register now so your account would tweet now and after you die." Similarly, Dur3a.org promises that "your account will tweet in your life and in your death." This is more than a mere detail. While the posthumous prayer apps resonate with traditions in many different religions, it is notable that such features have emerged within Islam. According to the Quran, one does not receive one's judgement immediately upon death. Instead, those who pass away must wait in their graves for Allah to end Earth and make His final judgment of each respective individual. In order to be eventually sent to paradise, a Muslim's sins must be outweighed by his or her good deeds. But in between the time of death and the final judgement day, a number of factors may posthumously increase their standing in the eyes of Allah. The Prophet Muhammed specifically mentioned three things that can improve one's afterlife reward this way: the continuous effects of charity; the provision of knowledge used by future generations; and virtuous descendants who pray for you (Sahih Muslim 1330, 42: 7064).

Islamic scholars debate about how this should be interpreted (see for instance Al-Halbali 2016). However, most interpreters agree that the reward of the afterlife, at least to some degree, is subject to posthumous improvement. Because contributing to the dissemination of Islam is considered an inherent good, it is less important whether this activity is performed by someone personally, or through knowledge that one helps disseminating. So, the hypothesis is that putting in motion an app to post supplications on one's behalf after one's death could help increase one's chances of a good afterlife.

With few exceptions, the Islamic community has thus far accepted the incorporation of new technologies into religious practice, especially when it comes to realizing their missionary potential. In her book *When Religion Meets New Media,* Campbell (2010, p. 96) describes this new emphasis on technology, noting that "Engagement with media is a religious imperative, one which Muslims will be held accountable for in the afterlife." Thus, it seems that the dissemination of Islam through media may have profound implications in both life and death. Yet, little is known about the attitudes of the larger Islamic community when it comes to Islamic Prayer Apps, which undoubtedly ties religious practice closer to online technologies. And research is needed to study the specific role that these apps may play within the lives and religious practices of their users. Likewise, little, if anything, is known about the stance of social media platforms themselves on this widespread phenomenon.

Unlike a platform such as Facebook, which now has a "memorialization" feature, Twitter handles deceased users by permanently taking down their accounts after some months of inactivity (Twitter n.d.). However, an account that has subscribed to a Islamic Prayer App will not go inactive after the user's death. It will keep tweeting, and will therefore not be identified as inactive or abandoned. This means that the huge presence of the Islamic Twitter supplications will likely continue to grow long after the account holders have died. An array of similar applications, albeit with a secular framing of "immortalizing one's social media presence", have been launched mainly targeting secular Western audiences (Ohman and Floridi 2017). However, such a project of social media immortalization still remains fringe in comparison to the Islamic Prayer Apps. Considering their popularity, it may, even within only a few decades, become increasingly difficult to differentiate between traffic generated by living and deceased users-and not because of the futurist community in Silicon Valley, but because of Islamic worshipers.

16.4 Broader Implications

Religion has always been one of the most significant aspects of human life, individually, socially, as well as technologically. It should not be surprising if it is now emerging as a possible key driver for the mainstream adoption of social technology. Islam is not the only religion incorporating creative automation technologies into worship. The iTunes App Store contains more than 6000 applications related to spirituality and religion (Buie and Blythe 2013, p. 2315). In early 2019, Pope Francis launched a new app called "Click to Pray,"[1] with which Catholics may participate in the Pope's prayers and share them on social media. Similarly, the Church of England's new voice activation feature for Amazon's Alexa allows owners of the device to ask it to read daily prayers (BBC 2018). This could be understood as part of a larger movement; after all, the offering of a wax candle in a church can now often be replaced by the turning on of an electric one. In a narrow capacity, technological

[1] https://clicktopray.org/

services can even substitute for priests, answering religious questions like "Who is God?" or what it means to believe in Jesus Christ. Similarly, Jewish communities have started employing new technologies to automate home facilities during the Sabbath (Woodruf et al. 2007), and members of diverse religious communities can now chat and make prayer requests to religious chatbots, such as those created by the Californian start-up Prayerbot. These examples, alongside the scale of the Islamic Prayer Apps, show that religion is far from a small or marginal force in contemporary social automation.

The case of the Islamic Prayer App also provides valuable insights into the complex, fast-evolving discourse on Twitter bots and online automation policy. Much recent literature on social bots focuses on false accounts set up during elections, and other political events, to inflate engagement metrics and help spread problematic political content (Ferrara et al. 2016). In contrast, Islamic Prayer Apps share no hashtags, and do not appear to try and influence the social network, instead remaining unobtrusively out of view. As well, because many of the observed accounts appear to be ordinary users who have partially "botified" or automated their accounts, they complicate the existing discourse, which is often methodologically and conceptually predicated on the assumption that "bots" and "not bots" exist as two distinct categories that can be easily separated (Gorwa and Guilbeault 2018; Stieglitz et al. 2017). Many Twitter users rely on a variety of publicly available services, from Twitter's own platform 'Tweetdeck' to 'If This Then That' to automate parts of their online activity. But how exactly should such behavior be understood? And what are the ethical ramifications?

There have been many positive uses of automated social media accounts, which have been deployed creatively by journalists (Lokot and Diakopoulos 2016), activists fighting corruption (Savage et al. 2016), and those promoting institutional transparency (Ford et al. 2016), but religious uses have been largely unexplored. The Islamic Prayer Apps we analysed here arguably represent one of the largest examples to date of Twitter automation being deployed organically in a creative and culturally significant way. The fact that a phenomenon of these unprecedented proportions has gone unnoticed by researchers shows the limitations of our current scope. To quote Bell (2006, p. 155): "We appear to be stubbornly secular in our imaginings of home and leisure contexts of computing." Indeed, now is the time to broaden the conversation.

Acknowledgements Our sincere thanks to Bence Kollanyi, for assistance with data collection.

References

Akma, N., and F.H. Abdul Razak. 2013. On the emergence of techno-spiritual: The concepts and current issues. In *Computer and mathematical sciences graduates national colloquium 2013 (SISKOM2013)*.

Al-Halbali, I.J. 2016. *The three that follow to the g rave*. Birmingham: Dar As-Sunnah Publishers.

Arab Social Media Report. 2014. *Twitter in the Arab Region*. Available at: http://arabsocialmediareport.com/Twitter/LineChart.aspx. Accessed 12 June 2018.

BBC. 2018. Church of England offers prayers read by Amazon's Alexa. *BBC.com*. https://www.bbc.co.uk/news/uk-44233053. Accessed 27 Mar 2019.

Bell, G. 2006. No more SMS from Jesus: Ubicomp, religion and techno-spiritual practices. In *UbiComp 2006: Ubiquitous Computing. UbiComp 2006*, Lecture Notes in Computer Science, ed. P. Dourish and A. Friday, vol. 4206. Berlin/Heidelberg: Springer. https://doi.org/10.1007/11853565_9.

Bessi, A., and E. Ferrara. 2016. Social bots distort the 2016 US Presidential election online discussion. *First Monday*. https://doi.org/10.5210/fm.v21ill.7090.

Bruns, A., T. Highfield, and J. Burgess. 2013. The Arab spring and social media audiences. *American Behavioral Scientist* 57 (7): 871–898.

Buie, E., and M. Blythe. 2013. *Spirituality: There's an app for that! (But not a lot of research).* CHI 2013 Extended Abstracts, April 27–May 2, 2013, Paris, France.

Campbell, H.A. 2010. *When religion meets new media*. London: Routledge.

Ferrara, E., O. Varol, C.B. Davis, F. Menczer, and A. Flammini. 2016. The rise of social bots. *Communications of the ACM* 59 (7): 96–104.

Ford, H., E. Dubois, and C. Puschmann. 2016. Keeping Ottawa honest-One tweet at a time? Politicians, journalists, wikipedians and their twitter bots. *International Journal of Communication* 10: 4891–4914. ISSN 1932-8036.

Gorwa, R., and D. Guilbeault. 2018. Unpacking the social media bot: A typology to guide research and policy. *Policy and Internet*. https://doi.org/10.1002/poi3.184.

Lokot, T., and N. Diakopoulos. 2016. News Bots: Automating news and information dissemination on Twitter. *Digital Journalism* 4 (6): 682–699. https://doi.org/10.1080/21670811.2015.1081822.

Öhman, C., and L. Floridi. 2017. The political economy of death in the age of information: A critical approach to the digital afterlife industry. *Minds and Machines*. https://doi.org/10.1007/s11023-017-9445-2.

———. 2018. An ethical framework for the digital afterlife industry. *Nature Human Behavior*. https://doi.org/10.1038/s41562-018-0335-2.

Rothenberg, M. 2017. Why the emoji recycling symbol is taking over Twitter. *Medium*. Available at: https://medium.com/@mroth/why-the-emoji-recycling-symbol-is-taking-over-twitter-65ad4b18b04b. Accessed 25 Apr 2018.

Sahih Muslim. 1330. *Sahih Muslim Vol.* 7, *Book of Zuhd and Softening of Hearts, Hadith 7064.* Retrieved from: http://www.iupui.edu/-msaiupui/042.smt.html. Accessed 27 Mar 2019.

Savage, S., A. Monroy-Hernandez, and T. Hollerer. 2016. Botivist: Calling volunteers to action using online bots. In *Proceedings of the 19th ACM conference on Computer-Supported Cooperative Work & Social Computing*, 813–822. ACM.

Stieglitz, S., F. Brachten, B. Ross, and A.-K. Jung. 2017. Do social bots dream of electric sheep? A categorisation of social media bot accounts. *arXiv: 1710.04044 [Cs]*. Retrieved from http://arxiv.org/abs/1710.04044.

Twitter. n.d. *Inactive account policy*. Available at: https://help.twitter.com/en/rules-and-policies/inactive-twitter-accounts. Accessed 27 Mar 2019.

Woodruf, A., S. Augustin, and B. Foucault. 2007. *Sabbath day home automation: "It's like mixing technology and religion"*. CHI 2007, April 28–May 3, 2007, San Jose, California, USA.

Wyche, S.P., K.E. Caine, B. Davison, M. Arteaga, and R.E. Grinter. 2008. *Sun dial: Exploring technospiritual design through a mobile Islamic call to prayer application*. CHI 2008, April 5–April 10, 2008, Florence, Italy ACM.

Chapter 17
Artificial Intelligence, Deepfakes and a Future of Ectypes

Luciano Floridi

Abstract In this chapter, I introduce the concept of "digital ectype". An ectype is a copy that has a special relation with its source (the origin of its creation), the archetype, like the impression left by a seal. It is not the real thing, but it is clearly linked in a significant, authentic way with the real thing itself. I argue that digital technologies are able to separate the archetypal source – what was in the mind of the artist, for example – from the process (style, method, procedure) that leads from the source to the artefact. Once this link is severed, one can have digital ectypes that are "authentic" in style and content, but not "original", in terms of archetypal source, and digital ectypes that are "original" in terms of archetypal source (they do come from where they purport to come) yet not "authentic" in terms of production, performance, or method (they are not the ones used by the source to deliver the artefact). In other words, digital ectypes can be authentic but unoriginal artefacts, or inauthentic but original artefacts.

Keywords Authenticity · Copy · Deepfake · Ectype · Reproduction

The art world is full of reproductions. Some are plain replicas, for example the Mona Lisa. Others are fakes or forgeries, like the "Vermeers" painted by Han van Meegeren that sold for $60 million (Kreuger and van Meegeren 2010). The distinction between a *replica* and a *fake* is based on the concept of *authenticity*.

Is this artefact what it claims to be?[1] The answer seems simple but, in reality, things are complicated. Today, the paintings of the forger John Myatt are so famous that they are valued at up to $40,000 each, as "genuine fakes" (Furlong 1986). They are not what they say they are, but they are authentically painted by him and not by

[1] I have discussed the nature of questions and epistemic relevance in (Floridi 2008).

L. Floridi (✉)
Oxford Internet Institute, University of Oxford, Oxford, UK
e-mail: luciano.floridi@oii.ox.ac.uk

L. Floridi (ed.), *Ethics, Governance, and Policies in Artificial Intelligence*, Philosophical Studies Series 144, https://doi.org/10.1007/978-3-030-81907-1_17

Fig. 17.1 A fake, the original is "Lodge on Lake Como" by Carl Frederik Peder Aagaard (1833–1895)

another forger. And they are beautiful. A bit as if one were to utter a beautiful lie, not any ordinary lie. And an artist like Magritte seems to have painted not only false Picassos and Renoirs during the Nazi occupation of Belgium (Marien 1983), but also faked his own work, so to speak, in the famous case of the two copies of the painting "The Flavour of Tears" (1948), both by Magritte, but one of which he passed off as false-partly as a surrealist act and partly to make money. In this mess, and as if things were not confusing enough, digital technologies further reshuffle what is possible and our understanding of it.

Thanks to digital technologies, today it is much easier to establish the authenticity of a work. There are databases where you can check authors' signatures, and millions of images that can be viewed with a few clicks. Selling a fake is more difficult. Figure 17.1 shows a reproduction of the "Lodge on Lake Como" by Carl Frederik Peder Aagaard (1833–1895), a Danish landscape painter and decorative artist. It was on sale in 2016 on eBay. The painting is very popular on the web, and there are plenty of good replicas. Nothing wrong with them. However, if you check Fig. 17.1 carefully, you will notice that this is sold as an unsigned "original", which is misleading to say the least. Both the quality of the painting and the price are suspicious, and a Google image search quickly reveals that this is a mere replica.

At the time of writing, the painting was no longer available and the seller did not seem to be active on eBay anymore.

Of course, fakes are not always reproductions; they can also be "new works" by a famous artist, like Pollock or Van Gogh. In this case, sophisticated scientific techniques to establish authenticity include tests run using AI. A research paper, published last November by Ahmed Elgammal, Yan Kang and Milko Den Leeuw (Elgammal et al. 2017) proposed "a computational approach for analysis of strokes in line drawings by artists", based on neural networks. The training collection consisted of a dataset of 300 digitised drawings with over 80,000 strokes, by Pablo Picasso, Henry Matisse and Egon Schiele, and a few works by other artists. By segmenting individual strokes, the system learned to quantify the characteristics of individual strokes in drawings, thus identifying the unique properties for each artist. The software managed to classify "individual strokes with accuracy 70%–90%, and aggregate over drawings with accuracy above 80%, while being robust to be deceived by fakes (with accuracy 100% for detecting fakes in most settings)". It turns out that the way in which individuals draw lines is as unique as their fingerprints or their gait, and AI can help one to discover it, as if it were a microscope.

But AI is not just for identifying fakes. Let us stay in the Netherlands, a very interesting project[2] by Microsoft, in collaboration with the Rembrandt House Museum, has led to the creation of a portrait of a gentleman, which both is and is not a Rembrandt (see Fig. 17.2).

Analysing the known works of Rembrandt, an algorithm identified the most common subject (a portrait of a Caucasian man, 30–40 years old), the most common traits (facial hair, facing to the right, wearing a hat, a collar and dark clothing, etc.), the most suitable style to reproduce these characterising properties, the brushstrokes, in short, all the information needed to produce a new painting by Rembrandt. Having created it, it was reproduced using a 3D printer, to ensure that the depth and layering of the colour would be as close as possible to Rembrandt's style and way of painting. The result is a masterpiece. A Rembrandt that Rembrandt never painted, but which challenges our concepts of "authenticity" and "originality", given the painting's strong link with Rembrandt himself. I do not know the value of the painting. My bet is that it would be quite expensive if it were auctioned as reliably authenticated as *that unique* Microsoft's Rembrandt.

We do not have a word to define an artefact such as Microsoft's Rembrandt. So let me suggest *ectype*. The word comes from Greek and it has a subtle meaning that is quite useful here: an ectype is a copy, yet not any copy, but rather a copy that has a special relation with its source (the origin of its creation), the archetype. In particular, an ectype is the impression left by a seal. It is not the real thing, but it is clearly linked in a significant, authentic way with the real thing itself. Locke used "ectypes" to refer to ideas or impressions that correspond, although somewhat inadequately, to some external realities (the archetypes) to which they refer (Locke 2008). Digital technologies are able to separate the archetypal source-what was in the mind of the artist, for

[2] See https://news.microsoft.com/europe/features/next-rembrandt/

Fig. 17.2 The Rembrandt that is not a Rembrandt. Microsoft Project with the Rembrandt House Musewn

example-from the process (style, method, procedure) that leads from the source to the artefact (Floridi 2017). Once this link is severed, one can have ectypes that are "authentic" in style and content, but not "original", in terms of archetypal source, like Microsoft's Rembrandt. But one can also have ectypes that are "original" in terms of archetypal source (they do come from where they purport to come) yet not "authentic" in terms of production, performance, or method (they are not the ones used by the source to deliver the artefact). In other words, ectypes can be authentic but unoriginal artefacts, like Microsoft's Rembrandt, or inauthentic but original artefacts. A great example of an inauthentic original ectype was provided in March by an audio recording of John F. Kennedy's last speech. Despite being an ordinary speech from a decades-old campaign trail, it suddenly made headline news. Because it was the Dallas Trade Mart speech of 22 November 1963, the text that JFK *would* have read, had he not been assassinated mere moments before, on his way to deliver it. The text is *original:* it comes from the source. But the voice that recites is *inauthentic,* because it was synthesised by software that analysed 831 recordings of Kennedy's speeches and interviews, in order to "learn" how to speak like him. The software finally gave voice to JFK's last speech 55 years late. So here is a Kennedy who is and is not a Kennedy, similar and yet different from the Rembrandt that is and is not a Rembrandt. They are both ectypes (see Table 17.1).

We saw that the production of ectypes does not stop at the work of art, but involves any artefact, from texts to photos, from audio recordings to videos. It is well known that the history of manuscripts, printing, photography, cinema and television

Table 17.1 Archetype, fake and ectypes

	Original source	Authentic production
Leonardo's Mona Lisa	Yes	Yes
Han van Meegeren's forged Vermeers	No	No
Microsoft's Rembrandt	No	(Qualified) Yes
JFK's Trade Mart speech	Yes	No

is paved with fakes. Expect more ectypes too. In particular, artists love to break boundaries and it is easy to imagine that, like Magritte faking his own painting, they will start producing their own ectypes. Imagine a painter using the software developed by Microsoft to produce her own new works. It would still be an ectype, and this would explain why (with qualifications) the process would capture some authenticity. The reproduction of the work of art by mechanical means will have acquired a new meaning (Benjamin 2008).

With ectypes, we usually know where things stand. But someone could cheat. Last May, Google presented Google Duplex, a version of its AI assistant that simulates being human to help users with simple interactive tasks, like booking a restaurant table. The company was quick to state that it will not intentionally mislead anyone, and that it will make sure always to clarify when a user is interacting with an artificial agent. But someone else could use these technologies for criminal or evil purposes. This is what happens with Deepfake, a set of techniques used to synthesise new visual products, for example by replacing faces in the originals. The typical cases involve porn movies in which the faces of famous actresses like Gal Gadot or Scarlett Johansson (this is regularly about women's faces) are used to replace the original faces. In this case too, large databases are needed to instruct the software (which is available for free, and there is also an app), so if you are not a public figure the risks are lower. Deepfake also concerns politicians, like President Obama, for example.

What is the future ahead of us? Digital technologies seem to undermine our confidence in the original, genuine, authentic nature of what we see and hear. But what the digital breaks it can also repair, not unlike the endless struggle between software virus and antivirus. In our case, in addition to educating people, acquiring new sensitivities and having the right legal framework, there are at least a couple of interesting digital strategies. For artefacts that are already available, it is easy to imagine.

AI systems that give us a hand. It would be interesting to analyse Microsoft's Rembrandt and Kennedy's speech with an artificial system to see whether it discovered them to be ectypes. Research is already available on methods to expose Deepfake videos generated with neural networks (Li et al. 2018). In short, let us remember the software developed to analyse drawings: there are plenty of sophisticated tools for detection of image forgery. And more are likely to be developed as the demand for them increases. Next, as regards new artefacts, because originality and authenticity are also a matter of provable historical continuity from the source to the product through the process of production, the much-vaunted blockchain, or a similar solution, could make a big difference. Blockchain is like a register that stores transactions in an accruable, safe, transparent and traceable way. As a secure and distributed register of transactions, blockchain is being explored as a means of reliably certifying the origins and history of particular products: whether in terms of securing food supply chains, or in recording the many linked acts of creation and ownership that define the provenance of an artwork. In the future, we may adopt the same solution wherever there is a need to ensure (or establish) the originality and authenticity of some artefact, be it a written document, a photo, a video or a painting. And of course, a future artist may want to ensure, through a blockchain, that her work of art as an ectype is really what it says it is. At that point we shall have travelled full circle, for we shall have "genuine ectypes", like the Microsoft's Rembrandt, or Kennedy's speech.

References

Benjamin, W. 2008. *The work of art in the age of mechanical reproduction*. London: Penguin.

Elgammal, A., Y. Kang, and M. Den Leeuw. 2017. Picasso, Matisse, or a fake? Automated analysis of drawings at the stroke level for attribution and authentication. *arXiv preprint arXiv: l711.03536*.

Floridi, L. 2008. Understanding epistemic relevance. *Erkenntnis* 69 (1): 69–92.

———. 2017. Digital's cleaving power and its consequences. *Philosophy & Technology* 30 (2): 123–129.

Furlong, M. 1986. *Genuinefake: A biography of Alan Watt5*. Portsmouth: Heinemann.

Kreuger, F.H., and H. van Meegeren. 2010. *Han van Meegeren revisited: His art & list of works*. Delft: F.H. Kreuger.

Li, Y., M.-C. Chang, H. Farid, and S. Lyu. 2018. In Ictu Oculi: Exposing AI generated fake face videos by detecting eye blinking. *arXiv preprint arXiv:l806.02877*.

Locke, J. 2008. *An essay concerning human understanding*. New York/Oxford: Oxford University Press.

Marien, M. 1983. *Le radeau de la memoire: souvenirs determines*. Paris: Pre-aux-Clercs.

Chapter 18
The Ethics of AI in Health Care: A Mapping Review

Jessica Morley, Caio C. V. Machado, Christopher Burr, Josh Cowls, Indra Joshi, Mariarosaria Taddeo, and Luciano Floridi

Abstract This article presents a mapping review of the literature concerning the ethics of artificial intelligence (AI) in health care. The goal of this review is to summarise current debates and identify open questions for future research. Five literature databases were searched (Scopus, Google Scholar, Philpapers, Web of Science, Pub Med), in April 2019, to support the following research question: "how can the primary ethical risks presented by AI-health be categorised, and what issues must policymakers, regulators and developers consider in order to be 'ethically mindful?'". A series of screening stages were carried out—for example, removing articles that focused on digital health in general (e.g. data sharing, data access, data privacy, surveillance/nudging, consent, ownership of health data, evidence of efficacy)—yielding a total of 156 papers that were included in the review.

We find that ethical issues can be (a) epistemic, related to misguided, inconclusive or inscrutable evidence; (b) normative, related to unfair outcomes and

The authors Jessica Morley and Caio C. V. Machado have contributed equally to the writing of this chapter.

J. Morley · C. C. V. Machado · L. Floridi (✉)
Oxford Internet Institute, University of Oxford, Oxford, UK
e-mail: Jessica.morley@kellogg.ox.ac.uk; jessica.morley@phc.ox.ac.uk; luciano.floridi@oii.ox.ac.uk

C. Burr
Alan Turing Institute, London, UK
e-mail: cburr@turing.ac.uk

J. Cowls · M. Taddeo
Oxford Internet Institute, University of Oxford, Oxford, UK

Alan Turing Institute, London, UK
e-mail: josh.cowls@oii.ox.ac.uk; mariarosaria.taddeo@oii.ox.ac.uk

I. Joshi
NHSX, London, UK
e-mail: indra.joshi@nhsx.nhs.uk

L. Floridi (ed.), *Ethics, Governance, and Policies in Artificial Intelligence*, Philosophical Studies Series 144, https://doi.org/10.1007/978-3-030-81907-1_18

transformative effectives; or (c) related to traceability. We further find that these ethical issues arise at six levels of abstraction: individual, interpersonal, group, institutional, and societal or sectoral. Finally, we outline a number of considerations for policymakers and regulators, mapping these to existing literature, and categorising each as epistemic, normative or traceability-related and at the relevant level of abstraction. This article contributes to the debate on AI in health care by offering a comprehensive analysis of the relevant literature, focusing on the ethical implications for individuals, interpersonal relationships, groups, institutions, societies and the health sector as a whole. Our goal is to inform policymakers, regulators and developers of what they must consider if they are to enable health and care systems to capitalise on the dual advantage of ethical AI; maximising the opportunities to cut costs, improve care, and improve the efficiency of health and care systems, whilst proactively avoiding the potential harms. We argue that if action is not swiftly taken in this regard, a new 'AI winter' could occur due to chilling effects related to a loss of public trust in the benefits of AI for health care.

Keywords Artificial intelligence · Ethics · Healthcare · Health policies · Machine learning

Research Highlights

- We provide a review of the literature, which covers over 156 scientific articles on the ethics of Artificial Intelligence in Healthcare, offering a typology for academics and policymakers seeking to advance research on the field or identify issues to be addressed according to their cause and stakeholders.
- We also highlight 11 key considerations identified from recurrent or overarching issues that were common to the literature covered.
- Although some technical solutions have been put forward for mitigating issues relating to data bias, data quality, and ensuring social inclusion in decision-making, these remain relatively untested. Unless a competitive advantage of taking such pro-ethical steps becomes clear without these approaches being made mandatory, it is unlikely that they will have a significant impact on the ethical impacts of AI-Health in the near future.
- Broader issues regarding the protection of equality of care, fair distribution of benefits, and the protection and promotion of societal values have scarcely been considered. Given that effective healthcare is a fundamental component of modern society this is concerning.
- Many different issues are at stake at every stage of the process, ranging from development all the way to the human interactions where these technologies will be introduced. This paper offers a common framework that allows specific discussions to fit into the bigger picture, based on several levels of abstration (i.e. individual, interpersonal, group, institutional, and sectoral or societal) and considering epistemic, normative and traceability ethical concerns at each level.

18.1 Introduction

Healthcare systems across the globe are struggling with increasing costs and worsening outcomes (Topol 2019). This presents those responsible for overseeing healthcare systems with a 'wicked problem', meaning that the problem has multiple causes, is hard to understand and define, and hence will have to be tackled from multiple different angles. Against this background, policymakers, politicians, clinical entrepreneurs and computer and data scientists increasingly argue that a key part of the solution will be Artificial Intelligence (AI), particularly Machine Learning (Chin-Yee and Upshur 2019). The argument stems not from the belief that all healthcare needs will soon be taken care of by "robot doctors" (Chin-Yee and Upshur 2019). Instead, the argument rests on the classic definition of AI as an umbrella term for a range of techniques that can be used to make machines complete tasks in a way that would be considered intelligent *were* they to be completed by a human. For example, as mapped by (Harerimana et al. 2018), decision tree techniques can be used to diagnose breast cancer tumours (Kuo et al. 2001); Support Vector Machine techniques can be used to classify genes (Brown et al. 2000) and diagnose Diabetes Mellitus (Barakat et al. 2010); ensemble learning methods can predict outcomes for cancer patients (Kourou et al. 2015); and neural networks can be used to diagnose stroke (Wang et al. 2017). From this perspective, AI represents a growing *resource* of *interactive*, *autonomous*, and often *self-learning* (in the machine learning sense) *agency*, that can be used on demand (Floridi 2019a, b), presenting the opportunity for potentially transformative cooperation between machines and doctors (Bartoletti 2019).

If harnessed effectively, such AI-clinician cooperation, where AI is used to provide comprehensive evidence-based clinical decision-support to the clinician (AI-Health), could offer great opportunities for the improvement of healthcare services and ultimately patients' health (Taddeo and Floridi 2018) by significantly improving human clinical capabilities in diagnosis (Arieno et al. 2019; De Fauw et al. 2018; Kunapuli et al. 2018), drug discovery (Álvarez-Machancoses and Fernández-Martínez 2019; Fleming 2018), epidemiology (Hay et al. 2013), personalised medicine (Barton et al. 2019; Cowie et al. 2018; Dudley et al. 2015) and operational efficiency (Lu and Wang 2019; Nelson et al. 2019). However, as Ngiam and Khor (2019) stress, if these AI solutions are to be embedded in clinical practice, then a clear governance framework is needed to protect people from harm, including harm arising from unethical conduct. We use the term 'cooperation' here and suggest that AI will be chiefly used for clinical decision support. This differentiates from arguments often made by the popular press which suggest that AI will be used to 'replace' clinicians.

To support policymakers, the task of the following pages is to classify the ethical risks presented by AI-health, align these with specific questions that must be answered by policymakers, and provide example actions that could be taken by healthcare governing bodies to develop the requisite governance framework. The intention is to ensure that the ethical challenges raised by implementing AI in healthcare settings are tackled *proactively* (Char et al. 2018). We seek to do this because if the ethical risks are not tackled proactively, by encouraging AI-health policymakers, developers and regulators to be ethically mindful, there is a potential risk of incurring significant opportunity costs (Cookson 2018). For instance, ethical mistakes or misunderstandings may lead to social rejection and/or distorted legislation and policies, which in turn cripple the acceptance and advancement of [the necessary] data science. Encouraging this kind of proactive ethical analysis is essential but also challenging because, although bioethical principles for clinical research and healthcare are well established, and issues related to privacy, effectiveness, accessibility and utility are clear (Nebeker et al. 2019), other issues are less obvious (Char et al. 2018). For example, AI processes may lack transparency, making accountability problematic, or may be biased, leading to unfair, discriminatory behaviour or mistaken decisions (Mittelstadt et al. 2016). Identification of these less obvious concerns requires input from the medical sciences, economics, computer sciences, social sciences, law, and policy-making. Yet, research in these areas is currently happening in siloes, is overly focused on individual level impacts (Redacted for anonymity), or does not consider the fact that the ethical concerns may vary depending on the stage of the algorithm development pipeline (Redacted for anonymity).

Whilst AI-Health remains in the early stages of development and relatively far away from having a major impact on frontline clinical care (Panch et al. 2019), there is still time to develop this framework. However, this window of opportunity is closing fast, as the pace at which AI-Health solutions are gaining approval for use in clinical care in the US is accelerating (Topol 2019). Both the Chinese (Zhang et al. 2018) and British governments (Department of Health and Social Care 2019) have made it very clear that they intend on investing heavily in the spread and adoption of AI-Health technologies. It is for these reasons that the goal of this article is to offer a cross-disciplinary mapping review of the potential ethical implications of the development of AI-Health in order to support policy discussion, which will in turn orient the development of better design practices, and transparent and accountable deployment strategies. We will do this in terms of digital ethics. That is, we will focus on the evaluation of moral problems related to data, algorithms and corresponding practices (Floridi and Taddeo 2016), with the hope of enabling governments and healthcare systems looking to adopt AI-Health to be ethically mindful (Floridi 2019a). Specifically, the review question is: "how can the primary ethical risks presented by AI-health be categorised, and what issues must policymakers, regulators and developers consider in order to be 'ethically mindful?'"

18.2 Methodology

A mapping review methodology (Grant and Booth 2009) was used to find literature from across disciplinary boundaries that highlighted ethical issues *unique* to the use of AI algorithms in healthcare. This type of review is used to map and categorise existing literature on a particular topic (in this case the ethics of AI) and contextualise the findings within broader literature. The mapping review methodology was developed by the Evidence for Policy and Practice Information and Co-ordinating Centre in London to offer policymakers, practitioners and researchers an explicit and transparent means of identifying narrower policy and practice-relevant review questions (Grant and Booth 2009). As our goal is to support the policy discussion and with these issues orient the development of better design practices, and transparent and accountable deployment strategies, this was the most appropriate methodology.

Our review question focused on "how can the primary ethical risks presented by AI-health be categorised, and what must policymakers, regulators and developers consider in order to be 'ethically mindful? We were concerned with categorising issues in order to facilitate future research and discussion. We chose five literature databases that are relevant to these issues and that are at the cross-section of the technical, medical, ethical and social science literature: Scopus, Google Scholar, Philpapers, Web of Science, Pub Med. Our literature review searches were conducted in April 2019, with references being added or removed throughout the drafting iterations. The search engines are not identical, so we used variations of the following generic search term string: ethic* AND algorithm* OR AI* OR "Artificial Intelligence"* OR "Machine Learning"* AND health* (see Appendix for details on results and search queries). Initial results were screened on title. Those that were deemed relevant were downloaded so that the abstracts and keywords could be reviewed for relevance. At this stage, we excluded any results that were focused on issues related to digital health in general (e.g. data sharing, data access, data privacy, surveillance/nudging, consent, ownership of health data, evidene of efficacy) to remain focused on mapping the current debate about the ethics of AI specifically. Recorrds that the authors had prior knowledge of, and which were relevant to the research question but not included in the initial database searches, were also added.

To ensure that the focus stayed on the *unique* ethical issues, the map, developed by (Mittelstadt et al. 2016), of the epistemic, normative, and overarching ethical concerns related to algorithms was used as a base. The typology offered by Mittelstadt et al. identifies problems pertaining to algorithmic decision making and their possible causes, such as error in input or discriminatory output. Traceability arises from the complexity of the system when all of the pieces are put together. This typology will be cross-referenced with each level of abstraction (LoA) we propose below.

First, the selected literature was reviewed to identify healthcare examples of each of the concerns highlighted in the original map, as shown in Table 18.1, and then reviewed more thoroughly to identify how the ethical issues may vary depending on whether the analysis was being conducted at: (i) individual, (ii) interpersonal, (iii) group (e.g. family or population), (iv) institutional, (v) sectoral, and/or (v) societal levels of abstraction (LoAs) (Floridi 2008). An LoA can be imagined as an interface that enables one to observe some aspects of a system analysed, while making other aspects opaque or indeed invisible. LoAs are common in computer science, where systems are described at different LoAs (computational, hardware, user-centred

Table 18.1 A summary of the epistemic, normative and overarching ethical concerns related to algorithmic use in healthcare based on Mittelstadt et al. (2016) from (Redacted for anonymity)

	Ethical Concern	Explanation	Medical example
Epistemic concerns	Inconclusive Evidence	Algorithmic outcomes (e.g. classification) are probabilistic and not infallible. They are rarely sufficient to posit the existence of a causal relationship.	*EKG readers in smartwatches may 'diagnose' a patient as suffering from arrhythmia when it may be due to a fault with the watch not being able to accurately read that user's heartbeat (for example due to the colour of their skin) or the 'norm' is inappropriately calibrated for that individual (Hailu* 2019*)*
	Inscrutable Evidence	Recipients of an algorithmic decision very rarely have full oversight of the data used to train or test an algorithm or the data points used to reach a specific decision.	*A clinical decision support system deployed in a hospital may make a treatment recommendation, but it may not be clear on what basis it has made that 'decision' raising the risk that it has used data that are inappropriate for the individual in question or that there is a bug in the system leading to issues with over or under prescribing (Wachter 2015).*
	Misguided Evidence	Algorithmic outcomes can only be as reliable (but also as neutral) as the data they are based on.	*Watson for Oncology is in widespread use in China for 'diagnosis' via image recognition but has primarily been trained on a Western data set leading to issues with concordance and poorer results for Chinese patients than their Western counterparts (Liu et al. 2018).*
Normative Concerns	Unfair outcomes	An action can be found to having more of an impact (positive or negative) on one group of people	An algorithm *'learns' to prioritise patients it predicts to have better outcomes for a particular disease. This turns out to have a discriminatory effect on people within the Black and minority ethnic communities (Garattini et al.* 2019*).*
	Transformative effects	Algorithmic activities, like profiling, re-conceptualise reality in unexpected ways.	*An individual using personal health app has limited oversight over what passive data it is collecting and how that is*

(continued)

Table 18.1 (continued)

	Ethical Concern	Explanation	Medical example
			being transformed into a recommendation to improve, limiting their ability to challenge any recommendations made and a loss of personal autonomy and data privacy (Kleinpeter 2017).
Overarching	Traceability	Harm caused by algorithmic activity is hard to debug (to detect the harm and find its cause), and it is hard to identify who should be held responsible for the harm caused.	*If a decision made by clinical decision support software leads to a negative outcome for the individual, it is unclear who to assign the responsibility and or liability to and therefore to prevent it from happening again (Racine et al. 2019).*

etc.). Note that LoAs can be combined in more complex sets, and can be, but are not necessarily hierarchical, with higher or lower 'resolution' or granularity of information. This helped the review avoid the narrow focus on individual-level impacts highlighted in the introduction. This approach is not intended to imply that there is no overlap between the levels.

18.3 Findings

What follows is a detailed discussion of the issues uncovered. A total of 223 titles were selected, duplicates were removed and, as reading commenced, relevant bibliography references were also added, resulting in approximately 147 papers to be read and included in the review. The flowchart below illustrates our methodology. Also, a summary map of our findings (Table 18.2) is provided at the end of the section (Fig. 18.1).

18.3.1 Epistemic Concerns: Inconclusive, Inscrutable, and Misguided Evidence

Many factors are encouraging the development of AI-Health (Chin-Yee and Upshur 2019). One of the main driving forces is the belief that algorithms can make more objective, robust and evidence-based clinical decisions (in terms of diagnosis, prognosis or treatment recommendations) than a human healthcare practitioner (HCP) can (Kalmady et al. 2019). This is not an unfounded position. Machine

Table 18.2 Summary of the epistemic, normative and overarching ethical concerns associated with AI-Health at the six different LoAs as identified by the literature review

	A. Individual	B. Interpersonal	C. Group	D. Institutional	E. Sectoral	F. Societal
1. Epistemic concern (inconclusive, inscrutable and misguided evidence)	Misdiagnosis or missed diagnosis	Loss of trust in HCP-Patient relationships, de-personalisation of care	Misdiagnosis or missed diagnosis at scale—some groups more affected than others	Waste of funds and resources not directed to areas of greater need	Excessively broad data sharing between public and private entities	Poorer public healthcare provision and worsening health outcomes for society
2. Normative (transformative effects and unfair outcomes)	Surveillance & undermining of autonomy and integrity of self	Deskilling of HCPs, overreliance on AI-tools, and undermining of consent practices and redefining roles in the healthcare system	Profiling and discrimination against certain groups seen as being less healthy or higher risk	Transformation of care pathways & imposing of specific values at scale—redefining 'good care'	Siloing of new AI tool development within private sector	Inequalities in outcomes
3. Overarching (traceability)	'Bad Users' could come to be blamed for their own ill-health	See institutional	Specific groups framed as being more morally irresponsible with regards to their health than others	Lack of clarity over liability with regards to issues with safety and effectiveness could halt adoption or result in certain groups being blamed more often than others	See institutional	Society must decide through regulation preferable answers to the questions regarding liability and risk allocation in healthcare provision. However, all groups in society may not be given an equal say in this process

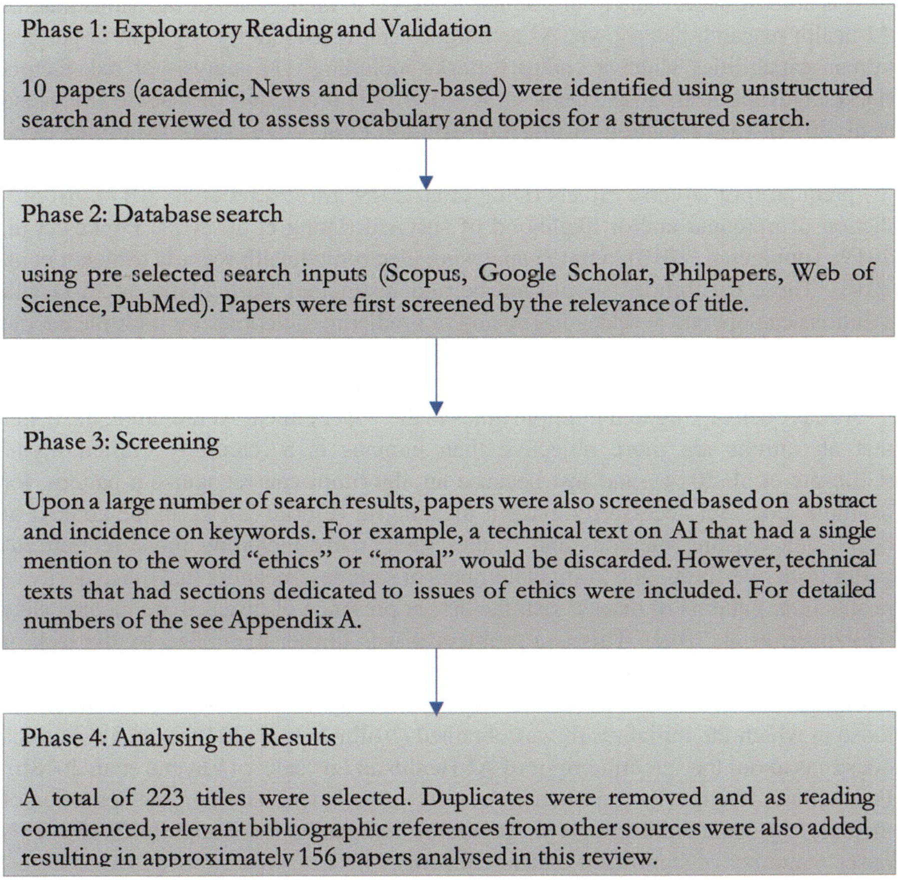

Fig. 18.1 Flowchart offering and overview of the steps taken in our literature review, filtering from several thousand titles to identified abstracts and selecting 156 papers to read

learning methods, especially ensemble and unsupervised methods (Harerimana et al. 2018), can take into account a far greater range of evidence (data) than a Healthcare Provider (HCP) when making a clinical decision, including five of the seven dimensions of healthcare data provided by the US Department of Health and Human services: (1) demographic and socioeconomic data; (2) symptom and existing diagnosis data; (3) treatment data; (4) outcome data; and (5) other omic data (Holzinger et al. 2019). If designed taking into account the multiple epistemic concerns, this ability enables clinical algorithms to act as digital companions (Redacted for anonymity), reducing the information asymmetry that exists between a HCP and the individual seeking care by making available information accessible to both parties and helping ensure that the most informed decision possible is made by the person who has the right to make it (Redacted for anonymity).

It is at least in part due to this ability to make ‘evidence-based’ decisions that, as AI-health research has shown, AI techniques can considerably augment or surpass human capabilities when it comes to tasks including: (1) analysis of risk factors (De Langavant et al. 2018; Deng et al. 2018); (2) prediction of disease (Moscoso et al. 2019); (3) prediction of infection (Barton et al. 2019; López-Martínez et al. 2019); (4) population health monitoring (Lu et al. 2019; Zacher and Czogiel 2019); (5) prediction of adverse effects (Ding et al. 2019; Mortazavi et al. 2017); (6) prediction of outcome and/or likelihood of survival (Dong et al. 2019; Popkes et al. 2019; Topuz et al. 2018); and (7) analysing electronic health records (Shickel et al. 2018). These capabilities should not be underestimated, particularly as AI-Health solutions can operate at scale, diagnosing or predicting outcomes for multiple people at once—something that an HCP could never do. Yet in many ways this almost unwavering faith in the truth-telling power of AI-Health is flawed.

As has been highlighted multiple times in the wider ethical AI literature, the belief that algorithms are more objective than humans is a ‘carefully crafted myth’ (Gillespie et al. 2014), and just because an algorithm can recognise a pattern, for example, does not necessarily make it meaningful (Floridi 2014). In the context of healthcare, existing methods and studies (potentially including those referenced) suffer from overfitting due to small numbers of samples, meaning that the majority of results (e.g. patterns of disease risk factors, or presence of disease) are inconclusive (Holzinger et al. 2019). This is a problem that is further magnified by the lack of reproducibility, and external validity, of results. AI-Health solutions are often untranslatable between different settings and rarely work in settings different to those in which the initial result was obtained (Vollmer et al. 2018b), raising serious questions about the scientific rigor of AI-Health and its safety (Vayena et al. 2018b). Furthermore, the results can often be heavily value-laden, based on the definition of ‘healthy’ by influential people or powerful companies (McLaughlin 2016). This raises a number of significant ethical concerns.

At the **individual LoA** there is considerable risk of misdiagnosis. This can happen in at least two ways: either by an individual using a wearable device that has a bug or is inappropriately calibrated for them (e.g. they could be ‘told’ that they are suffering from a health condition when they are not, or vice versa), or, an HCP relying on clinical decision support software (CDSS) (Ruckenstein and Schüll 2017) could be given an inaccurate diagnosis or recommendation which they do not question due to a tendency to uncritically accept the decisions of automated systems (Challen et al. 2019). Moreover, this can have impacts in medical practice, causing overreliance on the machine diagnostics and deskilling of practitioners (Cabitza et al. 2017). Not only is this a risk for individuals, but it also reverses the advantage of AI-Health solutions being able to operate at scale by introducing the **group LoA** ethical concern of misdiagnosis or missed diagnosis happening repeatedly. Whilst an HCP might give one person the wrong diagnosis and then be corrected, a faulty algorithm, based on the misguided, inscrutable or inconclusive evidence could give the same wrong diagnosis to hundreds or thousands of people at a time (Topol 2019). The scale of the problems is as large as the scale of the solutions.

Building on this, there are also ethical implications at the **interpersonal LoA**. HCP-patient relationships are primarily based on trust and empathy, and whilst AI-Health solutions can take over tasks that are more routine and standardised, they cannot reproduce the emotional virtues of which human HCPs are capable (Ngiam and Khor 2019). Consequently, an over-reliance on the 'quantitative' and objective evidence that fuels clinical algorithms (Cabitza et al. 2017) could discredit other forms of diagnosis and treatment (Rosenfeld et al. 2019)—a prominent concern in the case of clinical psychiatry (Burns 2015). This could lead to the de-humanisation or *im*personalisation of care provision (Juengst et al. 2016), from a service based on *listening* and *theory* to one based purely on *categorisation* (an issue that could again lead to a **group LoA** harm of group-profiling and associated discrimination by providers including insurers; see Sect. 18.3.2). Not only is this effectively 'paternalism in disguise' (Juengst et al. 2016) but it could also lead to poorer health outcomes due to the disconnect between pure medical evidence and actual behaviour change (Emanuel and Wachter 2019).

Finally, scaling up to the **institutional, sectoral and societal LoAs**, there is the concern that public health decisions are increasingly made on predictive AI-Health algorithms, which too often rely on the same flawed assumptions as outlined above. Regarding these assumptions, consider the example of Google Flu Trends monitoring the influenza virus. The initial algorithm distorted the spread of the virus in the US (Vayena et al. 2015) making it appear that there were a greater number of influenza cases than there were by mis-classifying influenza-like-illnesses as confirmed cases of influenza (Ortiz et al. 2011). This study carries obvious limitations: the healthcare-seeking behaviour from the population, for example searching for information on the outbreak of a Flu, make this research susceptible to distortion. Such distortion would be affected, for example, if there is large media coverage of an epidemic, or by demographic factors, such as digital divides.

If policy decisions about where to deploy health resources are based on such poor-quality evidence, this could result in the waste of public funds (e.g., promoting vaccination campaigns where they are not needed), damage local economies (e.g., scaring away tourists from a region)—which would result in a positive feedback loop of less money available for public expenditure—and lead to poorer quality public healthcare provision, and thus worse health outcomes for society at large. This worry is particularly paramount when it is considered that the ultimate ambition of AI-Health is to create a learning healthcare system where the 'system' is constantly learning from the data it receives on the performance of its interventions (Faden et al. 2013). Furthermore, it is worth noting that, at this juncture, the example offered above of Flu Trends does not represent the limits of Google's interest—and that of its subsidiaries and its siblings under parent company Alphabet—in public health. As we discuss below, the engagement between Alphabet's AI subsidiary DeepMind and a major UK hospital has attracted the attention of data protection regulators, the press, and academics (Information Commissioner 2018; Powles and Hodson 2017). The challenge of ensuring that AI-Health systems function accurately has in turn sparked debates about the appropriateness of sharing data between public and private entities. In response to claims that patient data transferred from the Royal

Free Hospital to DeepMind was "far in excess of the requirements of those publicly stated needs" (Powles and Hodson 2017), DeepMind representatives argued that "data processed in the application have been defined by and are currently being used by clinicians for the direct monitoring and care of AKI [acute kidney injury] patients" (King et al. 2018). Powles and Hodson responded in turn that it is a "statement of fact that the data transferred is broader than the requirements of AKI" (Powles and Hodson 2018). As this series of claims and counter-claims demonstrates, the quality and quantity of data required for a particular AI-Health application is likely to be a matter of dispute in the context of the collection and sharing of patient data in training AI-Health.

Ultimately, data is necessary for medical practice and thus so are AI-Health solutions that can take in greater volumes of data. But data collected and used in this way is insufficient to inform medical practice; it must be transformed to be useful (Car et al. 2019) and if this transformation process is flawed the results could be hugely damaging, resulting in either wasted funds and poorer health provision, or undue sharing of patient data with private sector actors under the guise of AI-Health.

18.3.2 Normative Concerns: Unfair Outcomes and Transformative Effects

As referenced in the introduction, healthcare systems across the globe are struggling with increasing costs and decreasing outcomes (Topol 2019) and their administrators increasingly believe that the answer may well lie in making healthcare systems more informationally mature and able to capitalise on the opportunities presented by AI-Health significantly to improve outcomes for patients, and to reduce the burdens on the system (Cath et al. 2017). Whilst it would be ethically remiss to ignore these opportunities (Floridi 2019a), it would be equally ethically problematic to ignore the fact that these opportunities are not created by AI-Health technologies *per se* but by their ability to fundamentally change the intrinsic nature of the ways in which healthcare is delivered by coupling, re-coupling and de-coupling different parts of the system. This changes the affordances and constraints of different governing bodies, regulators, and system agents, undermining the mechanisms in place to hold those delivering care accountable and thus introducing new risks (Floridi 2017a). For example(Redacted for anonymity):

- Coupling: patients and their data are so strictly and interchangeably linked that the patients *are* their genetic profiles, latest blood results, personal information, allergies etc. (Floridi 2017a). What the legislation calls "data subjects" become "data patients";
- Re-Coupling: research and practice have been sharply divided since the publication of the National Commission for the Protection of Human Subjects in the 1970s, but in the digital scenario described above, they are re-joined as one and the same again (Petrini 2015; Faden et al. 2013);

- De-Coupling: presence of Healthcare Provider (HCP) and location of Patient become independent, for example because of the introduction of online consultations (NHS England 2019).

As a result of these transformations a number of ethical concerns arise.

Starting once again with the **individual LoA**: as more diagnostic and therapeutic interventions become based on AI-Health solutions, individuals may be encouraged to share more and more personal data about themselves (Racine et al. 2019)—data that can then be used in opaque ways (Sterckx et al. 2016). This means that the ability for individuals to be meaningfully involved in shared decision making is considerably undermined (Vayena et al. 2018a, b). As a result, the increasing use of algorithmic decision-making in clinical settings can have negative implications for individual autonomy, as for an individual to be able to exert agency over the AI-Health derived clinical decision, they would need to have a good understanding of the underlying data, processes and technical possibilities that were involved in it being reached (DuFault and Schouten 2018) and be able to ensure their own values are taken into consideration (McDougall 2019). The vast majority of the population do not have the level of eHealth literacy necessary for this (Kim and Xie 2017), and those that do (including HCPs) are prevented from gaining this understanding due to the black-box nature of AI-Health algorithms (Watson et al. 2019). In extreme instances, this could undermine an individual's confidence in their ability to refuse treatment (Ploug and Holm 2019). Such issues pose a substantial threat to an individual's integrity of self (the ability of an individual to understand the forces acting on them) (Cheney-Lippold 2017). Given that damage to a person's psychological integrity can be perceived as a 'harm', not accounting for this potentiality poses the risk of creating a system that violates the first principle of medical ethics: *primum non nocere* ("first, do no harm") (Andorno 2004; Redacted for anonymity).

It is not necessarily the case that harmful impacts will primarily be felt by the patients. At the **interpersonal LoA**, HCPs may themselves feel increasingly left 'out of the loop' as decisions are made by patients and their 'clinical advice' algorithm in a closed digital loop (Nag et al. 2017). As a result, HCPs may too feel unable to exert their own agency over the decision-making capacity of AI-Health solutions. Though the use of algorithmic decision-making makes diagnostics seem like a straightforward activity of identifying symptoms and fitting them into textbook categories, medical practice is much less clear-cut than it seems (Cabitza et al. 2017). Clinical practice involves a series of evaluations, trial and error, and a dynamic interaction with the patient and the medical literature. As a result, formal treatment protocols should be seen more as evaluative guidelines than well-defined, isolated categories. AI-Health solutions may not be in accordance with current best practice, which is necessary to handle the great degree of uncertainty and can only be fully evaluated by physicians (Cabitza et al. 2017). Therefore, AI-Health solutions need to allow HCPs to exert influence in the decision-making process.

At the **group LoA** the concern is that AI-Health systems may well be able to better identify illnesses and injuries that have well-established and fairly set (and therefore automatable) treatment protocols. These are more likely to exist for afflictions most commonly suffered by white men as there is a greater volume of medical trials data for this group than there is for almost any other group. Algorithms trained on such biased datasets could make considerably poorer predictions for, for example, younger black women (Vollmer et al. 2018a, b). If HCPs are left out of the loop completely and learning healthcare systems primarily rely on automated decisions, there is considerable potential to exacerbate existing inequalities between the "haves" and the "have-nots" of the digital healthcare ecosystem, i.e. those that generate enough data on themselves to ensure accurately trained algorithms and those that do not (Topol 2019).

To mitigate these and associated risks, **institutions** need to be asking the crucial question: how much clinical decision-making should we be delegating to AI-Health solutions (Di Nucci 2019)? If it is known that algorithms which enable profiling (e.g. those that determine genetic risk profiles) can ignore outliers and provide the basis for discrimination (Garattini et al. 2019), deciding whether healthcare is also seen as a means of promoting social justice is crucial in order to establish: what type of data services will be embedded in the system (Voigt 2019); what data should be collected; and which values should be embedded in algorithmic decision-making services (McDougall 2019). This decision also determines what sort of population-level behavioural change the health system should be able to aim for depending on cost management, data collection and fairness in data-driven systems (Department of Health and Social Care 2018). If not carefully considered, this process of transforming the provision of care risks over-fitting the system to a specific set of values that may not represent those of society at large (McDougall 2019).

Another, more subtle yet pervasive transformative effect arises at the **sectoral** level. Powles and Hodson (2017) argue that one risk that may arise from collaboration between public and private sector entities such as that between the Royal Free London hospital and DeepMind is that the positive benefits of AI-Health "solutions" will be siloed within private entities. They note that in the Royal Free case, "DeepMind [was given] a lead advantage in developing new algorithmic tools on otherwise privately-held, but publicly-generated datasets" (Powles and Hodson 2017, p. 362). This, they suggest, may mean that the only feasible way that future advances may be developed is "via DeepMind on DeepMind's terms". This interpretation was contested by DeepMind, who called it "unevidenced and untrue" and claimed that the Information Commissioner agreed with their stance in her 2018 ruling (King et al. 2018). Whatever the circumstances of this particular case, the broader risk of privately held AI-Health solutions—trained on datasets that have been generated *about* the public *by* public actors but then (lawfully) shared with private companies—is worthy of caution going forward and a worthwhile topic of ongoing discussion in public health ethics.

As may now be clear, these transformative effects also have significant ethical implications at the **societal LoA**. Before institutions can establish where and how (and, from the sectoral perspective, whether) AI-Health solutions can improve care, society itself must make difficult decisions about what care *is* and what constitutes *good* care (Coeckelbergh 2014). To offer a simplistic example, does it mean purely providing a technical diagnosis and an appropriate prescription or does it involve contemplating a series of human necessities that revolve around well-being (Burr et al. 2020a)? If it is the former, then it is relatively easy to automate the role of non-surgical clinicians through AI (although this does not imply that doctors *should* be substituted by AI systems). However, if is the latter, then providing good healthcare means encompassing psychological wellbeing and other elements related to quality of life, which would make human interaction an essential part of healthcare provision, as a machine does not have the capability to make emotionally-driven decisions. Consequently, certain decisions may completely exceed the machine's capabilities and thus delegating these tasks to AI-Health would be ethically concerning (Matthias 2015).

Consider, for example, a situation where an AI-Health solution decides which patients are sent to the Intensive Care Unit (ICU). Intensive care is a limited resource and only people who are at risk of losing their lives or suffering grave harms are sent there. Triage decisions are currently made by humans with the aim of maximising well-being for the greatest number of people. Doctors weigh different factors when making this decision, including the likelihood of people surviving if they are sent to the ICU. These situations often involve practitioners (implicitly) taking moral stances, by prioritising individuals based on their age or health conditions. These cases are fundamentally oriented by legal constraints and medical norms (e.g. adherence to bioethical principles or codes of best practice), yet personal expertise, experience and values also inevitably play a role. Having the support of AI-Health in the ICU screening increases the number of agents and complicates the norms involved in these decisions, since the doctor may follow his or her professional guidelines, while the algorithm will be oriented by the values embedded in its code. Unless there is a transparent process for society to be involved in the weighing of values embedded in these decision-making tools (for instance, how is 'fair' provision of care defined?) (Cohen et al. 2014), then the use of algorithms in such scenarios could result in the overfitting of the health system to a specific set of values that are not representative of society at large.

In response to this risk, some attempts have already been made to involve the public at large in decisions over the design and deployment of AI systems. In early 2019, the UK's Information Commissioner's Office and the National Institute for Health Research staged a series of "citizens' juries" to obtain the opinions of a representative cross-section of British society regarding the use of AI in health (Information Commissioner 2019). The "juries" were presented with four scenarios, two relating to health—using AI to diagnose strokes, and using it to find potential matches for a kidney transplant—and another two relating to criminal justice. Notably, the juries "strongly favoured accuracy over explanation" in the two scenarios involving AI in health (National Institute for Health Research 2019). This is just one example of research which attempted to obtain public opinion data regarding AI in health, and there are reasons to suppose that the apparent preference among participants for accurate over explainable AI systems reflects the high-stakes and fast-moving scenarios that were presented (as opposed to, say, the more routine illnesses and injuries we are focusing on here). Nonetheless, it demonstrates the plausibility and preferability of involving the public in designing AI-Health systems.

To conclude this sub-section, the notion that AI-Health technologies are ethically neutral is unrealistic, and having them perform moral decision-making and enforcement may provoke immoral and unfair results (Rajkomar et al. 2018). The direct involvement of the public in the design of AI-Health may help mitigate these risks. This should be borne in mind by all those involved in the AI-driven transformation of healthcare systems.

18.3.3 Overarching Concerns: Traceability

The previous sub-section outlined how the increasing use of AI-Health is fundamentally transforming the delivery of healthcare and the ethical implications of this process, particularly in terms of potentially unfair outcomes. This transformation process means that healthcare systems now rely on a dynamic, cyclical and intertwined series of interactions between human, artificial and hybrid agents (Vollmer et al. 2018a; Turilli and Floridi 2009). This is making it increasingly challenging identify interaction-emerging risks and allocate liability, raising ethical concerns with regards to moral responsibility.

Moral responsibility involves both looking forward, where an individual, group or organisation is perceived as being in charge of guaranteeing a desired outcome, and looking backwards to appropriate blame and possibly redress, when a failure has occurred (Wardrope 2015). In a well-functioning healthcare system, this responsibility is distributed evenly and transparently across all nodes so that the causal chain of a given outcome can be easily replicated in the case of a positive outcome, or prevented from repeating in the case of a negative outcome (Floridi 2013, 2016). In an algorithmically-driven healthcare system, a single AI diagnostic tool might involve many people organising, collecting and brokering data, and performing analyses on it, making this transparent allocation of responsibility almost impossible. In essence, not only is the decision-making process of a single algorithm a black-box, but the entire chain of actors that participate in the end product of AI-Health solutions is extremely complex. This makes the entire AI-Health ecosystem inaccessible and opaque, making responsibility and accountability difficult.

To clearly outline the ethical implications of this at-scale lack of traceability, let us take the example of a digital heart-rate monitor that 'intelligently' processes biological and environmental data to signal to its user their risk of developing a heart condition.

At the **individual LoA** this process relies on what can be termed the 'digital medical gaze' (Redacted for anonymity) and is based on this micro-cycle of self-reflection adapted from (Garcia et al. 2014):

1. Gaining Knowledge: Algorithm reads multi-omic dataset to determine risk of heart attack and providers individual with a 'heart health score'
2. Gaining Awareness: on the advice of the algorithm, individual starts monitoring their activity level and becomes aware of how active they are

3. Self-reflection: as directed by the algorithm the individual reflects on how much high fat food they are eating in a day and compares this to their optimal diet based on their genomic profile and their level of activity
4. Action: individual takes the advice of the algorithm and takes specific actions to improve their heart-health score e.g. starts regular exercise.

If this process of self-reflection does not 'work' in the sense that it does not result in a person taking appropriate action to improve their heart-health, for any number of reasons, including data inaccuracy, and the individual still ends up experiencing heart failure, this process of algorithmic surveillance (Rich and Miah 2014) risks creating an elaborate mechanism for victim-blaming (Danis and Solomon 2013; McLaughlin 2016). The individual may be seen as being a 'bad user' for failing to act upon the allegedly objective and evidence-based advice of the algorithm (see Sect. 18.3.1), and may therefore be framed as being morally responsible for their poor health and not deserving of state-provided healthcare. Yet, due to the lack of traceability, there can be no certainty that the poor outcome was due to the lack of action by the individual: it could be a faulty device, buggy code, or the result of biased datasets (Topol 2019). Moreover, even if a negative outcome were to result purely from an individual disregarding the guidance, the adoption of digital infrastructure that enables failure to be ascribed to a morally 'culpable' individual is itself a matter of ethical concern. These new insights may enable lives to be saved and quality of life to be drastically improved, yet they also shift the ethical burden of 'living well' squarely onto newly accountable individuals. The ontological shift that this new infrastructure permits—from individuals-as-patients deserving quality healthcare, regardless of their prior choices as fallible humans, to individuals-as-agents expected to take active steps to pre-empt negative outcomes—raises stark questions for bioethics, which has traditionally been seen as an "ethics of the receiver" (Floridi 2008). Moreover, these technological changes might prompt a shift in the ethical framework, burdening the individuals, while not providing *de facto* means of behavioural change. Many concerns stem from socio-demographic issues which entail harmful habits, and cannot oversimplified to a matter of delivering the adequate information to the patient (Owens and Cribb 2019).

Due to issues of bias (discussed further in Sect. 18.3.2), there is, further, a **group LoA** ethical risk that some groups may come to be seen as being more morally irresponsible about their healthcare than others. Heart-rate monitors, for example, are notoriously less accurate for those with darker skin (Hailu 2019), meaning that they could give considerably less accurate advice to people of colour than to those with light skin. If this results in people of colour being less able to use AI-Health advice to improve their heart-health, then these groups of people may be seen as morally reprehensible when it comes to their health. Furthermore, the healthcare could then 'learn' to predict that people of colour have worse heart-health, potentially resulting in these groups of individuals being discriminated against by, for example, insurers (Martani et al. 2019).

At the **interpersonal**, **institutional** and **sectoral LoAs**, this moral responsibility translates into liability. If for example, instead of the heart-health algorithm providing the advice back to the individual, it provides the data to the individual's HCP and the HCP provides advice that either fails to prevent an adverse event or directly causes an adverse event, this could be the basis of a medical malpractice suit (Price 2018). In this scenario, it remains unclear where the liability will eventually sit (Ngiam and Khor 2019). Current law implies that the HCP would be at fault, and therefore liable, for an adverse event as the algorithm in this scenario would be considered a diagnostic support tool—just like a blood test—with no decision making capacity, so it is the HCP's responsibility to act appropriately based on the information provided (Price et al. 2019; Schönberger 2019; Sullivan and Schweikart 2019). However, the supply chain for any clinical algorithm is considerably more complex and less transparent than that of a more traditional diagnostic tool meaning that many are questioning whether this is actually how the law will be interpreted in the future. For example, does the liability really sit with the HCP for not questioning the results of the algorithm, even if they were not able to evaluate the quality of the diagnostic against other sources of information, including their own personal knowledge of the patient due to the black-box nature of the algorithm itself? And what about the role of the hospital or care facility: does it have a responsibility to put in place a policy allowing HCPs to overrule algorithmic advice when this seems indicated? Similarly, what role do commissioners or retailers of the device that contains the algorithm play? Do they not carry some responsibility for not checking its accuracy, or do they assume that this responsibility sits with the regulator (for example, MHRA in the UK, the FDA in the US or the CFDA in China) who should, therefore, carry the burden for not appropriately assessing the product before it was deployed in the market? What if the problem is further back in the chain, stemming from inaccurate coding or poor-quality training data? There is a clear lack of distributed responsibility (Floridi 2013, 2016)—a problem that is exacerbated by a lack of transparency—making it hard to hold individual parts of the chain accountable for poor outcomes which poses a significant ethical risk.

In their overview of patient-safety issues with AI in healthcare, He et al. (2019) state that those working in the field are trying to establish a systems-wide approach that does not attribute blame to individuals or individual companies, but conclude that where liability will ultimately rest remains to be seen. This is problematic because, as Hoffman et al. (2019) stress, uptake of algorithmic-decision-making tools by the clinical community is highly unlikely until this liability question is resolved (Vollmer et al. 2018a, b), which could result in the overarching ethical concern raised in the introduction—that of a significant missed opportunity. Many, including (Holzinger et al. 2019), believe that explainability is the answer to solving this problem and that, if HCPs can understand how a decision was reached, then reflecting on the output of an algorithm is no different from any other diagnostic tool. Indeed Schönberger (2019) argues that legally this is the case and that as long as it can be proven that the duty of care was met, then harm caused to a patient by an erroneous prediction of an AI-Health system would not yet constitute medical negligence but that it *might* in the near future constitute negligence to *not* rely on the algorithmic output, which brings us back to the issues outlined in Sect. 18.3.1.

Overall, this lack of clarity will continue to persist for some time (Schönberger 2019), making it once again a social issue. Society will ultimately dictate what the socially acceptable and socially preferable (Floridi and Taddeo 2016) answers are to these pressing questions. The ethical issue is whether all parts of society will have an equal say in this debate, as in the example of citizens' juries above, or whether it will be those individuals or groups with the loudest voices that get to set the rules. As (Beer 2017) attests, when thinking about the power of an algorithm, we need to think beyond the impact and consequences of the code, to the powerful ways in which notions and ideas about the algorithm circulate throughout the social world.

18.4 The Need for an Ethically-Mindful and Proportionate Approach

The literature surveyed in this review clearly indicates the need for an agreed standard for AI-Health ethical evaluation. While these issues are all connected, they cannot be treated under the blanket discussion of "Ethics of AI" when discussing specific recommendations and solutions. For example, handling privacy at the individual LoA, considering design issues, is different from handling privacy at a group level, where the concern is raised from the ways in which the aggregate data is treated. For these reasons, we need to consider the epistemic, normative and traceability ethical concerns at the six different LoAs to set the different fields of discussion. Protecting people from the harms of AI-Health goes beyond protecting data collection and ensuring that the algorithmic models have been validated. The discussion needs to discern how these issues present differently at the different stages of the algorithmic developmet lifecyle, the ethical issues present at the data collection stage are likely to be different to those present at the deployment stage.

An example of an issue that shows up often is the legal challenge of liability allocation in cases of medical error. This legislative and regulatory discussion is directly dependent on understanding the ethical issues of each stage an AI implementation (e.g. data collection, training, deployment) and making a normative decision on how these risks and burdens are going to distributed through society. Therefore, the ethical discussion needs to be plotted before engaging in any sort of policy, regulatory or legislative discussion.

Similarly, many challenges will not be addressed through rules. Much of the risk of handling data and algorithms stems from professionals not adopting measures to protect privacy and support cybersecurity. In these cases, policy-makers can use our framework to identify at which levels they can best tackle issues. For example, one issue can be *individual* capacitation, where a solution would be promoting doctors' and patients' understanding and control over AI tools; educating about how AI-Health produces predictions or recommendations that are used in treatment plans, and access to and protection of patient data (Ngiam and Khor 2019). The issue, however, could occur at an organisational (*group*) scale, so better control over how the interface and design of AI-Health products influences HCP-patient-artificial-agent interactions (Cohen et al. 2014) could address the issue. Finally, some cases could be handled at an institutional level, organising campaigns and creating certifications for professionals seeking to use AI-Health tools is also necessary for the adequate implementation and use of AI (Kluge et al. 2018).

To tackle these challenges, regulators will have to consider hard and soft mechanisms, meaning what *ought* to be done and what *may* be done based on the existing moral obligations (Floridi 2018). These mechanisms will have to consider the different stakeholders involved in each issue and LoA, to balance the need to protect individuals from harm, whilst still supporting innovation that can deliver genuine system and patient benefit (Redacted for anonymity). In short, healthcare systems should not be overly cautious about the adoption of AI-Health solutions, but should be mindful of the potential ethical impacts (Floridi 2019a) so that proportionate governance models can be developed (Sethi and Laurie 2013). These governance models can, in turn, help ensure that those responsible for ensuring that healthcare systems are held accountable for the delivery of high-quality equitable and safe care.

What these regulations, standards and policies should cover and how they should be developed remain open questions (Floridi 2017b), which will likely be 'solved' multiple times over by different healthcare systems operating in different settings. However, in order to lend a more systematic approach to addressing these outstanding questions, enabling greater coherence and speed in addressing these challenges, in Table 18.3 below we have assembled a list of essential cross-cutting considerations that emerge from our mapping review. The table indicates from which aspect of our mapping review (ethical concern × LoA, corresponding to a cell in Table 18.2) each consideration is assigned by an increasing Level of Abstraction: Individual (A), Interpersonal (B), Group (C), Institutional (D), Sectoral (E) and Societal (F).

There are steps being taken towards regulation and legislation, however, these discussions often fail to address broader ethical questions such as "what constitutes good healthcare?" (Coeckelbergh 2014), "what services should be contemplated in our standard of 'care'?", and others. Without addressing these larger questions, it is hard to orient greater normative frameworks and produce coherency across stakeholders in each LoA. For these reasons, their development is progressing slowly (which is why the relevant literature is unlikely to reflect all current developments) and almost all focus solely on interventions positioning themselves as being health-related in the medical sense, not in the wider, wellbeing sense (e.g., healthy exercise, diet, sleeping habits).

Awareness of the need to consider these questions is increasing, and efforts are being made at both a national and international level to adapt existing regulations so that they remain fit for purpose (The Lancet Digital Health 2019). The American Food and Drug Administration (FDA) is now planning on regulating Software as a Medical Devices (SaMD) (Food and Drug Administration (FDA) 2019) and in both the EU and the UK Regulation 2017/745 on medical devices comes into effect in April 2020 and significantly increases the range of software and non-medical products that will need to be classed (and assessed) as medical devices. This

Table 18.3 Eleven key considerations for policymakers that arose from the literature review, denoted by an increasing Level of Abstraction: Individual (A), Interpersonal (B), Group (C), Institutional (D), Sectoral (E) and Societal (F)

Consideration	Key supporting literature	Relevant aspects (ascending LoA[a])	Example body responsible for answering this question based on the English National Health System
What skills will the professional healthcare workforce require in order to make safe and effective use of AI-Health solutions in the future?	Kluge et al. (2018)	Epistemic (A, B, C, F) Normative (B, C, D, E) Overarching (A, C)	Health Education England should survey the skills currently available in the workforce and conduct a gap analysis of the skills that will be needed
Which tasks should be delegated to AI-Health solutions, and which should not?	Di Nucci (2019)	Epistemic (A, B, C, D, F) Normative (B, C, D, F) Overarching (A, C, D)	Department of Health and Social Care should conduct a multi-stakeholder engagement process to understand which tasks are socially acceptable to be delegated to AI-health and make this offficial policy,.
What evidence is needed to 'prove' clinical effectiveness of an AI-Health solution?	Greaves et al. (2018)	Epistemic (A, B, C, E, F) Normative (E) Overarching (A, C, D, F)	Medicines and Healthcare Regulators Medicines and Healthcare products Regulatory Agency should update the medical device regulations to include a minimum required standard of accuracy and a minimum standard of evidence to demonstrate that the AI-health product is genuinely capable of performing at this level
What mechanisms should be put in place to enable people to report and seek redress for AI-Health associated harms?	Schönberger (2019)	Epistemic (A, C, E, F) Normative (A, C, E, F) Overarching (A, C, D)	The Care Quality Commission should update its inspection framework to regularly check that AI-health products in use are continuing to operate safely. Medicines and Healthcare Regulators Medicines and

(continued)

Table 18.3 (continued)

Consideration	Key supporting literature	Relevant aspects (ascending LoA[a])	Example body responsible for answering this question based on the English National Health System
			Healthcare products Regulatory Agency should introduce a 'yellow card' scheme for AI-health products so that users can report errors and be assured that they are being taken care of.
What mechanisms should be put in place to ensure all relevant stakeholder views are included in the development of AI-Health solutions?	Aitken et al. (2019)	Epistemic (C, E, F) Overarching (A, C, D, F)	The Health Research Authority should update its guidance on ethical approval for AI-health research and product development to set out the minimum participation requirements for diverse stakeholders
How can the explainability of AI-Health solutions be guaranteed?	Watson et al. (2019)	Epistemic (A, C) Normative (A, C, E) Overarching (A, D)	Medicines and Healthcare Regulators Medicines and Healthcare products Regulatory Agency should update the regulations governing medical devices to set out the minimum standards for 'explainability' of AI-health products
What mechanisms can be put in place to ensure reliability, replicability and safety of AI-Health solutions?	Challen et al. (2019)	Epistemic (A, C, F) Normative (C, E, F) Overarching (A, C, D)	The National Institute for Health and Care Excellence should make make it a requirement of formal health technology assessment that, within the bounds of technical feasibility and respecting intellectual property, developers make code open to enable reproducibility and error checking.
How can transparency over how algorithmic tools are integrated into	Vayena et al. (2015)	Epistemic (A, B, C, D, E, F)	NHS England should make it a requirement of all NHS trusts, hospitals

(continued)

Table 18.3 (continued)

Consideration	Key supporting literature	Relevant aspects (ascending LoA[a])	Example body responsible for answering this question based on the English National Health System
the healthcare workflow, how it shapes decisions, and how it affects process optimization within medical services, be guaranteed?		Normative (A, B, D, F) Overarching (A, D, F)	and providers of care to declare when an AI-health solution is being used in a specific care pathway and to be clear about how its safety and quality is being regularly assessed.
How can traditional and non-traditional sources of health data be incorporated into AI-Health decision making? And how can it be appropriately protected and how can it be harmonised?	Maher et al. (2019), Ploug and Holm (2016), Richardson Milam et al. (2015) and Townend (2018)	Epistemic (A, C, D, E, F) Normative (A, C, D, E, F) Overarching (A, C, D, E)	The Health Research Authority and NHS Digital should update guidance and regulations governing secondary uses of health data to incorporate the specific considerations of AI-Health as we have outlined in this paper
How are bioethical concepts (beneficence, non-maleficence, autonomy and justice (Beauchamp and Childress 2013) challenged by AI-Health?	Mittelstadt (2019)	Epistemic (B, F) Normative (A, C, D, F) Overarching (A, F)	The Nuffield Council on Bioethics should update its guidance on the bioethical principles for data initiatives to incorporate AI-health specific considerations.
How can concepts such as fairness, accountability and transparency can be maintained at scale (redacted for anonymity)?	Rosenfeld et al. (2019)	Epistemic (C, D, E, F) Normative (D, E, F) Overarching (F)	The Care Quality Commission, should develop a mechanism for monitoring these impacts at scale as part of its regular review process that is designed to ensure safe and high-quality care. This may require the Department of Health and Social Care extending its regulatory powers.

[a]Denoted by an increasing Level of Analysis: Individual (A), Interpersonal (B), Group (C), Institutional (D), Sectoral (E) and Societal (F)

practical, normative debate necessarily needs to go through the discussion about what is expected of a medical device, and therefore what is considered to be treatment.

Similarly, there has been moves to pass ethical codes without considering this multi-layered interests and challenges. However, some changes are worth noting. The UK has published its Code of Conduct for data-driven health and care technologies, standards for evidence of clinical effectiveness for digital health technologies (Greaves et al. 2018)—a digital assessment questionnaire standards for apps—and is currently developing a 'regulation as a service' model to ensure that there are appropriate regulatory checks at all stages of the AI development cycle. The World Health Organisation has a number of projects under way to develop guidance for member states (Aicardi et al. 2016) (World Health Organisation 2019). In China, several norms provide specific and detailed instructions to ensure health data security and confidentiality (Wang 2019) to ensure that health and medical big data sets can be used as a national resource to develop algorithms (Zhang et al. 2018) for the improvement of public health (Li et al. 2019).

The ethical questions involved in the use of AI for healthcare trickle down to issues of which matters can or should be regulated within the scope of healthcare, against what is considered simply a wellbeing service. Therefore, thinking in the terms of the proposed framework helps policymakers also understand and delineate the scope of their regulation. For example, some algorithmic tools potentially enable people to bypass formal and well-regulated healthcare systems entirely by accessing technology directly, either by using a wearable device or consulting online databases (Burr et al. 2020b). There must be a discussion, considering the LoAs and concerns, on whether these services have *de facto* overstepped the boundaries into healthcare in any of those levels.

Similarly, although some technical solutions have been put forward for mitigating issues with data bias (Gebru et al. 2018; Holland et al. 2018) and data quality (Dai et al. 2018) and ensuring social inclusion in decision-making (Balthazar et al. 2018; Friedman et al. 2017; Rahwan 2018), these remain relatively untested. Unless a competitive advantage of taking such pro-ethical steps becomes clear without these approaches being made mandatory, it is unlikely that they will have a significant impact on the ethical impacts of AI-Health in the near future. As a result, there is still little control over the procedures followed and quality control mechanisms (Cohen et al. 2014) involved in the development, deployment and use of AI-Health.

As these comparatively easier to tackle problems do not yet have adequate solutions, it is unsurprising that the bigger issues regarding the protection of equality of care (Powell and Deetjen 2019), fair distribution of benefits (Balthazar et al. 2018) (Kohli and Geis 2018) and the protection and promotion of societal values (Mahomed 2018) have barely even been considered. Given that healthcare systems in many ways act as the core of modern societies this is concerning. If mistakes are made too early in the adoption and implementation of AI in healthcare, the fall-out could be significant enough to undermine public trust, resulting in significant opportunity costs, and potentially encouraging individuals to seek their healthcare from outside of the formal systems where they may be presented with even greater risks. A coherent approach is needed and urgently, hopefully this systematic overview of the issues to be considered can help speed up its development.

18.5 Conclusion

This thematic literature review has sought to map out the ethical issues around the incorporation of data-driven AI technologies into healthcare provision and public health systems. In order to make this overview more useful, the relevant topics have been organised into themes and six different levels of abstraction (LoAs) have been highlighted. The hope is that by encouraging a discussion of the ethical implications of AI-Health at individual, interpersonal, group, institutional and societal LoAs, policymakers and regulators will be able to segment a large and complex conversation into tractable debates around specific issues, stakeholders, and solutions. This is important, as Topol (2019) states 'there cannot be exceptionalism for AI in medicine,' especially not when there is potentially so much to gain (Miotto et al. 2018).

With this in mind, the review has covered a wide range of topics while also venturing into the specificity of certain fields. This approach has enabled a fuller and more nuanced understanding of the ethical concerns related to the introduction of AI into healthcare systems than has been previously seen in the literature. Inevitably, there are limitations to this approach. Firstly, it is important to note that the selection of articles and policy documents was restricted to those written in English. This means that some ethical issues will have been overlooked (e.g. those in Spanish-speaking countries or in China). Second, academic literature, much like regulation, tends to struggle to keep pace with technological development. This literature review did not seek to identify ethical issues associated with specific use cases of AI first-hand, for example, by reviewing recently published studies available on pre-print servers such as arXiv, but instead focused on providing an overview of the ethical issues already identified and becoming mature. As a result, there may well be ethical concerns that are associated with more emergent use cases of AI for healthcare that we have not identified as they have not yet been discussed in formal peer-reviewed publications.

To overcome these limitations, further research could seek to expand the literature review by including a wider range of search queries, and by taking a case-study approach to analysing the ethical issues of specific practices and then aggregating these. This could be complemented by a comprehensive review of the policies, standards and regulations in development in different healthcare systems across the globe to assess the extent to which these are likely to be effective at mitigating these ethical concerns.

In this article, we hope to have provided a sufficiently comprehensive and detailed analysis of the current debates on ethical issues related to the introduction of AI into healthcare systems. The aim is to help policymakers and legislators develop evidence-based and proportionate policy and regulatory interventions. In particular, we hope to encourage the development of a system of transparent and distributed responsibility, where all those involved in the clinical algorithm supply chain can be held proportionately and appropriately accountable for the safety of the patient at the end, not just the HCP. It is only by ensuring such a system is developed that policymakers and legislators can be confident that the inherent risks we have described are appropriately mitigated (as far as possible) and only once this is the case will the medical community at large feel willing and able to adopt AI technologies.

Appendix – Methodology

This review process resulted in 156 papers suitable for analysis and inclusion in the initial review. Subsequent relevant papers that met the criteria were added at a later date during the writing up of the results.

This literature review also included accessory readings and case studies that were encountered during the research process. This includes bibliography obtained from the references of the papers analysed, and case studies identified in the readings (e.g. the Deep Mind case study). It is our belief that these exploratory readings enrich our systematic approach by developing on interesting findings and topics identified throughout our investigation (Table 18.4).

Table 18.4 Showing the final results from all searches

Database	Search query	Results	Titles selected	Titles downloaded
SCOPUS	ethic* AND algorithm* AND health*	596	39	19
	(ethic* AND (“Artificial Intelligence” OR ai) AND health*)	239	37	15
	(moral* AND (“Artificial Intelligence” OR ai) AND health*)	46	2	0
	(fair* AND (“Artificial Intelligence” OR ai) AND health*)	122	6	3
	(moral* OR ethic*) AND “machine learning” AND health*	91	14	9
	(fair* AND “machine learning” AND health*)	70	5	3
Web of Science	((fair* OR moral* OR ethic*) AND (“machine learning” OR “Artificial Intelligence” OR “AI” OR algorithm*) AND health*)	668	45	26
Philpapers	“machine learning” AND health*	3	1	1
	Artificial Intelligence AND health* AND ethic*	1000+	–	–
	algorithm* AND health* AND ethic*	5	0	0
	ethics AND “artificial intelligence” AND health	3	2	2
	AI or Artificial Intelligence or Fair AND ethic or moral or health AND health[12]	9	0	0
Google Scholar	ethics algorithms health	15,400	18	18
	ethics of machine learning in health	21,300	11	10
		716,000	2	1

(continued)

Table 18.4 (continued)

Database	Search query	Results	Titles selected	Titles downloaded
	ETHICS & HEALTH and at least one of: algorithm OR machine learning OR artificial intelligence OR AI			
	ETHICS & HEALTH and at least one of: algorithm OR AI	105,000	2	2
	MORAL & HEALTH and at least one of: algorithm OR AI	26,900	2	1
	FAIR & HEALTH And at least one of: algorithm OR AI	38,000	0	0
PubMed	ETHICS & ARTIFICIAL INTELLIGENCE OR MACHINE LEARNING	34,193	37	37
Total		958,645	223	147

It is important to note that multiple search queries were made to cover all the combinations and the numbers in the table thus represent the sum of results, titles evaluated and downloaded (not all found files were accessible for download). It is also important to note that only the first 500 most relevant results from Google Scholar were reviewed and anything written before 2014 was excluded to make the number of results more manageable

Funding Taddeo and Floridi's work was partially supported by Privacy and Trust Stream – Social lead of the PETRAS Internet of Things research hub – PETRAS is funded by the UK Engineering and Physical Sciences Research Council (EPSRC), grant agreement no. EP/N023013/1. Caio's, Taddeo's and Floridi's work was also partially supported by a Microsoft grant and a Google grant.

References

Aicardi, C., L. Del Savio, E.S. Dove, F. Lucivero, N. Tempini, and B. Prainsack. 2016. Emerging ethical issues regarding digital health data. On the World Medical Association Draft Declaration on Ethical Considerations Regarding Health Databases and Biobanks. *Croatian Medical Journal* 57 (2): 207–213. https://doi.org/10.3325/cmj.2016.57.207.

Aitken, M., M.P. Tully, C. Porteous, S. Denegri, S. Cunningham-Burley, N. Banner, C. Black, M. Burgess, L. Cross, J. Van Delden, E. Ford, S. Fox, N. Fitzpatrick, K. Gallacher, C. Goddard, L. Hassan, R. Jamieson, K.H. Jones, M. Kaarakainen, et al. 2019. Consensus statement on public involvement and engagement with data-intensive health research. *International Journal of Population Data Science* 4 (1). https://doi.org/10.23889/ijpds.v4i1.586.

Álvarez-Machancoses, Ó., and J.L. Fernández-Martínez. 2019. Using artificial intelligence methods to speed up drug discovery. *Expert Opinion on Drug Discovery* 14 (8): 769–777. https://doi.org/10.1080/17460441.2019.1621284.

Andorno, R. 2004. The right not to know: An autonomy based approach. *Journal of Medical Ethics* 30 (5): 435–439. https://doi.org/10.1136/jme.2002.001578.

Arieno, A., A. Chan, and S.V. Destounis. 2019. A review of the role of augmented intelligence in breast imaging: From automated breast density assessment to risk stratification. *American Journal of Roentgenology* 212 (2): 259–270. https://doi.org/10.2214/AJR.18.20391.

Balthazar, P., P. Harri, A. Prater, and N.M. Safdar. 2018. Protecting your patients' interests in the era of big data, artificial intelligence, and predictive analytics. *Journal of the American College of Radiology* 15 (3): 580–586. https://doi.org/10.1016/j.jacr.2017.11.035.

Barakat, N., A.P. Bradley, and M.N.H. Barakat. 2010. Intelligible support vector machines for diagnosis of Diabetes mellitus. *IEEE Transactions on Information Technology in Biomedicine* 14 (4): 1114–1120. https://doi.org/10.1109/TITB.2009.2039485.

Bartoletti, I. 2019. AI in healthcare: Ethical and privacy challenges. In *Lecture notes in computer science (Including subseries lecture notes in artificial intelligence and lecture notes in bioinformatics), 11526 LNAI*, 7–10. https://doi.org/10.1007/978-3-030-21642-9_2.

Barton, C., U. Chettipally, Y. Zhou, Z. Jiang, A. Lynn-Palevsky, S. Le, J. Calvert, and R. Das. 2019. Evaluation of a machine learning algorithm for up to 48-hour advance prediction of sepsis using six vital signs. *Computers in Biology and Medicine* 109: 79–84. https://doi.org/10.1016/j.compbiomed.2019.04.027.

Beauchamp, T.L., and J.F. Childress. 2013. *Principles of biomedical ethics*. 7th ed. Oxford University Press.

Beer, D. 2017. The social power of algorithms. *Information, Communication & Society* 20 (1): 1–13. https://doi.org/10.1080/1369118X.2016.1216147.

Brown, M.P.S., W.N. Grundy, D. Lin, N. Cristianini, C.W. Sugnet, T.S. Furey, M. Ares, and D. Haussler. 2000. Knowledge-based analysis of microarray gene expression data by using support vector machines. *Proceedings of the National Academy of Sciences* 97 (1): 262–267. https://doi.org/10.1073/pnas.97.1.262.

Burns, T. 2015. *Our necessary shadow: The nature and meaning of psychiatry*. Pegasus Books.

Burr, C., M. Taddeo, and L. Floridi. 2020a. The ethics of digital well-being: A thematic review. *Science and Engineering Ethics*. https://doi.org/10.1007/s11948-020-00175-8.

Burr, C., J. Morley, M. Taddeo, and L. Floridi. 2020b. Digital psychiatry: Risks and opportunities for public health and well-being. *IEEE Transactions on Technology and Society*. https://doi.org/10.1109/TTS.2020.2977059.

Cabitza, F., R. Rasoini, and G.F. Gensini. 2017. Unintended consequences of machine learning in medicine. *JAMA* 318 (6): 517. https://doi.org/10.1001/jama.2017.7797.

Car, J., A. Sheikh, P. Wicks, and M.S. Williams. 2019. Beyond the hype of big data and artificial intelligence: Building foundations for knowledge and wisdom. *BMC Medicine* 17 (1). https://doi.org/10.1186/s12916-019-1382-x.

Cath, C., S. Wachter, B. Mittelstadt, M. Taddeo, and L. Floridi. 2017. Artificial Intelligence and the 'Good Society': The US, EU, and UK approach. *Science and Engineering Ethics*. https://doi.org/10.1007/s11948-017-9901-7.

Challen, R., J. Denny, M. Pitt, L. Gompels, T. Edwards, and K. Tsaneva-Atanasova. 2019. Artificial intelligence, bias and clinical safety. *BMJ Quality & Safety* 28 (3): 231–237. https://doi.org/10.1136/bmjqs-2018-008370.

Char, D.S., N.H. Shah, and D. Magnus. 2018. Implementing machine learning in health care – Addressing ethical challenges. *The New England Journal of Medicine* 378 (11): 981–983. https://doi.org/10.1056/NEJMp1714229.

Cheney-Lippold, J. 2017. *We are data: Algorithms and the making of our digital selves*. New York University Press.

Chin-Yee, B., and R. Upshur. 2019. Three problems with big data and artificial intelligence in medicine. *Perspectives in Biology and Medicine* 62 (2): 237–256. https://doi.org/10.1353/pbm.2019.0012.

Coeckelbergh, M. 2014. Good healthcare is in the "how": The quality of care, the role of machines, and the need for new skills. In *Machine medical ethics*, vol. 74. Springer.

Cohen, I.G., R. Amarasingham, A. Shah, B. Xie, and B. Lo. 2014. The legal and ethical concerns that arise from using complex predictive analytics in health care. *Health Affairs* 33 (7): 1139–1147. https://doi.org/10.1377/hlthaff.2014.0048.

Cookson, C. 2018, September 6. Artificial intelligence faces public backlash, warns scientist. *Financial Times*. https://www.ft.com/content/0b301152-b0f8-11e8-99ca-68cf89602132

Cowie, J., E. Calveley, G. Bowers, and J. Bowers. 2018. Evaluation of a digital consultation and self-care advice tool in primary care: A multi-methods study. *International Journal of Environmental Research and Public Health* 15 (5). https://doi.org/10.3390/ijerph15050896.

Dai, W., K. Yoshigoe, and W. Parsley. 2018. Improving data quality through deep learning and statistical models. *ArXiv:1810.07132 [Cs]* 558: 515–522. https://doi.org/10.1007/978-3-319-54978-1_66.

Danis, M., and M. Solomon. 2013. Providers, payers, the community, and patients are all obliged to get patient activation and engagement ethically right. *Health Affairs* 32 (2): 401–407. https://doi.org/10.1377/hlthaff.2012.1081.

De Fauw, J., J.R. Ledsam, B. Romera-Paredes, S. Nikolov, N. Tomasev, S. Blackwell, H. Askham, X. Glorot, B. O'Donoghue, D. Visentin, G. van den Driessche, B. Lakshminarayanan, C. Meyer, F. Mackinder, S. Bouton, K. Ayoub, R. Chopra, D. King, A. Karthikesalingam, et al. 2018. Clinically applicable deep learning for diagnosis and referral in retinal disease. *Nature Medicine* 24 (9): 1342–1350. https://doi.org/10.1038/s41591-018-0107-6.

De Langavant, L.C., E. Bayen, and K. Yaffe. 2018. Unsupervised machine learning to identify high likelihood of dementia in population-based surveys: Development and validation study. *Journal of Medical Internet Research* 20 (7). https://doi.org/10.2196/10493.

Deng, X., Y. Luo, and C. Wang. 2018. Analysis of risk factors for cervical cancer based on machine learning methods. In *2018 5th IEEE International Conference on Cloud Computing and Intelligence Systems (CCIS)*, 631–635. https://doi.org/10.1109/CCIS.2018.8691126.

Department of Health and Social Care. 2018. *Annual Report of the Chief Medical Office 2018: Health 2040—Better Health Within Reach.* https://assets.publishing.service.gov.uk/government/uploads/system/uploads/attachment_data/file/767549/Annual_report_of_the_Chief_Medical_Officer_2018_-_health_2040_-_better_health_within_reach.pdf

———. 2019. *Health Secretary announces £250 million investment in artificial intelligence* [Gov.uk]. Retrieved August 8, 2019, from https://www.gov.uk/government/news/health-secretary-announces-250-million-investment-in-artificial-intelligence

Di Nucci, E. 2019. Should we be afraid of medical AI? *Journal of Medical Ethics*. https://doi.org/10.1136/medethics-2018-105281.

Ding, Y., J. Tang, and F. Guo. 2019. Identification of drug-side effect association via multiple information integration with centered kernel alignment. *Neurocomputing* 325: 211–224. https://doi.org/10.1016/j.neucom.2018.10.028.

Dong, R., X. Yang, X. Zhang, P. Gao, A. Ke, H. Sun, J. Zhou, J. Fan, J. Cai, and G. Shi. 2019. Predicting overall survival of patients with hepatocellular carcinoma using a three-category method based on DNA methylation and machine learning. *Journal of Cellular and Molecular Medicine* 23 (5): 3369–3374. https://doi.org/10.1111/jcmm.14231.

Dudley, J.T., J. Listgarten, O. Stegle, S.E. Brenner, and L. Parts. 2015. *Personalized medicine: From genotypes, molecular phenotypes and the quantified self, towards improved medicine*, 342–346.

DuFault, B.L., and J.W. Schouten. 2018. Self-quantification and the datapreneurial consumer identity. *Consumption Markets & Culture*: 1–27. https://doi.org/10.1080/10253866.2018.1519489.

Emanuel, E.J., and R.M. Wachter. 2019. Artificial intelligence in health care: Will the value match the hype? *JAMA* 321 (23): 2281–2282. https://doi.org/10.1001/jama.2019.4914.

Faden, R.R., N.E. Kass, S.N. Goodman, P. Pronovost, S. Tunis, and T.L. Beauchamp. 2013. An ethics framework for a learning health care system: A departure from traditional research ethics and clinical ethics. *Hastings Center Report* 43 (s1): S16–S27. https://doi.org/10.1002/hast.134.

Fleming, N. 2018. How artificial intelligence is changing drug discovery. *Nature* 557 (7707): S55–S57. https://doi.org/10.1038/d41586-018-05267-x.

Floridi, L. 2008. The method of levels of abstraction. *Minds and Machines* 18 (3): 303–329. https://doi.org/10.1007/s11023-008-9113-7.

———. 2013. Distributed morality in an information society. *Science and Engineering Ethics* 19 (3): 727–743. https://doi.org/10.1007/s11948-012-9413-4.

———. 2014. *The 4th revolution: How the infosphere is reshaping human reality.* Oxford University Press.

———. 2016. Faultless responsibility: On the nature and allocation of moral responsibility for distributed moral actions. *Philosophical Transactions of the Royal Society A: Mathematical, Physical and Engineering Sciences* 374 (2083): 20160112. https://doi.org/10.1098/rsta.2016.0112.

———. 2017a. Digital's cleaving power and its consequences. *Philosophy & Technology* 30 (2): 123–129. https://doi.org/10.1007/s13347-017-0259-1.

———. 2017b. The logic of design as a conceptual logic of information. *Minds and Machines* 27 (3): 495–519. https://doi.org/10.1007/s11023-017-9438-1.

———. 2018. Soft ethics, the governance of the digital and the General Data Protection Regulation. *Philosophical Transactions. Series A, Mathematical, Physical, and Engineering Sciences* 376 (2133). https://doi.org/10.1098/rsta.2018.0081.

———. 2019a. AI opportunities for healthcare must not be wasterd. *Health Management* 19.

———. 2019b. What the near future of artificial intelligence could be. *Philosophy & Technology* 32 (1): 1–15. https://doi.org/10.1007/s13347-019-00345-y.

Floridi, L., and M. Taddeo. 2016. What is data ethics? *Philosophical Transactions of the Royal Society A: Mathematical, Physical and Engineering Sciences* 374 (2083): 20160360. https://doi.org/10.1098/rsta.2016.0360.

Friedman, B., D.G. Hendry, and A. Borning. 2017. A survey of value sensitive design methods. *Foundations and Trends® in Human–Computer Interaction* 11 (2): 63–125. https://doi.org/10.1561/1100000015.

Garattini, C., J. Raffle, D.N. Aisyah, F. Sartain, and Z. Kozlakidis. 2019. Big data analytics, infectious diseases and associated ethical impacts. *Philosophy & Technology* 32 (1): 69–85. https://doi.org/10.1007/s13347-017-0278-y.

Garcia, J., N. Romero, D. Keyson, and P. Havinga. 2014. Reflective healthcare systems: Mirco-cylce of self-reflection to empower users. *Interaction Design and Architecture(s)* 23 (1): 173–190.

Gebru, T., J. Morgenstern, B. Vecchione, J.W. Vaughan, H. Wallach, H. Daumeé III, and K. Crawford. 2018. Datasheets for datasets. *ArXiv:1803.09010 [Cs]*. http://arxiv.org/abs/1803.09010.

Gillespie, T., P.J. Boczkowski, and K.A. Foot. 2014. *Media technologies: Essays on communication, materiality, and society*. The MIT Press.

Grant, M.J., and A. Booth. 2009. A typology of reviews: An analysis of 14 review types and associated methodologies: A typology of reviews, Maria J. Grant & Andrew Booth. *Health Information & Libraries Journal* 26 (2): 91–108. https://doi.org/10.1111/j.1471-1842.2009.00848.x.

Greaves, F., I. Joshi, M. Campbell, S. Roberts, N. Patel, and J. Powell. 2018. What is an appropriate level of evidence for a digital health intervention? *The Lancet* 392 (10165): 2665–2667. https://doi.org/10.1016/S0140-6736(18)33129-5.

Hailu, R. 2019. Fitbits and other wearables may not accurately track heart rates in people of color. *STAT*. https://www.statnews.com/2019/07/24/fitbit-accuracy-dark-skin/

Harerimana, G., B. Jang, J.W. Kim, and H.K. Park. 2018. Health big data analytics: A technology survey. *IEEE Access* 6: 65661–65678. https://doi.org/10.1109/ACCESS.2018.2878254.

Hay, S. I., George, D. B., Moyes, C. L., & Brownstein, J. S. (2013). Big data opportunities for global infectious disease surveillance.

He, J., S.L. Baxter, J. Xu, J. Xu, X. Zhou, and K. Zhang. 2019. The practical implementation of artificial intelligence technologies in medicine. *Nature Medicine* 25 (1): 30–36. https://doi.org/10.1038/s41591-018-0307-0.

Hoffman, L., E. Benedetto, H. Huang, E. Grossman, D. Kaluma, Z. Mann, and J. Torous. 2019. Augmenting mental health in primary care: A 1-year study of deploying smartphone apps in a multi-site primary care/behavioral health integration program. *Frontiers in Psychiatry* 10: 94. https://doi.org/10.3389/fpsyt.2019.00094.

Holland, S., A. Hosny, S. Newman, J. Joseph, and K. Chmielinski. 2018. The dataset nutrition label: A framework to drive higher data quality standards. *ArXiv:1805.03677 [Cs]*. http://arxiv.org/abs/1805.03677.

Holzinger, A., B. Haibe-Kains, and I. Jurisica. 2019. Why imaging data alone is not enough: AI-based integration of imaging, omics, and clinical data. *European Journal of Nuclear Medicine and Molecular Imaging*. https://doi.org/10.1007/s00259-019-04382-9.

Information Commissioner. 2018, June 6. *Royal Free—Google DeepMind trial failed to comply with data protection law*. https://ico.org.uk/about-the-ico/news-and-events/news-and-blogs/2017/07/royal-free-google-deepmind-trial-failed-to-comply-with-data-protection-law/

———. 2019, June 3. *Project ExplAIn interim report*. https://ico.org.uk/about-the-ico/research-and-reports/project-explain-interim-report/

Juengst, E., M.L. McGowan, J.R. Fishman, and R.A. Settersten. 2016. From "personalized" to "precision" medicine: The ethical and social implications of rhetorical reform in genomic medicine. *Hastings Center Report* 46 (5): 21–33. https://doi.org/10.1002/hast.614.

Kalmady, S.V., R. Greiner, R. Agrawal, V. Shivakumar, J.C. Narayanaswamy, M.R.G. Brown, A.J. Greenshaw, S.M. Dursun, and G. Venkatasubramanian. 2019. Towards artificial intelligence in mental health by improving schizophrenia prediction with multiple brain parcellation ensemble-learning. *NPJ Schizophrenia* 5 (1): 2. https://doi.org/10.1038/s41537-018-0070-8.

Kim, H., and B. Xie. 2017. Health literacy in the eHealth era: A systematic review of the literature. *Patient Education and Counseling* 100 (6): 1073–1082.

King, D., A. Karthikesalingam, C. Hughes, H. Montgomery, R. Raine, G. Rees, and On behalf of the DeepMind Health Team. 2018. Letter in response to Google DeepMind and healthcare in an age of algorithms. *Health and Technology* 8 (1): 11–13. https://doi.org/10.1007/s12553-018-0228-4.

Kluge, E.-H., P. Lacroix, and P. Ruotsalainen. 2018. Ethics certification of health information professionals. *Yearbook of Medical Informatics* 27 (01): 037–040. https://doi.org/10.1055/s-0038-1641196.

Kohli, M., and R. Geis. 2018. Ethics, artificial intelligence, and radiology. *Journal of the American College of Radiology* 15 (9): 1317–1319. https://doi.org/10.1016/j.jacr.2018.05.020.

Kourou, K., T.P. Exarchos, K.P. Exarchos, M.V. Karamouzis, and D.I. Fotiadis. 2015. Machine learning applications in cancer prognosis and prediction. *Computational and Structural Biotechnology Journal* 13: 8–17. https://doi.org/10.1016/j.csbj.2014.11.005.

Kunapuli, G., B.A. Varghese, P. Ganapathy, B. Desai, S. Cen, M. Aron, I. Gill, and V. Duddalwar. 2018. A decision-support tool for renal mass classification. *Journal of Digital Imaging* 31 (6): 929–939. https://doi.org/10.1007/s10278-018-0100-0.

Kuo, W.-J., R.-F. Chang, D.-R. Chen, and C.C. Lee. 2001. Data mining with decision trees for diagnosis of breast tumor in medical ultrasonic images. *Breast Cancer Research and Treatment* 66 (1): 51–57. https://doi.org/10.1023/A:1010676701382.

Li, B., J. Li, Y. Jiang, and X. Lan. 2019. Experience and reflection from China's Xiangya medical big data project. *Journal of Biomedical Informatics* 93. https://doi.org/10.1016/j.jbi.2019.103149.

López-Martínez, F., E.R. Núñez-Valdez, J. Lorduy Gomez, and V. García-Díaz. 2019. A neural network approach to predict early neonatal sepsis. *Computers & Electrical Engineering* 76: 379–388. https://doi.org/10.1016/j.compeleceng.2019.04.015.

Lu, H., and M. Wang. 2019. RL4health: Crowdsourcing reinforcement learning for knee replacement pathway optimization. *ArXiv:1906.01407 [Cs, Stat]*. http://arxiv.org/abs/1906.01407.

Lu, F.S., M.W. Hattab, C.L. Clemente, M. Biggerstaff, and M. Santillana. 2019. Improved state-level influenza nowcasting in the United States leveraging internet-based data and network approaches. *Nature Communications* 10 (1). https://doi.org/10.1038/s41467-018-08082-0.

Maher, N., J. Senders, A. Hulsbergen, N. Lamba, M. Parker, J.-P. Onnela, A. Bredenoord, T. Smith, M. Broekmann, et al. 2019. Passive data collection and use in healthcare: A systematic review of ethical issues. *International Journal of Medical Informatics* 129: 242–247.

Mahomed, S. 2018. Healthcare, artificial intelligence and the fourth industrial revolution: Ethical, social and legal considerations. *South African Journal of Bioethics and Law* 11 (2): 93. https://doi.org/10.7196/SAJBL.2018.v11i2.00664.

Martani, A., D. Shaw, and B.S. Elger. 2019. Stay fit or get bit—Ethical issues in sharing health data with insurers' apps. *Swiss Medical Weekly* 149: w20089. https://doi.org/10.4414/smw.2019.20089.

Matthias, A. 2015. Robot lies in health care: When is deception morally permissible? *Kennedy Institute of Ethics Journal* 25 (2): 169–162. https://doi.org/10.1353/ken.2015.0007.

McDougall, R.J. 2019. Computer knows best? The need for value-flexibility in medical AI. *Journal of Medical Ethics* 45 (3): 156–160. https://doi.org/10.1136/medethics-2018-105118.

McLaughlin, K. 2016. *Empowerment: A critique*. http://public.eblib.com/choice/publicfullrecord.aspx?p=4332655

Miotto, R., F. Wang, S. Wang, X. Jiang, and J.T. Dudley. 2018. Deep learning for healthcare: Review, opportunities and challenges. *Briefings in Bioinformatics* 19 (6): 1236–1246. https://doi.org/10.1093/bib/bbx044.

Mittelstadt, B. 2019. The ethics of biomedical 'big data' analytics. *Philosophy & Technology* 32 (1): 17–21. https://doi.org/10.1007/s13347-019-00344-z.

Mittelstadt, B.D., P. Allo, M. Taddeo, S. Wachter, and L. Floridi. 2016. The ethics of algorithms: Mapping the debate. *Big Data & Society* 3 (2): 205395171667967. https://doi.org/10.1177/2053951716679679. Redacted for anonymity.

Mortazavi, B.J., N. Desai, J. Zhang, A. Coppi, F. Warner, H.M. Krumholz, and S. Negahban. 2017. Prediction of adverse events in patients undergoing major cardiovascular procedures. *IEEE Journal of Biomedical and Health Informatics* 21 (6): 1719–1729. https://doi.org/10.1109/JBHI.2017.2675340.

Moscoso, A., J. Silva-Rodríguez, J.M. Aldrey, J. Cortés, A. Fernández-Ferreiro, N. Gómez-Lado, Á. Ruibal, and P. Aguiar. 2019. Prediction of Alzheimer's disease dementia with MRI beyond the short-term: Implications for the design of predictive models. *NeuroImage: Clinical* 23: 101837. https://doi.org/10.1016/j.nicl.2019.101837.

Nag, N., V. Pandey, H. Oh, and R. Jain. 2017. Cybernetic health. *ArXiv:1705.08514* [Cs]. http://arxiv.org/abs/1705.08514

National Institute for Health Research. 2019, June 14. *Involving the public in complex questions around artificial intelligence research*. https://www.nihr.ac.uk/blog/involving-the-public-in-complex-questions-around-artificial-intelligence-research/12236

Nebeker, C., J. Torous, and R.J. Bartlett Ellis. 2019. Building the case for actionable ethics in digital health research supported by artificial intelligence. *BMC Medicine* 17 (1). https://doi.org/10.1186/s12916-019-1377-7.

Nelson, A., D. Herron, G. Rees, and P. Nachev. 2019. Predicting scheduled hospital attendance with artificial intelligence. *Npj Digital Medicine* 2 (1): 26. https://doi.org/10.1038/s41746-019-0103-3.

Ngiam, K.Y., and I.W. Khor. 2019. Big data and machine learning algorithms for health-care delivery. *The Lancet Oncology* 20 (5): e262–e273. https://doi.org/10.1016/S1470-2045(19)30149-4.

NHS England. 2019. The NHS long term plan. *NHS*. https://www.longtermplan.nhs.uk/wp-content/uploads/2019/01/nhs-long-term-plan.pdf

Ortiz, J.R., H. Zhou, D.K. Shay, K.M. Neuzil, A.L. Fowlkes, and C.H. Goss. 2011. Monitoring influenza activity in the United States: A comparison of traditional surveillance systems with Google flu trends. *PLoS One* 6 (4): e18687. https://doi.org/10.1371/journal.pone.0018687.

Owens, J., and A. Cribb. 2019. 'My Fitbit thinks I can do better!' Do health promoting wearable technologies support personal autonomy? *Philosophy & Technology* 32 (1): 23–38. https://doi.org/10.1007/s13347-017-0266-2.

Panch, T., H. Mattie, and L.A. Celi. 2019. The "inconvenient truth" about AI in healthcare. *Npj Digital Medicine* 2 (1): 77. https://doi.org/10.1038/s41746-019-0155-4.

Petrini, C. 2015. On the 'pendulum' of bioethics. *Clinica Terapeutica* 166 (2): 82–84. https://doi.org/10.7417/CT.2015.1821.
Ploug, T., and S. Holm. 2016. Meta consent – A flexible solution to the problem of secondary use of health data. *Bioethics* 30 (9): 721–732.
———. 2019. The right to refuse diagnostics and treatment planning by artificial intelligence. *Medicine, Health Care, and Philosophy*. https://doi.org/10.1007/s11019-019-09912-8.
Popkes, A.-L., H. Overweg, A. Ercole, Y. Li, J.M. Hernández-Lobato, Y. Zaykov, and C. Zhang. 2019. Interpretable outcome prediction with sparse Bayesian neural networks in intensive care. *ArXiv:1905.02599 [Cs, Stat]*. http://arxiv.org/abs/1905.02599.
Powell, J., and U. Deetjen. 2019. Characterizing the digital health citizen: Mixed-methods study deriving a new typology. *Journal of Medical Internet Research* 21 (3): e11279.
Powles, J., and H. Hodson. 2017. Google DeepMind and healthcare in an age of algorithms. *Health and Technology*: 1–17. https://doi.org/10.1007/s12553-017-0179-1.
———. 2018. Response to DeepMind. *Health and Technology* 8 (1): 15–29. https://doi.org/10.1007/s12553-018-0226-6.
Price, W.N. 2018. Medical malpractice and Black-box medicine. In *Big data, health law, and bioethics*, ed. I.G. Cohen, H.F. Lynch, E. Vayena, and U. Gasser, 1st ed., 295–306. Cambridge University Press. https://doi.org/10.1017/9781108147972.027.
Price, W.N., S. Gerke, and I.G. Cohen. 2019. Potential liability for physicians using artificial intelligence. *Journal of the American Medical Association*. https://doi.org/10.1001/jama.2019.15064.
Racine, E., W. Boehlen, and M. Sample. 2019. Healthcare uses of artificial intelligence: Challenges and opportunities for growth. *Healthcare Management Forum*. https://doi.org/10.1177/0840470419843831.
Rahwan, I. 2018. Society-in-the-loop: Programming the algorithmic social contract. *Ethics and Information Technology* 20 (1): 5–14. https://doi.org/10.1007/s10676-017-9430-8.
Rajkomar, A., M. Hardt, M.D. Howell, G. Corrado, and M.H. Chin. 2018. Ensuring fairness in machine learning to advance health equity. *Annals of Internal Medicine* 169 (12): 866. https://doi.org/10.7326/M18-1990.
Rich, E., and A. Miah. 2014. Understanding digital health as public pedagogy: A critical framework. *Societies* 4 (2): 296–315. https://doi.org/10.3390/soc4020296.
Richardson, V., S. Milam, and D. Chrysler. 2015. Is sharing de-identified data legal? The state of public health confidentiality laws and their interplay with statistical disclosure limitation techniques. *The Journal of Law, Medicine & Ethics* 43 (s1): 83–86.
Rosenfeld, A., D. Benrimoh, C. Armstrong, N. Mirchi, T. Langlois-Therrien, C. Rollins, M. Tanguay-Sela, J. Mehltretter, R. Fratila, S. Israel, E. Snook, K. Perlman, A. Kleinerman, B. Saab, M. Thoburn, C. Gabbay, and A. Yaniv-Rosenfeld. 2019. Big data analytics and AI in mental healthcare. *ArXiv:1903.12071 [Cs]*. http://arxiv.org/abs/1903.12071.
Ruckenstein, M., and N.D. Schüll. 2017. The Datafication of health. *Annual Review of Anthropology* 46 (1): 261–278. https://doi.org/10.1146/annurev-anthro-102116-041244.
Schönberger, D. 2019. Artificial intelligence in healthcare: A critical analysis of the legal and ethical implications. *International Journal of Law and Information Technology* 27 (2): 171–203. https://doi.org/10.1093/ijlit/eaz004.
Sethi, N., and G.T. Laurie. 2013. Delivering proportionate governance in the era of eHealth: Making linkage and privacy work together. *Medical Law International* 13 (2–3): 168–204. https://doi.org/10.1177/0968533213508974.
Shickel, B., P.J. Tighe, A. Bihorac, and P. Rashidi. 2018. Deep EHR: A survey of recent advances in deep learning techniques for electronic health record (EHR) analysis. *IEEE Journal of Biomedical and Health Informatics* 22 (5): 1589–1604. https://doi.org/10.1109/JBHI.2017.2767063.
Sterckx, S., V. Rakic, J. Cockbain, and P. Borry. 2016. "You hoped we would sleep walk into accepting the collection of our data": Controversies surrounding the UK care.Data scheme and

their wider relevance for biomedical research. *Medicine, Health Care and Philosophy* 19 (2): 177–190. https://doi.org/10.1007/s11019-015-9661-6.

Sullivan, H.R., and S.J. Schweikart. 2019. Are current tort liability doctrines adequate for addressing injury caused by AI? *AMA Journal of Ethics* 21 (2): 160–166. https://doi.org/10.1001/amajethics.2019.160.

Taddeo, M., and L. Floridi. 2018. How AI can be a force for good. *Science* 361 (6404): 751–752. https://doi.org/10.1126/science.aat5991.

Topol, E.J. 2019. High-performance medicine: The convergence of human and artificial intelligence. *Nature Medicine* 25 (1): 44–56. https://doi.org/10.1038/s41591-018-0300-7.

Topuz, K., F.D. Zengul, A. Dag, A. Almehmi, and M.B. Yildirim. 2018. Predicting graft survival among kidney transplant recipients: A Bayesian decision support model. *Decision Support Systems* 106: 97–109. https://doi.org/10.1016/j.dss.2017.12.004.

Townend, D. 2018. Conclusion: Harmonisation in genomic and health data sharing for research: An impossible dream? *Human Genetics* 137 (8): 657–664.

Turilli, M., and L. Floridi. 2009. The ethics of information transparency. *Ethics and Information Technology* 11 (2): 105–112. https://doi.org/10.1007/s10676-009-9187-9.

Vayena, Effy, M. Salathé, L.C. Madoff, and J.S. Brownstein. 2015. Ethical challenges of big data in public health. *PLoS Computational Biology* 11 (2): e1003904. https://doi.org/10.1371/journal.pcbi.1003904.

Vayena, E., H. Tobias, A. Afua, and B. Allesandro. 2018a. Digital health: Meeting the ethical and policy challenges. *Swiss Medical Weekly* 148 (34). https://doi.org/10.4414/smw.2018.14571.

Vayena, E., A. Blasimme, and I.G. Cohen. 2018b. Machine learning in medicine: Addressing ethical challenges. *PLoS Medicine* 15 (11): e1002689. https://doi.org/10.1371/journal.pmed.1002689.

Voigt, K. 2019. Social justice, equality and primary care: (how) can 'big data' help? *Philosophy & Technology* 32 (1): 57–68. https://doi.org/10.1007/s13347-017-0270-6.

Vollmer, S., B.A. Mateen, G. Bohner, F.J. Király, and R. Ghani. 2018a. Machine learning and AI research for patient benefit: 20 critical questions on transparency. *Replicability, Ethics and Effectiveness*. 25.

Vollmer, S., B.A. Mateen, G. Bohner, F.J. Király, R. Ghani, P. Jonsson, S. Cumbers, A. Jonas, K.S.L. McAllister, P. Myles, D. Granger, M. Birse, R. Branson, K.G. Moons, G.S. Collins, J.P.A. Ioannidis, C. Holmes, and H. Hemingway. 2018b. Machine learning and AI research for patient benefit: 20 critical questions on transparency, replicability, ethics and effectiveness. *ArXiv:1812.10404 [Cs, Stat]*. http://arxiv.org/abs/1812.10404.

Wang, Z. 2019. Data integration of electronic medical record under administrative decentralization of medical insurance and healthcare in China: A case study. *Israel Journal of Health Policy Research* 8 (1). https://doi.org/10.1186/s13584-019-0293-9.

Wang, S., X. Jiang, S. Singh, R. Marmor, L. Bonomi, D. Fox, M. Dow, and L. Ohno-Machado. 2017. Genome privacy: Challenges, technical approaches to mitigate risk, and ethical considerations in the United States: Genome privacy in biomedical research. *Annals of the New York Academy of Sciences* 1387 (1): 73–83.

Wardrope, A. 2015. Relational autonomy and the ethics of health promotion. *Public Health Ethics* 8 (1): 50–62. https://doi.org/10.1093/phe/phu025.

Watson, D.S., J. Krutzinna, I.N. Bruce, C.E. Griffiths, I.B. McInnes, M.R. Barnes, and L. Floridi. 2019. Clinical applications of machine learning algorithms: Beyond the black box. *BMJ* 364: l886. https://doi.org/10.1136/bmj.l886.

World Health Organisation. 2019. *Big data and artificial intelligence*. Retrieved June 29, 2019, from https://www.who.int/ethics/topics/big-data-artificial-intelligence/en/

Zacher, B., and I. Czogiel. 2019. Supervised learning improves disease outbreak detection. *ArXiv:1902.10061 [Cs, Stat]*. http://arxiv.org/abs/1902.10061.

Zhang, L., H. Wang, Q. Li, M.-H. Zhao, and Q.-M. Zhan. 2018. Big data and medical research in China. *BMJ*: j5910. https://doi.org/10.1136/bmj.j5910.

Chapter 19
Autonomous Vehicles: From Whether and When to Where and How

Luciano Floridi

Abstract Mobility is an essential component of life in any society, so a transformation of mobility will affect the foundations of any society, and it is hard to imagine a more profound transformation of mobility than autonomous driving. This is why understanding attitudes towards the benefits and shortcomings of autonomous vehicles means being able to address societal welfare and individual well-being more successfully. In this chapter I argue that digital technologies have made it possible to detach the journey from the trip. It seems that, in the near future, we may be increasingly able to enjoy trips rather than journeys, with more freedom to choose to travel because we want to rather than because we need to.

Keywords Artificial Intelligence · Autonomous driving · Driverless vehicles · Mobility · Travelling

19.1 Introduction

When *Death of a Salesman* by Arthur Miller premiered in 1949, the job of selling merchandise by peddling wares in a designated area was common. It was a very different age-intermodal containers were being standardised, and consumerism was becoming rampant. In the play, the salesman, Willy Loman, is unhappy about all his travelling, and rightly so. The disappearance of his kind of job reminds us that, until recently, we thought that digital technologies were just going to decrease the need to move around. Today, we shop online, and the new selling agents are recommender

L. Floridi (✉)
Oxford Internet Institute, University of Oxford, Oxford, UK
e-mail: luciano.floridi@oii.ox.ac.uk

L. Floridi (ed.), *Ethics, Governance, and Policies in Artificial Intelligence*, Philosophical Studies Series 144, https://doi.org/10.1007/978-3-030-81907-1_19

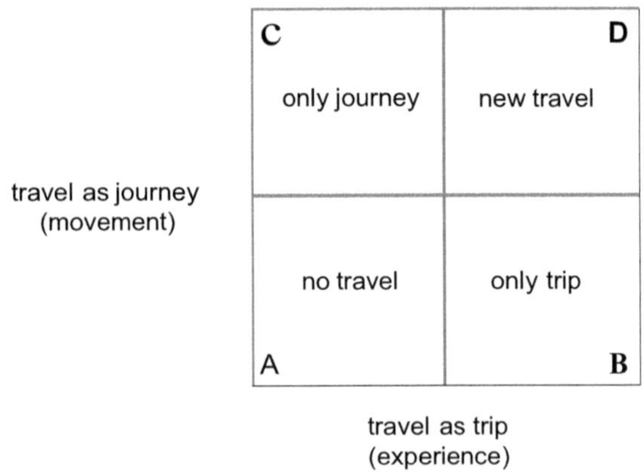

Fig. 19.1 The two components of travel: the physical journey and the experiential trip

systems (Milano et al. 2019), which canvass the space of information, or *irifosphere.* It is true that we move a lot of bytes and not just atoms around, yet this is not the whole story. It is really too simplistic to conclude that the only impact that the digital revolution has, and is still having, on mobility has been that of reducing it. One still hears some driverless car evangelists arguing this (sometimes hijacking a green rhetoric) but they are clearly mistaken, because more people who cannot drive today will be able to do so in the future, thanks to increased levels of automation. It is far more accurate to say that digital technologies are changing the very essence of mobility (they are – re-ontologising it), in four different ways (see Fig. 19.1).

19.2 Travelling, Journeys and Trips

Every act of travel, understood as the movement of people from one geographical location to another, has two main components, which may be mapped on two axes (Fig. 19.1).

One is the physical process of movement, say from home to office and back. Let's call this y-axis, the physical *journey.* A journey has many aspects that are easily *quantifiable,* in terms of medium employed, costs, time, distance, speed, and so forth. The other component of travelling is the experience. Let's call this *x*- axis, the experiential *trip.* A trip has *qualitative* aspects that are more subjective and hence more difficult to specify, such as perceived duration, novelty, comfort, enjoyability, and so forth. The digital revolution has deeply transformed human mobility by changing both the journey and the trip component of travelling.

Consider *A* first, in Fig. 19.1. This is the zero-travelling option I mentioned in Sect. 19.1. In some cases, digital technologies have eliminated both the journey and the trip because everything happens inside the infosphere. Telepresence[1] and the

[1]For an epistemological analysis of telepresence, see Floridi (2005).

remotization of jobs, for example, have hugely contributed to a lowering of the need for mobility. At the same time, option *A* may simply shift the occurrence of travelling; our shopping online has caused a huge increase in delivery services, hence of journeys (the courier's), if not of trips (ours). This explains why lower human mobility leads to higher artificial mobility, that is, the replacement of people with artificial agents whenever possible , from delivery drones in the sky and luggage-sized vehicles navigating about town, to ships without crews. In other words, the more *A* you have, the more C you are likely to need. If this is correct, the future of drones lies not so much in transporting humans but in delivering goods, in tandem with AI solutions that can handle tasks such as navigation, orientation, scheduling, safety, and delivery.

At the opposite comer of the quadrant, in *D,* we find new forms of mobility made possible by the infosphere. Driverless motorbikes are a bit of an oxymoron not because motorbikes need us on board for balancing purposes, for this engineering problem can easily be solved (a self-balancing motorbike can park itself), but because motorbikes are often about the trip and the journey at least in equal measure, that is, they are often about the trip experience of the physical journey. Likewise, if you rent a sport car for a day, it is because you want to enjoy (trip) the driving (journey). This is why I doubt we shall see a driverless Ferrari. What the digital revolution has done is to make some mobility cheaper and safer (journey), and more enjoyable (trip). Thus, sustainable mobility trends made possible by ICTs, especially in urban contexts, such as 3rd generation bicycle-sharing systems, are a form of mobility that relies on a new combination of less impactful journeys that are, at least ethically speaking, more satisfactory trips.

Looking to the bottom right, we find in *B* that digital technologies are trying to make any travel just a matter of enjoyable trip, eliminating as much as possible the tedious aspects of the journey. Cruises are a classic example; the journey is entirely absorbed by the trip. Consider on-board entertainments of all kinds in all sorts of vehicles, geolocation, digital maps, navigators, and other digital features: increasingly autonomous cars performing more and more functions independently of the driver, and these are all trends in the transformation of human mobility into a purely experiential phenomenon. In the future, one may imagine forms of *virtual* mobility becoming a reality in this context, e.g., replacing holiday tours as digital trips without physical journeys.

Finally, in the top left, in C, we find more efficient mobility, which tends to exclude more and more the human component: better and safer performance, lower costs, less time, better routes, just-in-time re-routing, re-scheduling, monitoring of consumption, better logistics, 24/7 services: these are some of the many aspects of a deep transformation of travel into unmanned journeys with no trip component, with AI systems in charge and humans at most 'on' the loop, placed in the *A* square. Here, one of the great successes has been freight transport, which is increasingly automated, especially at sea. The foreseeable future includes the further automation of public transport in public spaces (e.g., dedicated lanes) and environment-bounded technology-friendly, local mobilities, such as airport buses, and robots in industrial logistics, as in warehouses . In this context, we should be careful not to confuse the *logically possible* with the *actually feasible* . *In theory,* level-5 autonomous vehicles-that is, those that are completely autonomous and require no driver-are perfectly (i.e. logically) possible because there is nothing intrinsically contradictory in assuming that, one day, all potential difficulties will be resolved by the right kind of technology. *In practice* though, what we are likely to witness will be deep transformation (re-ontologisation again) of whole environments to ensure that the available technologies will be successful (we are 'enveloping' the world around the capacities of our digital technologies, see Floridi (2014)). Think of the difference between (the logical possibility of) developing totally reliable visual systems to enable an autonomous vehicle to identify and recognise road signs in any weather condition in any context, from the snowy and foggy countryside to a rainy and traffic-bound city at night, to (the actual feasibility of) re-engineering all road signs in a given environment (say an airport) to make them communicate with the vehicle wirelessly and seamlessly, through radio signals, rather than visually.

19.3 Not 'When' or Even 'Where' But 'How' Is the Question

Mobility is an essential component of life in any society. Every day, all over the world, billions of vehicles of all kinds (in particular, it is estimated that, since 2010, more than 1 billion passenger cars travel the streets and roads of the world) play key functions (e.g., in transport or leisure) and social roles (e.g., as statements or status symbols). They are part of our history and culture. A transformation of mobility will affect the foundations of any society. And it is hard to imagine a more profound transformation of mobility than autonomous driving. This is why understanding attitudes towards the benefits and shortcomings of autonomous vehicles means being able to address societal welfare and individual well-being more successfully (Floridi et al. 2018). We saw in Sect. 19.2 that digital technologies have made it possible to detach the journey from the trip. It seems that, in the near future, we may be increasingly able to enjoy trips rather than journeys, with more freedom to choose to travel because we want to rather than because we need to . Willy Loman would have liked this. Here, I would like to introduce one more distinction that is equally important. A recent study (AUDI 2019) is very valuable because it provides a wealth of information and insights about people's attitudes to autonomous driving in China, France, Germany, Italy, Japan, South Korea, Spain, the UK and the USA. Studying its findings, it becomes clear that one should not confuse technological *novelty* with *change.*

The majority of those surveyed expressed *interest* (82%) and *curiosity* (62%) about autonomous driving. However, a majority also raised *concerns* about loss of control (70%), technically unavoidable residual risks (66%) and the lack of a legal framework (65%). This is not as odd as it seems. Appreciating a *novelty* requires only an open mind but involves no actual risks or costs. Embracing a *change* implies a commitment that raises concerns about risks and costs (only 28% of people are willing to pay more for autonomous vehicles). Autonomous driving is both a realistic novelty and an unprecedented change. To translate high levels of interest and curiosity into low levels of concerns, one needs to provide better technology, more safety, and robust ethical and legal frameworks. Thus, high expectations about these latter variables are understandable.

Consider next that only a minority (8%) 'feel able to explain the subject'. This may seem worryingly low and even cast doubts on the value of the survey. It is not, however, and it should not worry us. Take cars with automatic transmission. In 2018, only 3.7% of the vehicles sold by CarMax (the largest used car retailer in the USA) had manual transmission. Cars with automatic transmission are by far the default option in the USA. Yet, arguably, only a very small percentage of drivers may 'feel able to explain' the difference between constantly variable transmission, dual clutch transmission, and simple automatic transmission. Attitudes are usually based on beliefs and experience rather than scientific knowledge. It would be a mistake to conclude that people's attitudes about something they cannot explain are insignificant or unreliable. What matters is that 90% of the people surveyed 'have heard of the technology' and 30% 'know it well'. A general conclusion that emerges from the survey may be summarised by a single word: *variety*. The question about the future of autonomous driving is not *when* or even *where,* but *how* it will take place. It will be a matter of what options, choices, and degrees of autonomous driving are offered to customers. Their needs, preferences, attitudes and circumstances differ. They are best addressed by a flexible variety of alternatives.

19.4 Conclusion

Inbad sci-fi movies, there are only new cars and a handful of models. Reality, however, is greasy and sticky, like a real engine. Public policies and business strategies about autonomous driving will need to make *variety* a feature, not a bug, and concentrate on re-engineering (enveloping) whole environments to make autonomous vehicles an ordinary reality. Hopefully, all this innovation will also ensure that autonomous vehicles will be environmentally more sustainable than the ones we drive today.

Acknowledgements I am very grateful to AUDI and the Audi initiative for the opportunity of collaborating on their study and for our many, informative meetings.

References

AUDI. 2019. *The pulse of autonomous driving-an international user typology and an emotional landscape of autonomous driving*. https://www.audi.com/enlcompany/researchland-audi-initiativelstudy-autonomous-driving.html.

Floridi, L. 2005. The philosophy of presence: From epistemic failure to successful observation. *Presence Teleoperators and Virtual Environments* 14 (6): 656–667.

———. 2014. *The fourth revolution-how the infosphere is reshaping human reality*. Oxford: Oxford University Press.

Floridi, L., J. Cowls, M. Beltrametti, R. Chatila, P. Chazerand, V. Dignum, C. Luetge, R. Madelin, U. Pagallo, F. Rossi, B. Schafer, P. Valcke, and E. Vayena. 2018. AI4People – An ethical framework for a good AI society: Opportunities, risks, principles, and recommendations. *Minds and Machines* 28 (4): 689–707. https://doi.org/10.1007/s11023-018-9482-5.

Milano, Silvia, Mariarosaria Taddeo, and Luciano Floridi. 2019. *Recommender systems and their ethical challenges*. Available at SSRN 3378581.

Chapter 20
Innovating with Confidence: Embedding AI Governance and Fairness in a Financial Services Risk Management Framework

Michelle Seng Ah. Lee, Luciano Floridi, and Alexander Denev

Abstract An increasing number of financial services (FS) companies are adopting solutions driven by artificial intelligence (AI) to gain operational efficiencies, derive strategic insights, and improve customer engagement. However, the rate of adoption has been low, in part due to the apprehension around its complexity and self-learning capability, which makes auditability a challenge in a highly regulated industry. There is limited literature on how FS companies can implement the governance and controls specific to AI-driven solutions. AI auditing cannot be performed in a vacuum; the risks are not confined to the algorithm itself, but rather permeates the entire organization. Using the risk of unfairness as an example, this paper will introduce the overarching governance strategy and control framework to address the practical challenges in mitigating risks AI introduces. With regulatory implications and industry use cases, this framework will enable leaders to innovate with confidence.

Keywords AI · Innovation · Fairness · Ginance · Governance · Risk management

Originally published in Berkeley Technology Law Journal Commentaries (2020)

M. S. A. Lee (✉)
University of Cambridge, Cambridge, UK
e-mail: Michelle.sengah.lee@cl.cam.ac.uk

L. Floridi
Oxford Internet Institute, University of Oxford, Oxford, UK
e-mail: luciano.floridi@oii.ox.ac.uk

A. Denev
Deloitte, London, UK
e-mail: adenev@deloitte.co.uk

L. Floridi (ed.), *Ethics, Governance, and Policies in Artificial Intelligence*,
Philosophical Studies Series 144, https://doi.org/10.1007/978-3-030-81907-1_20

20.1 Introduction

Adoption of AI in the FS sector is still in its infancy, according to a recent survey of more than 3000 C-suite executives conducted by Deloitte and the European Financial Management Association (EFMA). The survey results show that 11% had not started any activities in AI, and 40% were still learning how AI could be deployed in their organizations.[1]

For the purpose of this discussion, we use the term AI generally to refer to the collection of techniques that leverage machine learning to perform tasks that normally require human intelligence, including natural language processing, speech recognition, and decision-making under uncertainty. Traditional approaches to tasks were either a people-based process or a systemic rules-based process. Loans were either granted at the discretion of the bank manager or by using a scorecard to calculate a customer's risk level. The unprecedented availability of affordable computer power and the rise in volume and variety of data gave rise to new and advanced algorithms to analyze more information faster. These AI tools can be static and periodically updated, e.g. a revenue forecasting model that is updated per fiscal quarter, or live and continuously evolving with a real-time feedback cycle, e.g. a chatbot that learns in real-time from the user's input.

Despite the slow adoption rate, FS firms are exploring how to leverage AI to drive cost efficiencies and maintain competitiveness. Most banking executives (65%) see the highest potential impact of AI in customer service, while most insurance executives (78%) view back office and operations as the best part of the value chain for AI use.[2]

FS is a highly regulated industry, comprising a wide variety of complex business lines and products. Given the history of regulatory penalties levied for non-compliance or misconduct in the FS industry and the growing regulatory scrutiny around the use of AI, the conservatism in its adoption is understandable.

In the past year, the U.K. Financial Conduct Authority (FCA) and Information Commissioner's Office (ICO) have been actively issuing opinions on AI and machine learning. While regulators are not proactively designing regulation for AI, they are formulating their expectations with their recent publications on algorithmic trading,[3] supervision of internal models,[4] and Senior Managers and

---

[1] Louise Brett et al., *AI and You: Perceptions of Artificial Intelligence from the EMEA financial services industry*, Deloitte 9 (Apr. 2017), https://www2.deloitte.com/content/dam/Deloitte/cn/Documents/technology/deloitte-cn-tech-ai-and-you-en-170801.pdf [https://perma.cc/R688-FSQS]

[2] *Id.* at 7, 12.

[3] *See Algorithmic Trading Compliance in Wholesale Markets*, Fin. Conduct Authority (Feb. 2018), https://www.fca.org.uk/publication/multi-firm-reviews/algorithmic-trading-compliance-wholesale-markets.pdf [https://perma.cc/WWS2-UERJ] [hereinafter *Algorithmic Trading Compliance*].

[4] *See ECB guide to internal models*, European Cent. Bank (Mar. 2018), https://www.bankingsupervision.europa.eu/legalframework/publiccons/pdf/internal_models/ssm.guidegeneraltopics.en.pdf [https://perma.cc/HV3T-HC6K]

Certification Schemes (SM&CS).[5] Conscious of the lag between the pace at which new technologies evolve and the speed at which new regulations can be developed, regulators have historically adopted the principle of "technological neutrality." Therefore, the same regulatory principles in the aforementioned publications apply to firms regardless of the technology they use to perform a regulated activity. They can also be seen as indicators as to how AI may be regulated in the future.

Past literature on the use of AI has focused on the techniques, tools, and methodologies to ensure the fairness, accountability, and transparency of AI algorithms. However, there has been little effort to contextualize these findings within regulatory limitations, and the connection between the technical frameworks and the governance process of an organization has largely been overlooked. Despite the numerous competing mathematical formalizations of fairness, the practical implications for industry on how to implement fair algorithms are uncertain.

This paper will use the risk of discrimination as an example to discuss the practical FS challenges of managing risks introduced by AI. We will walk through an AI product lifecycle and reveal the process by which risks can be identified, assessed, controlled, and monitored in an FS company by deriving recommended practices and principles from past publications by regulators. While it may refer to external regulations, most examples will be drawn from the European Union and United Kingdom.

20.2 Fairness in the Financial Services Industry

Machine learning is increasingly being used to make or aid decisions that are consequential to FS customers, from evaluating their credit worthiness to recommending investment products to pricing their insurance premiums. It also impacts employees, with CV screening algorithms and performance tracking measures.

Historically, FS companies have focused on limited types of data that directly relate to the desired outcome. For auto insurance, such metrics included past driving convictions and number of years of driving experience.[6] For credit risk, they included debt-to-income ratio and past payment histories.[7] With the advent of big data analytics, firms are beginning to incorporate non-traditional types of data into their algorithms as proxies of risk. Controversially, insurance pricing has been found

---

[5] *See Senior Managers Regime*, Fin. Conduct Authority 3 (Mar. 2019), https://www.fca.org.uk/publication/corporate/applying-smr-to-fca.pdf [https://perma.cc/E95F-FPVE]

[6] *See for example; How Is My Insurance Premium Calculated, Think Insurance,* https://www.thinkinsurance.co.uk/personal/young-driver-insurance/how-is-my-insurance-premium-calculated

[7] Bank of England, *What risks do banks take,* https://www.bankofengland.co.uk/knowledgebank/what-risks-do-banks-take

to be influenced by an applicant's email domain name and surname,[8] and credit lending decisions can depend on an individual's Internet browsing history.[9]

Prior to the availability of big data and machine learning algorithms, companies could avoid liability by showing that any unequal treatment of protected class was unintentional because the protected attributes were not considered in the decision-making process. AI-driven processes are less transparent than traditional systemic rules-based processes due to their ability to extract patterns from complex feature relationships. Recent legal rulings, however, have transferred the emphasis from discriminatory intent to discriminatory impact. The U.S. Supreme Court upheld "disparate impact" claims under the Fair Housing Act in *Texas Department of Housing and Community Affairs v. Inclusive Communities Project, Inc.*[10] The case found unintentional discrimination to be illegal if the plaintiff can show a disproportionate impact on a protected group.[11] In the United Kingdom, *Essop v Home Office* similarly found indirect discrimination to be unlawful in hiring practices.[12]

As discrimination gets embedded in such complex relationships in social data within "black box" algorithms, and as governments increasingly focus on impact rather than intent of discrimination, new approaches to identifying the harm in these automated decision tools are required. Given a bias, people-based processes may arrive at different decisions. AI, by contrast, can replicate an identical bias at-scale, crystalizing the bias and removing the outcome ambiguity associated with human decision-making. This is especially concerning in domain areas with documented historical discrimination, as AI can exacerbate any underlying societal problems and inequalities. Even if AI is designed to augment human decision-making rather than completely replace it, the business users may not comprehend the confidence intervals provided and may not feel comfortable overriding the algorithm in practice, given the complexity of how it reached the decision.

On the other hand, the rulings stipulate that if the accused can prove a legitimate business necessity, this treatment can be deemed lawful; however, the required evidence for this justification is unclear and has not yet been studied. In the United States, the "business necessity clause" states disparate impact can be justified to meet

---

[8] John Leonard, *Admiral Insurance found to give higher quotes to Hotmail users and people called Mohammed*, COMPUTING (Jan. 24, 2018), https://www.computing.co.uk/ctg/news/3025139/admiral-insurance-found-to-give-higher-quotes-to-hotmail-users-and-people-called-mohammed [https://perma.cc/7793-U9SX]

[9] James Rufus Koren, *What does that Web search say about your credit?*, L.A. TIMES (July 17, 2016), https://www.latimes.com/business/la-fi-zestfinance-baidu-20160715-snap-story.html [https://perma.cc/T2M3-WZ5M]

[10] Deborah B. Baum et al., *Supreme Court Affirms FHA Disparate Impact Claims*, PILLSBURY WINTHROP SHAW PITTMAN LLP (July 21, 2015), https://www.pillsburylaw.com/en/news-and-insights/supreme-court-affirms-fha-disparate-impact-claims.html [https://perma.cc/7J85-7AMP]

[11] *Id.*

[12] Tom Lowenthal, *Essop v Home Office: Proving Indirect Discrimination*, OXFORD HUM. RTS. HUB (Apr. 6, 2017), http://ohrh.law.ox.ac.uk/essop-v-home-office-proving-indirect-discrimination [https://perma.cc/VN5K-Q6XP]

performance-related constraints, provided the least possible disparate impact is incurred given the constraints.[13] In the United Kingdom, following the Supreme Court ruling of *Essop v Home Office*, a provision, criterion, or practice (PCP) can be justified by showing it is a "proportionate means of achieving a legitimate aim."[14]

In July 2018, the FCA wrote that while traditionally they have focused on procedural fairness in assessing firms' conduct, there are cases for intervention to ensure distributive fairness in pricing discrimination.[15] The FCA lists six evidential questions to assess whether an intervention is required:

- customer vulnerability;
- scale of adverse effect;
- number of people affected;
- lack of transparency in pricing methodologies;
- essential nature of product or service; and
- societal views of unfairness.[16]

This suggests a step further in the regulators' focus on impact over intent, and organizations will need to shift to an outcome-based analysis of whether their processes are fair.

This paper will use the risk of unlawful discrimination as an example in exploring how an FS company would manage this risk throughout an AI solution's product lifecycle.

20.3 Managing Risks of AI Through Its Lifecycle

Academic research has focused on model and algorithmic risks, such as bias and accuracy, in isolation. In reality, model design and performance must also consider non-model risk domains, such as: regulatory and compliance risk, technology risk, people risk, supplier risk, conduct risk, and market risk.

For example, assessing a model for fairness is not a purely mathematical or computational problem. The appropriate definition, assessment, and remediation has to consider the regulations guiding the use case, the potential technological limitations in its implementation, and alignment with the company's risk appetite, ethics, and core values.

The adoption of AI does not require an overhaul of the existing enterprise Risk Management Framework (RMF), but rather an awareness of how AI may complicate

---

[13] Baum, *supra* note 8.

[14] Lowenthal, *supra* note 10.

[15] Mary Starks et al., *Price discrimination in financial services*, FIN. CONDUCT AUTHORITY 1 (July 2018), https://www.fca.org.uk/publication/research/price_discrimination_in_financial_services.pdf [https://perma.cc/3WK8-LT34]

[16] *Id.* at 6.

the detection of risks as they manifest themselves in unfamiliar ways. The volume and speed of data processed may require a much faster reaction speed for any errors, and the complexity of a machine learning algorithm may hinder its explainability and auditability.

Supervisors will expect firms to have robust and effective governance in place, including RMF, to identify, reduce, and control any of the risks associated with the development and ongoing use of each AI application across the business. The RMF should be approved by the board.[17]

20.4 Design

20.4.1 Definition of Scope

A recent FCA report[18] outlines the requirement for firms to define algorithmic trading, with the objective to ensure that firms establish an appropriate process to identify and manage its usage. The FCA can require firms to provide a description of their algorithmic trading strategies within 14 days.[19] Similarly, FS firms will need to define the scope for what constitutes an AI technology or solution.[20] The difference between AI and rules-based systems, robotic process automation, and static mathematical models, should be clear to both management and employees.[21]

The scope should reflect the regulatory implications around the firm's use of AI.[22] Increasingly, AI solutions in industry leverage third-party machine learning algorithms as accelerators for development. While the build process may have been outsourced, the FS firm is still liable for all associated risks.[23] A retail banking chatbot powered by Natural Language Processing application programming interface (API) provided by a third-party company should still fall under the scope of AI RMF because accountability for any legal or regulatory breach still lies with the firm. The FCA advises, where there is technical outsourcing, "the firm remains fully responsible for its regulatory obligations."[24]

---

[17] Tom Bigham et al., *AI and risk management*, DELOITTE 18 (2018), https://www2.deloitte.com/content/dam/Deloitte/global/Documents/Financial-Services/deloitte-gx-ai-and-risk-management.pdf [https://perma.cc/D3BT-3VP5]

[18] *See* Algorithmic Trading Compliance, *supra* note 3.

[19] *Id.* at 8.

[20] *Id.* at 8–9.

[21] *See* Bigham et al., *supra* note 17.

[22] *Id.*

[23] See Algorithmic Trading Compliance, supra note 3 at 5, 16, 26.

[24] *Id.* at 5.

20.4.2 Risk Identification and Assessment

For firms to identify and assess the impact of AI use cases on their risk appetite, they should first develop a set of clear and consistent assessment criteria to apply to all such cases. The firm should identify relevant risk domains for a solution as well as specific product risks and then assess whether the level of residual risk is acceptable given the existing controls. It is critical that the risk assessment and management process do not constrict creativity. The main objective is to ensure the risks are identified early and properly managed to create a safe setting for innovation. A few of relevant considerations include:

- **External vs. internal:** The intended audience of the AI solution will determine the conduct risk implications, as well as the threshold confidence level and performance the solution is required to reach prior to deployment. For example, an insurance pricing model with a customer user interface has higher risk of unfair outcomes than an income validation model being used by employees.
- **Use of personal information:** Under the General Data Protection Regulation (GDPR) in Europe, Privacy Impact Assessment should be performed if the organization plans to process personal data. An algorithm using personal information for decision-making should be assessed for fairness. In addition, GDPR gives consumers additional rights to understand and take control of how firms are using their personal data. The UK Information Commissioner pointed out that "where a decision has been made by a machine that has a significant impact on an individual, the GDPR requires that they have the right to challenge the decision and a right to have it explained to them."[25] While there have been disagreements among academics on the definition of and the legal basis for this "right to explanation,"[26] firms should nonetheless have a process in place to respond to customers' inquiries in a meaningful, transparent, and understandable manner and be able to demonstrate that an algorithm is compliant with data protection requirements.
- **Data accuracy and quality:** All input and training data into the machine learning model should be high quality and fit for its intended purpose. This includes a review of the data collection methodology for potential selection bias and an evaluation of the distribution of outcomes for possible biases against protected classes.

Societal views of unfairness, aside from being an FCA criterion for intervention, can lead to reputational damage. When an investigation revealed that motorists

---

[25] Science and Technology Committee, *Oral evidence: Algorithms in decision-making, HC 351*, House of Commons (Jan. 23, 2018), http://data.parliament.uk/writtenevidence/committeeevidence.svc/evidencedocument/science-and-technology-committee/algorithms-in-%20decisionmaking/oral/77536.html [https://perma.cc/W4SY-WXYQ]

[26] Sandra Wachter et al., *Why a Right to Explanation of Automated Decision-Making Does Not Exist in the General Data Protection Regulation*, 7 Int'l Data Privacy L. 76, 76–99 (2017).

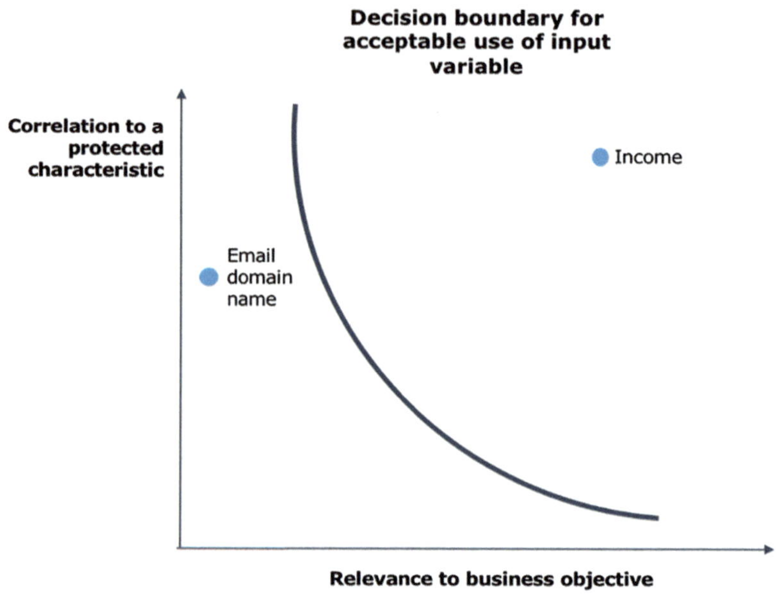

Fig. 20.1 Decision boundary for acceptable use of input variables

named Mohammed are being charged up to £919 more in car insurance than men with typically white, English names, it led to public outcry and calls for a boycott.[27] To avoid such scrutiny, features that are input into the model should be assessed for appropriateness.

Figure 20.1 visualizes a possible decision boundary for whether or not an input variable should be used in a model. Given the decisions in *Essop v Home Office* and *Texas Department of Housing and Community Affairs v. Inclusive Communities Project, Inc.*, even if a feature is correlated to a protected characteristic, there may be reasonable grounds to use it for business objectives. For example, among loan applicants, income levels may differ between men and women because more women work part-time. Income may still be used in a lending decision due to its high relevance to the risk of default. In contrast, an email domain name may be predictive of risk, but if it is highly correlated to race, it may need to be removed from the model due to the lack of foundation of a causal link to risk (Fig. 20.2).

This decision boundary may shift depending on the conditions outlined by the FCA for possible intervention. The drivers of decision-making in providing essential products, such as checking account, car insurance, or mortgage, may be subject to higher scrutiny than the rationale for offering premium credit cards. This is also

[27] Lester Holloway, *Boycott car insurance firms that discriminate*, Operation Black Vote (Jan. 25, 2018), https://www.obv.org.uk/news-blogs/boycott-car-insurance-firms-discriminate [https://perma.cc/9QWW-G7FN]

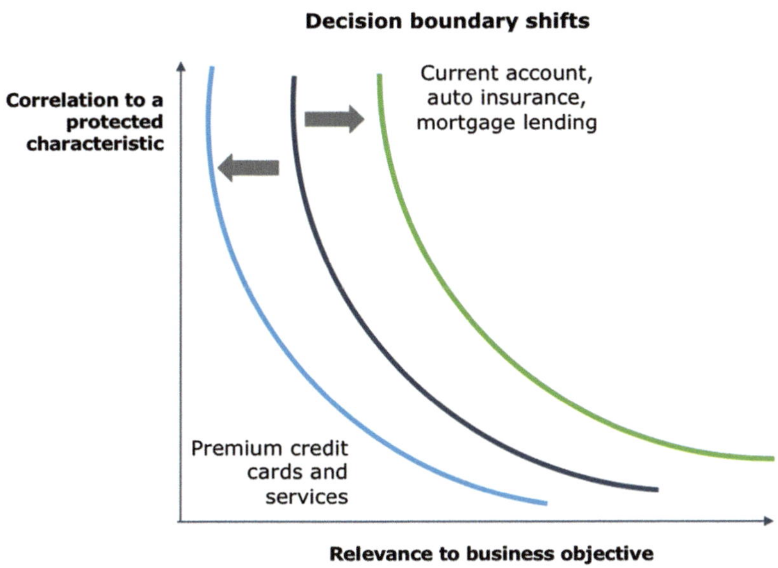

Fig. 20.2 Decision boundary shifts

related to the greater number of people and a higher proportion of vulnerable customers in essential financial products.

This pre-processing step ensures that the decision to include features correlated to protected characteristics is carefully considered within the context of the regulated domain and the potential impact on consumers.

20.4.3 Risk Management Plan and Control Design

Risk management plans should mitigate the risks identified in the assessment, and the residual risk should be in line with the given overall risk appetite. This includes the appropriate controls and testing methodologies, which may vary depending on the FS domain.

Considering the example of the risk of unlawful discrimination, the appropriate control would be to test the algorithm for fairness. Yet, the numerous competing mathematical definitions of fairness only obfuscate its criteria, hindering the ability of business leaders to enforce its implementation. In order to formulate a risk management plan, an appropriate and actionable definition of fairness should be assigned for each use case.

Fairness Through Unawareness This model attempts to avoid discrimination by excluding protected attributes from the model build. Given the power of machine learning algorithms to deduce complex patterns from other features, this does not guarantee a fair outcome. One example of this is the impact of the controversial EU ruling to prohibit car insurance companies from discriminating based on gender in order to counteract the fact that men paid more for insurance than women. Rather than the gap between men and women's insurance premiums narrowing, it has widened from £27 to £101, as insurance companies have turned to gender-correlated proxies for risk measurement, such as occupation and average length of driving history.[28]

While this may be considered more fair if we believe the new prices are reflective of true risk differences between men and women, it is less equitable and would not meet some of the constraints of other definitions of fairness. The model may still be discriminating based on gender through its proxies. In a 2018 study of one million insurance quotes in the United Kingdom, the median price was the highest for laborers (e.g. construction workers) and barbers—stereotypically male jobs—and the lowest for personal assistants and secretaries—stereotypically female jobs.[29]

Defining A as the protected attribute, Y as the actual outcome, and $\widehat{Y}$ as the predicted outcome, other fairness metrics in existing statistics literature include:

Demographic Parity[30] Demographic parity (group fairness) is a population-level metric where the outcome is independent of the protected attribute. Formally:

$$P\left(\widehat{Y}|A=0\right)=P\left(\widehat{Y}|A=1\right)$$

As Gajane and Pechenizkiy argue, this metric is feasible where there is no reliable "ground truth" data, such as in credit risk and employment where historical discrimination against protected groups is well-documented.[31] They are, on the other hand, ineffective where disproportionality in outcomes can be justified by non-protected, non-proxy attributes, as this can lead to reverse discrimination and inaccurate predictions.[32] It is also not stipulated to select the most optimal outcome.[33] In these cases, the tradeoff between accuracy and demographic parity may be too significant for application in business-critical usage, such as pricing. In employment, where there is an additional interest in increasing the diversity of the workforce,

---

[28] Patrick Collinson, *How an EU gender equality ruling widened inequality*, GUARDIAN (Jan. 14, 2017), https://www.theguardian.com/money/blog/2017/jan/14/eu-gender-ruling-car-insurance-inequality-worse [https://perma.cc/6BV8-334Z]

[29] Rebecca Rutt, *How much does your job cost in car insurance*, THIS IS MONEY (Apr. 26, 2018), http://www.thisismoney.co.uk/money/bills/article-5637979/The-jobs-expensive-car-insurance.html [https://perma.cc/YE2F-PJV3]

[30] Nina Grgic-Hlaca et al., *The case for process fairness in learning: Feature selection for fair decision making*, NIPS SYMP. ON MACHINE LEARNING & L. (2016) [https://perma.cc/D5XT-FZJV]

[31] Pratik Gajane & Mykola Pechenizkiy, *On formalizing fairness in prediction with machine learning*, ARXIV (May 28, 2018) https://arxiv.org/pdf/1710.03184.pdf [https://perma.cc/7TPK-HKFM]

[32] *Id.*

[33] *Id.*

demographic parity may be a useful metric to ensure an equitable representation of all protected classes.

Counterfactual Fairness[34] This model posits that given a causal model (U, V, F) with a set of observable variables (V), a set of latent background variables (U) not caused by V, and a set of functions (F), the counterfactual of belonging to a protected class is independent of the outcome. Where X represents the remaining attributes, A represents the binary protected attribute, and Y is the actual outcome, and $\widehat{Y}$ is the predicted outcome, formally:

$$P\left(\widehat{Y}_{A\leftarrow a}(U) = y|X = x, A = a\right) = P\left(\widehat{Y}_{A\leftarrow a}{'}(U) = y|X = x, A = a\right)$$

While the methodology of causal inference is robust, causal links are often difficult to hypothesize in complex FS domains. An FCA report acknowledges the challenge of disentangling the differences in actuarial risk (cost-based pricing) from different willingness to pay (price discrimination).[35] The metric is additionally prone to hindsight bias and outcome bias.[36]

Individual Fairness[37] Individual fairness states that similar individuals get similar outputs. Formally, for similar individuals i and j:

$$\widehat{Y}\left(X^{(i)}, A^{(i)}\right) \approx \widehat{Y}\left(X^{(j)}, A^{(j)}\right)$$

This criterion has a high dependency on the measurement of "similarity" between individuals that does not correlate to the protected characteristics. It is also more computationally intensive than population-level metrics, which could be a challenge for any real-time solutions with Big Data.

Equalized Odds/Equalized Opportunity[38] Equalized odds imply that predicted outcome given actual outcome is independent on predicted protected attribute given actual outcome. This guarantees that the predicted outcome has equal true positive rates across protected characteristics. Equalized opportunity focuses on the true positives: given a positive outcome, the prediction is independent of the protected attribute. Defining A as the protected attribute, Y as the actual outcome, and $\widehat{Y}$ as the predicted outcome, formalization of equalized odds is:

[34] Matt Kusner et al., *Counterfactual Fairness*, ARXIV (Mar. 8, 2018) https://arxiv.org/pdf/1703.06856.pdf [https://perma.cc/82PV-BF6Z]

[35] Starks et al., *supra* note 13.

[36] Gajane & Pechenizkiy, *supra* note 27.

[37] Grgic-Hlaca et al., *supra* note 26.

[38] Moritz Hardt et al., *Equality of Opportunity in Supervised Learning*, ARXIV (Oct. 7, 2016), https://arxiv.org/pdf/1610.02413.pdf [https://perma.cc/3C7W-YESH]

$$P\left(\widehat{Y}=1 \middle| A=0, Y=y\right)=P\left(\widehat{Y}=1 \middle| A=1, Y=y\right), y \epsilon \{0,1\}$$

Similarly, equalized opportunity meets the following condition:

$$P\left(\widehat{Y}=1 \middle| A=0, Y=1\right)=P\left(\widehat{Y}=1 \middle| A=1, Y=1\right)$$

The relative importance of the accuracy metrics can differ across FS use cases. For example, a mortgage lending company may be most concerned about the algorithm's false positive rates (i.e. approving loans that lead to default). A retail bank with an algorithm to predict expected churn may focus on the false negative rates (i.e. was offered a better rate but left anyway). The equalized odds and equalized opportunity metrics fail to address discrimination that may already be embedded in the data.[39]

For any AI with the risk of discrimination against protected classes, appropriate definition of fairness and justification should be required, taking into consideration the strengths and weaknesses of each option and the regulatory implications of its implementation in the FS domain. Once selected, the metric can be used to test the predictions for fairness as a part of the control process. The firm should also bring in stakeholders from legal risk and ethics teams to ensure the definitions are aligned to the companies' ethical values and risk appetite.

20.4.4 Defined Roles and Responsibilities

Given the potentially far-reaching implications of AI use on a business, FS firms may need to involve a wider set of stakeholders from first, second, and third lines of defense throughout the product lifecycle. Under SM&CS, senior management should be prepared to evidence an effective governance and risk framework for AI solutions. Good practice involves senior management's participation throughout the testing and development process and understanding of potential market conduct consequences.

The risk and compliance functions should be involved at each stage of the development, testing, and implementation process. In the publication on algorithmic trading, the FCA particularly noted that compliance staff should aim to have the required knowledge and skills to provide sufficient challenge to the development of algorithms, which may initially involve conducting a gap analysis of their ability to supervise algorithmic trading activity and establishing new roles and responsibilities where required.[40]

---

[39] Gajane & Pechenizkiy, *supra* note 27.

[40] *Algorithmic Trading Compliance*, *supra* note 3.

There should be close collaboration with the technical owner of the AI model and the business owner of the model outcome, with a gradual hand-off of accountability through the lifecycle from design to productionization.

20.5 Build

The AI development process can pose a challenge to traditional risk managers due to the agile approach often adopted by data science and AI teams. It is important to bridge the gap between traditional risk functions and technical teams, as the technical team is not always aware of business and risk challenges. For example, a team may use an open source tool without reviewing whether the license allows for its commercial use. Thus, controls for risks should be embedded into the development process. For example, use of an open source tool should trigger a required process to obtain approval from the legal team to proceed after reviewing the terms of the open source license. Below are analogous regulatory principles for other technologies that apply equally to AI.

20.5.1 Development and Testing Process

By maintaining a robust, consistent, and well-understood development and testing framework, firms need to ensure that their development of algorithms is consistent with the risk appetite and behavioral expectations of the firm. The requirements are similar to those proposed by the FCA for algorithmic trading. Before sign-off, firms need to complete a comprehensive review and approval process, and all stakeholders need to confirm that their assigned tasks are completed, verified, and documented.

20.5.2 Governance and Oversight

Firms should aim to have an independent multi-disciplinary governance committee to review the documentation and completion of testing procedures and to verify that the algorithm is consistent with the original specifications. Its members should be trained to understand the risks associated with AI applications. The issue of fairness, for example, requires both domain knowledge and understanding of the mathematical trade-offs. These committees should establish the testing and assurance process and regularly review the performance to identify emerging issues.

20.5.3 Documented Change Management, Testing, and Approval

Throughout the development and testing process, firms should ensure they have adequate documentation and a comprehensive audit trail for all AI applications deployed throughout their organization, including the relevant owners and key compliance and risk controls in place. Should there be a change in the definition of fairness, for example, this falls under the category of material change and approvals by relevant stakeholders should be recorded.

20.5.4 Transparency and Explainability

An analysis of model drivers should reveal any features that should not be impacting the model. If a person's preferred email address provider is highlighted as a potential driver for insurance pricing, this should feed into the algorithm's fairness analysis to ensure this feature is not being used as a proxy for a protected characteristic. Methodological transparency was explicitly listed by the FCA[41] and the GDPR as a requirement for algorithmic decision-making. As the UK Information Commissioner stated earlier this year, "[w]e may need, as a regulator, to look under the hood or behind the curtain to see what data were used, what training data were used, what factors were programmed into the system, and what question the AI system was trained to answer."[42] GDPR will require a shift in relationships with regulators, requiring appropriately funded regulatory affairs teams to discuss any planned high-risk automated data processing.

20.6 Productionize

Unlike robotic process automation and other rules-based and deterministic systems, risks in AI-driven solutions are dynamic and more challenging to detect. This requires a shift in mindset for risk managers, who will need to remain involved in the risk monitoring process. Prior to productionization, the solution should be safe to deploy at-scale by embedding automated controls.

---

[41] Starks et al., *supra* note 13.

[42] Bigham et al., *supra* note 15.

20.6.1 *Ensuring Solution Is Safe to Scale*

High data processing volume and speed may require a much faster reaction speed for any errors because risk events can propagate much faster. There should be sufficient controls in place prior to go-live, with rules and thresholds programmed for when human intervention is required. The FCA, in its publication on algorithmic trading compliance, mandated a clear explanation of the conditions that need to be met before being implemented into a live environment.[43]

20.6.2 *Review the Feedback Mechanism*

For machine learning algorithms with live incoming data, there should be a control to flag unsuitable input. A chatbot, for example, should not learn from inflammatory or profane user comments. A pricing algorithm should not react erratically to external shocks.

The appropriateness of the feedback loop should also be considered. In a credit risk algorithm, a bank is likely to lack data on the individuals who were denied a loan, even if they proceeded to get a loan elsewhere. The counterfactual of the decision, i.e. whether they would have paid back the loan had they been approved, is unknown. This missingness should be considered when evaluating the accuracy of the model. In the decision boundaries of the model, continuous experimentation to grant credit to those who were just outside the cut-off point can provide the business with evidence on whether the policy is appropriate.

20.6.3 *"Kill Switch" and Business Continuity*

Firms should document procedures and controls for a manual "kill switch" to stop an algorithm from operating once a critical error or abnormal behavior is detected. Business continuity plans may need to be redefined to provide a contingency plan for roll-back to manual processes with minimal disruption to critical business processes.

20.7 Monitor

Due to the continuously-evolving nature of AI, a more dynamic monitoring approach will be required to ensure a model is still performing as intended for its specific use case. The Key Performance Indicators (KPIs), including non-functional

[43] *Algorithmic Trading Compliance*, *supra* note 3.

requirements such as fairness, need to be continuously monitored for appropriateness, relevance, and accuracy. In addition, real-time measures of risk (KRIs) can help inform the second and third lines of defense. An example of this would be the number of complaints and appeals against an AI credit decision on the basis of perceived unfairness.

20.7.1 Automated Monitoring and Testing

AI-driven solutions can be leveraged for AI risk monitoring. For example, a machine learning-driven solution can monitor phone conversations between an insurance agent and a customer to predict the probability of mis-selling. In this tool called TrueVoice, subject matter experts in both insurance and conduct risk have developed and trained custom metrics such as customer vulnerability, dominance, and loss aversion, all of which indicate a higher likelihood of mis-selling.[44]

20.7.2 Vulnerable Customers

Another potential post-processing step may be needed to ensure fairness. If the model results in high variability in outcomes between protected classes, especially if vulnerable customers are involved, an organization may implement a rules-based approach to limit the variation. If a customer is rated as high risk due to the unusual circumstances surrounding his or her vulnerability, some flexibility is required. The FCA defines a vulnerable customer as "someone who, due to their personal circumstances, is especially susceptible to detriment, particularly when a firm is not acting with appropriate levels of care."[45] In particular, the FCA lists "lack of suitable affordable products for people in some non-standard situations" as a potential conduct risk and recommends that "[f]lexibility in the application of terms and conditions of products and services play[] a significant role [to] ensur[e] the needs of consumers in vulnerable circumstances are met."[46] An FS organization may put guardrails in place to limit the level of variability if a customer is deemed to be vulnerable.

[44] *TrueVoice*, Deloitte UK, (2019), https://www2.deloitte.com/uk/en/pages/risk/solutions/truevoice.html [https://perma.cc/QEJ7-EKTV] (last visited Sept 18, 2019).

[45] *Consumer Vulnerability*, Fin. Conduct Authority (Feb. 2015), https://www.fca.org.uk/publication/occasional-papers/occasional-paper-8-exec-summary.pdf [https://perma.cc/GK77-WAVK]

[46] *Id.*

20.7.3 Periodic Re-Validation

External and internal events can result in a change to the organization's risk profile. New legal and regulatory developments may require a change in the design of the model. Media scrutiny of a use case may make a solution non-viable. Legal teams should communicate any changes and their implications to business owners.

20.7.4 Internal Audit Planning

Internal Audit (IA) functions should receive training to acquire adequate expertise to properly understand the risks associated with each AI solution. AI components should be explicitly considered in their audit planning process, independent of the larger systems in which they sit. They should understand and handle compliance breaches and determine the frequency of the review required for each AI solution.

20.8 Conclusion

While adoption rates have been slow, AI will increasingly become an integral component of FS firms' strategies to achieve operational efficiency, improve customer service, and gain insights for competitive advantage. It is imperative that organizations understand the implications of this adoption from a risk perspective, such that appropriate governance and controls are put in place to mitigate the new and exacerbated risks.

This paper explored the practical implications of risk management throughout an AI solution's product lifecycle. With a particular focus on the United Kingdom and the European Union, suggested approaches were coupled with regulatory principles and precedents. The primary highlighted example use case was the risk of discrimination against protected classes. While there has been a wide array of studies on the technical and theoretical definitions of fairness, further work is required to devise a framework to determine which definitions are most appropriate in the practical implementation of fairness metrics in FS industry.

Risks of AI are not confined to the algorithm itself, but rather affect the entire organization. AI-specific considerations should be integrated into existing RMFs to ensure they remain fit for purpose. Only then will FS firms feel empowered to use AI, having the confidence that AI-related risks can be effectively identified and managed.

20.9 Funding

This research was partially supported by Deloitte (MSAL and AD); and by Privacy and Trust Stream – Social lead of the PETRAS Internet of Things research hub – PETRAS is funded by the Engineering and Physical Sciences Research Council (EPSRC), grant agreement no. EP/N023013/1, and Google UK Limited (LF).

Bibliography

Baum, Deborah B., Julia E. Judish, David J. Stute, and John Scalia. *Supreme court affirms FHA disparate impact claims*. Pillsbury Winthrop Shaw Pittman LLP, July 21, 2015. https://www.pillsburylaw.com/en/news-and-insights/supreme-court-affirms-fha-disparate-impact-claims.html (last visited Jun 12, 2019).

Bigham, Tom, Valeria Gallo, Suchitra Nair, Michelle Seng Ah Lee, Sulabh Soral, Tom Mews, Alan Tua, and Morgane Fouche. 2018. *AI and risk management*. https://www2.deloitte.com/content/dam/Deloitte/global/Documents/Financial-Services/deloitte-gx-ai-and-risk-management.pdf (last visited Jun 12, 2019).

Brett, Louise, Patrick Laurent, Paolo Gianturco, and Tiago Pereira Durao. 2017. *AI and You: Perceptions of Artificial Intelligence from the EMEA financial services industry*. https://www2.deloitte.com/content/dam/Deloitte/cn/Documents/technology/deloitte-cn-tech-ai-and-you-en-170801.pdf (last visited Jun 12, 2019).

Collinson, Patrick. How an EU gender equality ruling widened inequality. *The Guardian*, January 14, 2017. https://www.theguardian.com/money/blog/2017/jan/14/eu-gender-ruling-car-insurance-inequality-worse (last visited Jun 12, 2019).

Deloitte UK. 2019. *TrueVoice*. https://www2.deloitte.com/uk/en/pages/risk/solutions/truevoice.html (last visited Jun 12, 2019).

European Central Bank. 2018. *ECB guide to internal models*. https://www.bankingsupervision.europa.eu/legalframework/publiccons/pdf/internal_models/ssm.guidegeneraltopics.en.pdf (last visited Jun 12, 2019).

Financial Conduct Authority. 2015. *Consumer vulnerability*. https://www.fca.org.uk/publication/occasional-papers/occasional-paper-8-exec-summary.pdf (last visited Jun 12, 2019).

———. 2018a. *Algorithmic trading compliance in wholesale markets*. https://www.fca.org.uk/publication/multi-firm-reviews/algorithmic-trading-compliance-wholesale-markets.pdf (last visited Jun 12, 2019).

———. 2018b. *Price discrimination in financial services*. https://www.fca.org.uk/publication/research/price_discrimination_in_financial_services.pdf (last visited Jun 12, 2019).

———. 2019. *Senior managers regime*. https://www.fca.org.uk/publication/corporate/applying-smr-to-fca.pdf (last visited Jun 12, 2019).

Gajane, Pratik, and Mykola Pechenizkiy. 2017. *On formalizing fairness in prediction with machine learning*. arXiv preprint arXiv:1710.03184.

Grgic-Hlaca, Nina, Muhammad Bilal Zafar, Krishna Gummadi, and Adrian Weller. 2016. The case for process fairness in learning: Feature selection for fair decision making. In *NIPS symposium on machine learning and the law*.

Hardt, Moritz, Eric Price, and Nathan Srebro. 2016. Equality of opportunity in supervised learning. *CoRR* arXiv:1610.02413.

Holloway, Lester. *Boycott car insurance firms that discriminate*. Operation Black Vote, January 25, 2018. https://www.obv.org.uk/news-blogs/boycott-car-insurance-firms-discriminate (last visited Jun 12, 2019).

Koren, James Rufus. What does that Web search say about your credit. *Los Angeles Times*, July 17, 2016., https://www.latimes.com/business/la-fi-zestfinance-baidu-20160715-snap-story.html (last visited Jun 12, 2019).

Kusner, Matt, Joshua Loftus, Chris Russell, and Ricardo Silva. 2017. *Counterfactual fairness*. arXiv:1703.06856v2.

Leonard, John. 2018. *Admiral Insurance found to give higher quotes to Hotmail users and people called Mohammed*. https://www.computing.co.uk/ctg/news/3025139/admiral-insurance-found-to-give-higher-quotes-to-hotmail-users-and-people-called-mohammed (last visited Jun 12, 2019).

Lowenthal, Tom. *Essop v home office: Proving indirect discrimination*. Oxford Human Rights Hub, April 6, 2017. http://ohrh.law.ox.ac.uk/essop-v-home-office-proving-indirect-discrimination (last visited Jun 12, 2019).

Rutt, Rebecca. *How much does your job cost in car insurance*. This is Money, April 26, 2018. http://www.thisismoney.co.uk/money/bills/article-5637979/The-jobs-expensive-car-insurance.html (last visited Jun 12, 2019).

Science and Technology Committee. *Oral evidence: Algorithms in decision-making, HC 351*. January 23, 2018. http://data.parliament.uk/writtenevidence/committeeevidence.svc/evidencedocument/science-and-technology-committee/algorithms-in-%20decisionmaking/oral/77536.html (last visited Jun 12, 2019).

Wachter, Sandra, Brent Mittelstadt, and Luciano Floridi. 2017. Why a right to explanation of automated decision-making does not exist in the general data protection regulation. *International Data Privacy Law* 7: 76–99.

Chapter 21
Robots, Jobs, Taxes, and Responsibilities

Luciano Floridi

Abstract In this chapter I argue that the point is not to decide whether robots will qualify someday as a kind of persons, but to realise that we are stuck within the wrong conceptual framework. The digital is forcing us to rethink new solutions for new forms of agency. We must keep in mind that the debate is not about robots but about us, who will have to live with them, and about the kind of infosphere and societies we want to create.

Keywords Agency · Autonomy · Artificial Intelligence · Digital Ethics · Robots

AI and robots continue to make news. Alarmist headlines used to be about some kind of Terminator developing in the future to dominate and enslave us, like an inferior species. They are now about tireless machines that, like enslaved persons, will make us redundant, replacing and outperforming us more efficiently and cheaply than we can ever be. This master-slave dialectics is not science fiction. On the 16th of February 2017, the plenary session of the European Parliament voted in favour[1] of a resolution[2] to create a new ethical-legal framework according to which robots may qualify as "electronic persons". The Commission does not have to follow the Parliament's recommendations but, if it refuses, it will have to explain why. The following day, on the 17th of February, in an interview with *Quartz*,[3] Bill Gates, Microsoft co-founder, suggested that there should be a tax on robots.[4]

[1] http://www.europarl.europa.eu/news/en/news-room/20170210IPR61808/robots-and-artificial-intelligence-meps-call-for-eu-wide-liability-rules

[2] http://www.europarl.europa.eu/sides/getDoc.do?type=REPORT&reference=A8-2017-0005&language=EN

[3] https://qz.com/911968/bill-gates-the-robot-that-takes-your-job-should-pay-taxes/

[4] https://www.ft.com/content/d04a89c2-f6c8-11e6-9516-2d969e0d3b65

L. Floridi (✉)
Oxford Internet Institute, University of Oxford, Oxford, UK
e-mail: luciano.floridi@oii.ox.ac.uk

L. Floridi (ed.), *Ethics, Governance, and Policies in Artificial Intelligence*, Philosophical Studies Series 144, https://doi.org/10.1007/978-3-030-81907-1_21

Regulating robots is a very reasonable idea. Today, we live *onlife,* spending increasing amount of time inside the infosphere. In this digital ocean, robots are the real natives: we scuba dive, they are like fish. So robots of all kinds are going to multiply and proliferate, making the infosphere even more their own space. These smart, autonomous, and social agents perform an increasing number of tasks better than we can. Some of them are already among us. Others are discernible on the horizon, while later generations are still unforeseeable. The solutions that have already arrived come in soft forms, such as apps, webbots, algorithms, and software of all kinds; and hard forms, such as robots, house appliances, personal assistants, smart watches, and other gadgets. In health care, for example, robots and AI solutions are joining nurses, doctors, social workers, technicians, and experts, such as radiologists, by helping perform functions that, just a few years ago, were considered off-limits for technological disruption: cataloguing images, suggesting diagnosis, monitoring and even moving patients, interpreting radiographies, controlling insulin pumps, extracting new medical information from huge data sets, and so forth. Many trivial, routine tasks will be performed automatically either by AI or by people aided by AI. This is good news. We need AI to deal with increasing levels of complexity and difficulty. With an analogy, we need to remember that the best chess player is neither a human nor a computer, but a human using a computer.

While we can only guess at the scale of the coming disruption, everybody expects it to be profound. Any job in which people serve as menial interfaces – e.g. adjusting the dose of a medication for a patient-is now at risk. Yet new jobs will appear because we will need to manage and coordinate AI solutions. For example, someone will need to ensure that the data collected by insulin pumps and by smart apps are properly combined in order to improve the health care provided and the technologies of the future. What is more, many tasks will not be cost-effective for AI applications. The world never changes at the same pace. In some places, nurses will be irreplaceable for many routine tasks while in others they may coordinate and direct semi-autonomous robots through smart tablets and apps. And some old jobs will survive, even when a machine is doing most of the work: a doctor who delegates some routine tasks to a smart digital assistant will simply have more time to focus on other things, such as prevention. Jobs that were economically not viable until yesterday will become avail- able. Finally, other tasks will be delegated back to us-the patients-to perform them as users, such as testing for blood pressure, something trivial and routine in many countries but still impossible in others.

Another source of uncertainty concerns the point at which AI will no longer be controlled only by a guild of scientists, technicians, and managers. Still relying on the health care example, what will happen when AI becomes "democratised" and a "digital doctor" is available to millions of people on their smartphones or some other device? As Elena Bonfiglioli and Mathias Ekman recently wrote[5]: "As you think

[5]Bonfiglioli Elena and Ekman Mathias. 2016. "Innovation race for improved cancer outcomes". https://enterprise.microsoft.com/en-us/industries/health/innovation-race-for-improved-cancer-outcomes/

innovation in health, you want to think about how to scale the adoption of systems of intelligence making them accessible in more intuitive ways. The vision of AI as "conversations" will empower intelligent health experiences that mirror the way people collaborate and interact with one another, and the way machines proactively understand our intent. [...] Systems of intelligence will endemically transform the way we innovate for improved cancer outcomes, the way we optimise clinical and operational processes, and the way we think and do prevention. So, what if people across the healthcare continuum could collaborate and use machine learning to come up with ways to catch cancer earlier and improve outcomes for patients?".

We should investigate how we are going to socialise such systems of intelligence and how we shall best adopt them and adapt to them, from an ethical perspective, because many solutions are far from inevitable, and some may be preferable to others and should be privileged. There is no dystopian science-fiction scenario. *Brave New World* is not coming to life, and the Terminator is not lurking just beyond the horizon, either. There is a good chance that Satya Nadella, Microsoft CEO, may be right when he remarked: "humans and machines will work together - not against one another. Computers may win at games, but imagine what's possible when human and machine work together to solve society's greatest challenges like beating disease, ignorance, and poverty."[6] But there are of course risks and challenges in how we shall develop and socialise AI systems and we should tackle them now, to ensure that individual and social benefits are maximised. Quoting Nadella once more: "The most critical next step in our pursuit of A.I. is to agree on an ethical and empathic framework for its design". Add machine learning to artificial intelligence and robotics, mix these ingredients with the Internet, the Web, smart phones and apps, cloud computing, big data, and the Internet of Things, and it becomes obvious that there is no time to waste. We are laying down the foundations of the mature information societies of the near future, so we need new ethical solutions for the infosphere, to determine which forms of artificial agency and interactions we like to see flourishing in it. Against this background, one can look at the normative initiative taken by the European Parliament or the debate that has followed Gate's suggestion with a mixed sense of excitement for the aspiration but disappointment for the implementation. For there is too much fantasy about speculative scenarios and too little imagination in designing realistic solutions that could work well. Consider two key issues: *jobs* and *responsibilities.*

Robots replace human workers. Retraining unemployed people was never easy, but it is more challenging now that technological disruption is spreading so rapidly, widely, and unpredictably. Today, a bus driver replaced by a driverless bus is unlikely to become a web master, not least because even that job is at risk of automation. There will be many new forms of employment in other comers of the infosphere. Think of how many people have opened virtual shops on eBay. But these will require new and different skills. So more education and a universal basic income

[6]Nadella Satya. 2016. "The Partnership of the Future", *Slate,* http://www.slate.com/blogs/future_tense/20171011/did_a_federal_surveillance_court_really_reject_an_application_to_monitor.html

may be needed to mitigate the impact of robotics on the labour market, while ensuring a more equitable redistribution of its economic benefits. This means that society will need more resources. Unfortunately, robots do not pay taxes. And it is unlikely that more profitable companies may pay enough more taxes to compensate for the loss of revenues. So robots cause a higher demand for taxpayers' money but also a lower supply of it. The problem is exacerbated by the fact that people with low income purchase cheap goods, those produced more efficiently by increasingly roboticised processes. How can one get out of this tailspin? The report correctly identifies the problem . But its original recommendation [7] of a robotax on companies that employ robots may be unfeasible-for what exactly counts as a robot, if you need to pay a tax on it?-and counterproductive, for a robotax would disincentive innovation. The final text[8] approved by the European Parliament shuns the recommendation but does not offer an alternative solution to the revenue problem.

Consider next the allocation of responsibilities. If a robot breaks the window of my neighbour, who is responsible? The company who produced it, the shop who sold it, I the owner, or the robot itself, if the robot has become completely autonomous through a learning process and is now capable of intelligent-looking actions? In this case too, the report identifies the issue. It rightly recommends forms of risk management (insurance and compensation). But it also suggests the creation of a "specific legal status" for more advanced robots, as "electronic persons responsible for making good any damage they may cause". This has been approved in the final document. As a result, we may see a future in which companies do not pay a robotax and are not even liable for some kinds of robots. This is probably a mistake. There is no need to adopt science fiction solutions to solve practical problems of legal liability with which jurisprudence has been dealing successfully for a long time. If robots become one day as good as human agents-think of Droids in Star Wars-we may adapt rules as old as Roman law, according to which the owner of an enslaved person was responsible for any damage caused by that person *(respondeat superior).* As the Romans already knew, attributing some kind of legal personality to robots would deresponsabilise those who should control them. Not to speak of the counterintuitive attribution of rights. For example, do robots as "electronic persons" have the right to own the data they produce (machine- generate data)? Should they be "liberated"? It may be fun to speculate about such questions, but it is also distracting and irresponsible, given the pressing issues we have at hand. The point is not to decide whether robots will qualify some day as a kind of persons, but to realise that we are stuck within the wrong conceptual framework. The digital is forcing us to rethink new solutions for new forms of agency. While doing so we must keep in mind that the debate is not about robots but about us, who will have to live with them, and about

[7] http://www.europarl.europa.eu/sides/getDoc.do?type=REPORT&mode=XML&reference=A8-2017-0005&language=EN

[8] http://www.europarl.europa.eu/sides/getDoc.do?pubRef=-//EP//NONSGML+TA+P8-TA-2017-0051+O+DOC+PDF+VO//EN

the kind of infosphere and societies we want to create. We need less science fiction and more philosophy. [9]

References

Floridi, L., ed. 2008. *Philosophy of computing and information: 5 questions*. Automatic Press NIP.

———. 2012. *The road to the philosophy of information, Luciano Floridi's philosophy of technology*, 245–271. New York: Springer.

———. 2013. Distributed morality in an information society. *Science and Engineering Ethics* 19 (3): 727–743.

Pagallo, U. 2013. *The laws of robots: crimes, contracts, and torts*. Dordrecht: Springer.

[9]On the debate about a normative framework for robotics see (Pagallo 2013). In terms of philosophy of information see (Floridi 2008, 2012, 2013).

Chapter 22
What the Near Future of Artificial Intelligence Could Be

Luciano Floridi

Abstract In this chapter, I look into the possible developments of Artificial Intelligence (AI) in the near future and identify two likely trends: (a) a shift from historical to synthetic data; and (b) a translation of difficult tasks (in terms of abilities) into complex ones (in terms of computation). I then argue that (a) and (b) will be pursued as development strategies of AI solutions whenever and as far as they are feasible.

Keywords Artificial intelligence · Complexity · Foresight analysis · Synthetic data · Digital innovation

22.1 Introduction

Artificial intelligence (Al) has dominated recent headlines, with its promises, challenges, risks, successes, and failures. What is its foreseeable future? Of course, the most accurate forecasts are made with hindsight. But if some cheating is not acceptable, then smart people bet on the uncontroversial or the untestable. On the uncontroversial side, one may mention the increased pressure that will come from law-makers to ensure that AI applications align with socially acceptable expectations. For example, everybody expects some regulatory move from the EU, sooner or later. On the untestable side, some people will keep selling catastrophic forecasts, with dystopian scenarios taking place in some future that is sufficiently distant to ensure that the Jeremiahs will not be around to be proven wrong. Fear always sells

L. Floridi (✉)
Oxford Internet Institute, University of Oxford, Oxford, UK
e-mail: luciano.floridi@oii.ox.ac.uk

L. Floridi (ed.), *Ethics, Governance, and Policies in Artificial Intelligence*,
Philosophical Studies Series 144, https://doi.org/10.1007/978-3-030-81907-1_22

well, like vampire or zombie movies. Expect more. What is difficult, and may be quite embarrassing later on, is to try to "look into the seeds of time, and say which grain will grow and which will not" *(Macbeth,* Act I, Scene III), that is, to try to understand where AI is more likely to go and hence where it may not be going. This is what I will attempt to do in the following pages, where I shall be cautious in identifying the paths of least resistance, but not so cautious as to avoid any risk of being proven wrong.

Part of the difficulty is to get the level of abstraction right (Floridi 2008a, b), i.e. to identify the set of relevant observables ("the seeds of time") on which to focus because those are the ones that will make the real, significant difference. In our case, I shall argue that the best observables are provided by an analysis of the *nature of the data* used by AI to achieve its performance, and of the *nature of the problems* that AI may be expected to solve.[1] So, my forecast will be divided into two, complementary parts. In Sect. 22.2, I will discuss the nature of the data needed by AI; and in Sect. 22.3, I will discuss the scope of the problems AI is more likely to tackle successfully. I will conclude with some more general remarks about tackling the related ethical challenges. But first, let me be clear about what I mean by AI.

22.2 AI: A Working Definition

AI has been defined in many ways. Today, it comprises several techno-scientific branches, well summarised in Corea (Aug. 29, Corea 2018) in Fig. 22.1.

Altogether, AI paradigms still satisfy the classic definition provided by John McCarthy, Marvin Minsky, Nathaniel Rochester, and Claude Shannon in their seminal "Proposal for the Dartmouth Summer Research Project on Artificial Intelligence", the founding document and later event that established the new field of AI in 1955:

> For the present purpose the artificial intelligence problem is taken to be that of making a machine behave in ways that would be called intelligent if a human were so behaving. (Quotation from the 2006 re-issue in McCarthy et al. 2006)

As I have argued before (Floridi 2017), this is obviously a counterfactual: *were* a human to behave in that way, that behaviour *would* be called intelligent. It does not mean that the machine is intelligent or even *thinking*. The latter scenario is a fallacy and smacks of superstition. Just because a dishwasher cleans the dishes as well as, or even better than I do, it does not mean that it cleans them *like* I do, or needs any intelligence in achieving its task. The same counterfactual understanding of AI

---

[1] For a reassuringly converging review based not on the nature of data or the nature of problems, but rather on the nature of technological solutions, based on a large scale review of the forthcoming literature on AI, see "We analysed 16,625 papers to figure out where AI is headed next" https://www.technologyreview.com/s/612768/we-analyzed-I6625-papers-to-figure-out-where-ai-is-headed-next/

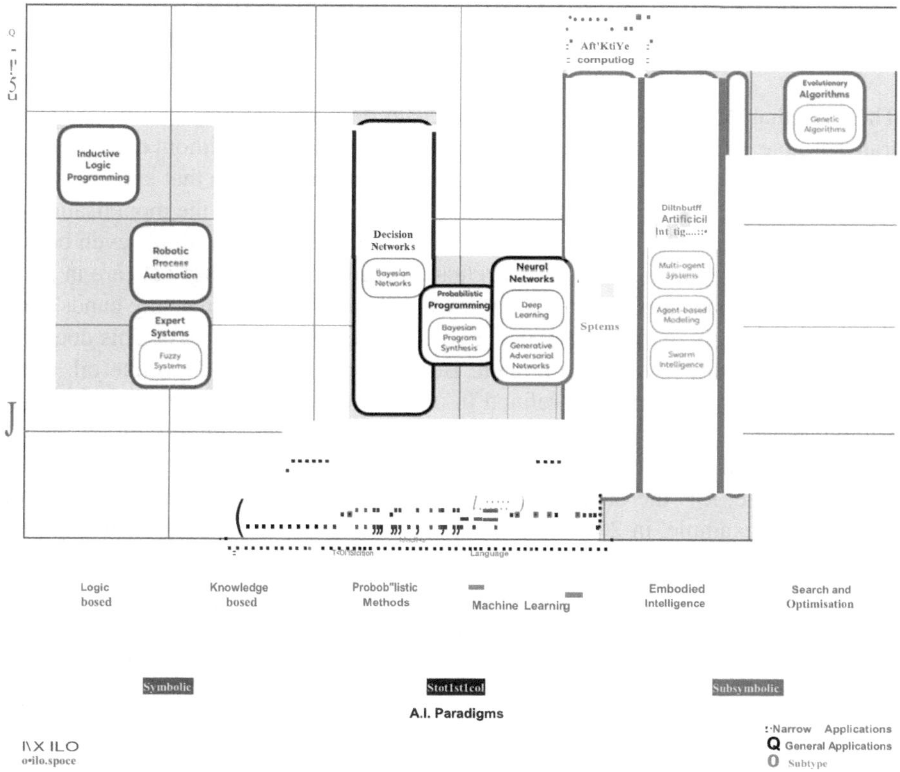

Fig. 22.1 AI knowledge map (AIKM). (Source: Corea Aug. 29, Corea 2018)

underpins the Turing test (Floridi et al. 2009), which, in this case, checks the ability of a machine to perform a task in such a way that the *outcome* would be indistinguishable from the outcome of a human agent working to achieve the same task (Turing 1950).

The classic definition enables one to conceptualise AI as a growing resource of interactive, autonomous, and often self-learning (in the machine learning sense, see Fig. 22.1) *agency,* that can deal with tasks that would otherwise require human intelligence and intervention to be performed successfully. This is part of the ethical challenge posed by Al, because artificial agents are

> sufficiently informed, 'smart', autonomous and able to perform morally relevant actions independently of the humans who created them [. . .]. (Floridi and Sanders 2004)

Although this aspect is important, it is not a topic for this article, and I shall return to it briefly only in the conclusion.

In short, AI is defined on the basis of outcomes and actions and so, in what follows, I shall treat AI as a *reservoir of smart agency on tap.* The question I wish to address is: what are the foreseeable ways in which such a technology will evolve and be used successfully? Let us start from the data it needs.

22.3 AI's Future: From Historical Data to Hybrid and Synthetic Data, and the Need for Ludification

They say that data are the new oil. Maybe. But data are durable, reusable, quickly transportable, easily duplicable, and simultaneously shareable without end, while oil has none of these properties. We have gigantic quantities of data that keep growing, but oil is a finite resource. Oil comes with a clear price, whereas the monetisation of the same data depends on who is using them and for what. And all this even before introducing the legal and ethical issues that emerge when *personal* data are in play, or the whole debate about ownership ("my data" is much more like "my hands" and much less like "my oil"). So, the analogy is a stretch, to say the least. This does not mean that is entirely worthless though. Because it is true that data, like oil, are a valuable resource and must be refined in order to extract their value. In particular, without data, algorithms-AI included-go nowhere, like an engine with an empty tank. AI needs data to *train,* and then data to *apply* its training. Of course, AI can be hugely flexible; it is the data that determine its scope of application and degree of success. For example, in 2016, Google used DeepMind's machine learning system to reduce its energy consumption:

> Because the algorithm is a general-purpose framework to understand complex dynamics, we plan to apply this to other challenges in the data centre environment and beyond in the coming months. Possible applications of this technology include improving power plant conversion efficiency (getting more energy from the same unit of input), reducing semiconductor manufacturing energy and water usage, or helping manufacturing facilities increase throughput.[2]

It is well known that AI learns from the data it is fed and progressively improves its results. If you show an immense number of photos of dogs to a neural network, in the end, it will learn to recognise dogs increasingly well, including dogs it never saw before. To do this, usually, one needs huge quantities of data, and it is often the case that the more the better. For example, in recent tests, a team of researchers from the University of California in San Diego trained an AI system on 101.6 million electronic health record (EHR) data points (including text written by doctors and laboratory test results) from 1,362,559 paediatric patient visits at a major medical centre in Guang-zhou, China. Once trained, the AI system was able to demonstrate:

> [. . .] high diagnostic accuracy across multiple organ systems and is comparable to experienced pediatricians in diagnosing common childhood diseases. Our study provides a proof of concept for implementing an AI-based system as a means to aid physicians in tackling large amounts of data, augmenting diagnostic evaluations, and to provide clinical decision support in cases of diagnostic uncertainty or complexity. Although this impact may be most evident in areas where healthcare providers are in relative shortage, the benefits of such an AI system are likely to be universal. (Liang et al. 2019)

However, in recent times, AI has improved so much that, in some cases, we are moving from an emphasis on the *quantity* of large masses of data, sometimes

[2] https://deepmind.com/blog/deepmind-ai-reduces-google-data-centre-cooling-bill-40/

improperly called Big Data (Floridi 2012), to an emphasis on the *quality* of data sets that are well curated. For example, in 2018, DeepMind, in partnership with Moorfields Eye Hospital in London, UK, trained an AI system to identify evidence of sight-threatening eye diseases using optical coherence tomography (OCT) data, an imaging technique that generates 3D images of the back of the eye. In the end, the team managed to

> demonstrate performance in making a referral recommendation that reaches or exceeds that of experts on a range of sight-threatening retinal diseases after training on *only 14,884 scans* [my italics]. (De et al. 2018, p. 1342)

I emphasise "only 14,884 scans" because "small data" of high quality is one of the futures of AL AI will have a higher chance of success whenever well-curated, updated, and fully reliable data sets become available and accessible to train a system in a specific area of application. This is quite obvious and hardly a new forecast. But it is a solid step forward, which helps us look further ahead, beyond the "Big Data" narrative. If *quality* matters, then *provenance* is crucial. Where do the data come from? In the previous example, they were provided by the hospital. Such data are sometimes known as *historical, authentic,* or *real-life* (henceforth I shall call them simply historical). But we also know that AI can generate its own data. I am not talking about *metadata* or *secondary data* about its uses (Floridi 2010). I am talking about its primary input. I shall call such *entirely* AI-generated data *synthetic.* Unfortunately, the term has an ambiguous etymology. It began to be used in the 1990s to refer to historical data that had been anonymised before being used, often to protect privacy and confidentiality. These data are synthetic only in the sense that they have been *synthesised* from historical data, e.g. through "masking".[3] They have a lower resolution, but their genesis is not an artificial source. The distinction between the historical data and those synthesised from them is useful, but this is not what I mean here, where I wish to stress the completely and exclusively *artificial provenance* of the data in question. It is an ontological distinction, which may have significant implications in terms of epistemology, especially when it comes to our ability to explain the synthetic data produced, and the training achieved by the AI using them (Watson et al. Forthcoming). A famous example can help explain the difference.

In the past, playing chess against a computer meant playing against the best human players who had ever played the game. One of the features of Deep Blue, the IBM's chess program that defeated the world champion Garry Kasparov, was

> an effective use of a Grandmaster game database. (Campbell et al. 2002, p. 57)

But AlphaZero, the last version of the AI system developed by DeepMind, learnt to play better than anyone else, and indeed any other software, by relying only on the *rules* of the game, with no data input at all from any external source. It had no historical memory whatsoever:

---

[3]https://www.tcs.com/blogs/the-masking-vs-synthetic-data-debate

> The game of chess represented the pinnacle of artificial intelligence research over several decades. State-of-the-art programs are based on powerful engines that search many millions of positions, *leveraging handcrafted domain expertise and sophisticated domain adaptations.* [my italics, these are the non-synthetic data]. AlphaZero is a generic reinforcement learning and search algorithm – originally devised for the game of Go-that achieved superior results within a few hours [...] *given no domain knowledge except the rules of chess* [my italics]. (Silver et al. 2018, p. 1144)

AlphaZero learnt by playing against itself, thus generating its own chess-related, synthetic data. Unsurprisingly, Chess Grandmaster Matthew Sadler and Women's International Master Natasha Regan,

> who have analysed thousands of AlphaZero's chess games for their forthcoming book Game Changer (New in Chess, January 2019), say its style is unlike any traditional chess engine. "It's like discovering the secret notebooks of some great player from the past," says Matthew.[4]

Truly synthetic data, as I am defining them here, have some wonderful properties. Not only do they share those listed at the beginning of this section (durable, reusable, quickly transportable, easily duplicable, simultaneously shareable without end, etc.). They are also clean and reliable (in terms of curation), they infringe no privacy or confidentiality at the *development* stage (though problems persist at the *deployment* stage, because of the predictive privacy harms (Crawford and Schultz 2014)), they are not immediately sensitive (sensitivity during the deployment stage still matters), if they are lost, it is not a disaster because they can be recreated, and they are perfectly formatted to be used by the system that generates them. With synthetic data, AI never has to leave its digital space, where it can exercise complete control on any input and output of its processes. Put more epistemologically, with synthetic data, AI enjoys the privileged position of a maker's knowledge, who knows the intrinsic nature and working of something because it made that something (Floridi 2018). This explains why they are so popular in security contexts, for example, where AI is deployed to stress-test digital systems. And sometimes synthetic data can also be produced more quickly and cheaply than historical data. AlphaZero became the best chess player on earth in 9 hours (it took 12 hours for shogi, and 13 days for Go).

Between historical data that are more or less masked (impoverished through lower resolution, e.g. through anonymisation) and purely synthetic data, there is a variety of more or less *hybrid* data, which you can imagine as the offspring of historical and synthetic data. The basic idea is to use historical data to obtain some new synthetic data that are not merely impoverished historical data. A good example is provided by Generative Adversarial Networks (GANs), introduced by Goodfellow et al. (2014):

> Two neural networks-a Generator and a Discriminator [my capitals in the whole text]-compete against each other to succeed in a game. The object of the game is for the Generator to fool the Discriminator with examples that look similar to the training set. [...] When the

[4]https://deepmind.com/blog/alphazero-shedding-new-light-grand-games-chess-shogi-and-go/

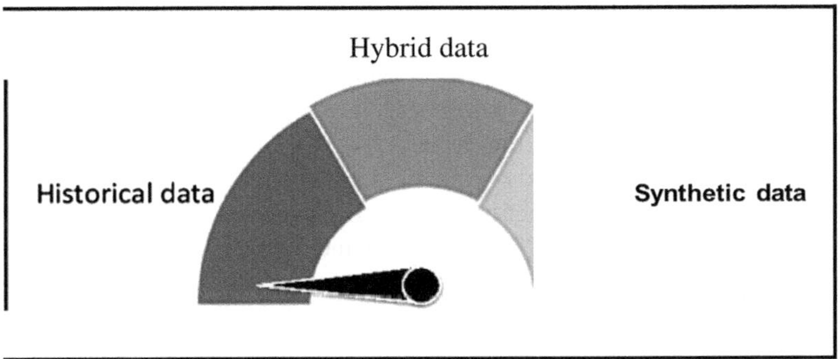

Fig. 22.2 Shifting from entirely historical to truly synthetic data

> Discriminator rejects an example produced by the Generator, the Generator learns a little more about what the good example looks like. [. . .] In other words, the Discriminator leaks information about just how close the Generator was and how it should proceed to get closer. [. . .] As time goes by, the Discriminator learns from the training set and sends more and more meaningful signals back to the Generator. As this occurs, the Generator gets closer and closer to learning what the examples from the training set look like. *Once again, the only inputs the Generator has are an initial probability distribution (often the normal distribution) and the indicator it gets back from the Discriminator. It never sees any real examples* [my italics].[5]

The Generator learns to create synthetic data that are like some known input data. So, there is a bit of a hybrid nature here, because the Discriminator needs to have access to the historical data to "train" the Generator. But the data generated by the Generator are new, not merely an abstraction from the training data. So, not a case of parthenogenesis, like AlphaZero giving birth to its own data, but close enough to deliver some of the very appealing features of synthetic data nevertheless. For example, synthetic human faces created by a Generator pose no problems in terms of privacy, consent, or confidentiality at the development stage.[6]

Many methods to generate hybrid or synthetic data are already available or being developed, often sector specific. There are also altruistic trends to make such data sets publicly and freely available (Howe et al. 2017). Clearly, the future of AI lies not just in "small data" but also, or perhaps mainly, in its increasing ability to generate its own data. That would be a remarkable development, and one may expect significant efforts to be made in that direction. The next question is: what factors can make the dial in Fig. 22.2 move from left to right?

The difference is made by the genetic process, i.e. by the rules used to create the data. *Historical data* are obtained by *recording rules,* as they are the outcome of some observation of a system behaviour. *Synthesised data* are obtained by *abstracting rules* that eliminate, mask or obfuscate some degrees of resolution

[5]https://securityintelligence.com/generative-adversarial-networks-and-cybersecurity-part-1/

[6]https://motherboard.vice.com/en_us/article/7xn4wy/this-website-uses-ai-to-generate-the-faces-of-people-who-dont-exist

from the historical data, e.g. through anonymisation. *Hybrid* and truly *synthetic data* can be generated by *constraining rules* or *constitutive rules*. There is no one-to-one mapping, but it is useful to consider hybrid data as the data on which we have to rely, using constraining rules, when we do not have constitutive rules that can generate synthetic data from scratch. Let me explain.

The dial moves easily towards synthetic data whenever AI deals with "games"-understood as any formal interactions in which players compete according to rules and in view of achieving a goal-the rules of which are *constitutive* and not merely *constraining*. The difference is obvious if one compares chess and football. Both are games, but in chess, the rules establish the legal and illegal moves before any chess-like activity is possible, so they are generative of all and only the acceptable moves. Whereas in football, a previous activity-let us call it kicking a ball-is "regimented" or structured by rules that arrive *after* the activity. The rules do not and cannot determine the moves of the players, they simply put boundaries to what moves are "legal". In chess, as in all board games whose rules are constitutive (draughts, Go, Monopoly, shogi...), AI can use the rules to play any possible legal move that it wants to explore. In 9 hours, AlphaZero played 44 million training games. To have a sense of the magnitude of the achievement consider that the *Opening Encyclopedia 2018* contains approximately 6.3 million games, selected from the whole history of chess. But in football, this would be meaningless because the rules do not make the game, they only shape it. This does not mean that AI cannot play virtual football; or cannot help identifying the best strategy to win against a team whose data about previous games and strategies are recorded; or cannot help with identifying potential players, or training them better. Of course, all these applications are now trivially feasible and already occur. What I mean is that when (1) a process or interaction can be transformed into a game, and (2) the game can be transformed into a *constitutive-rule* game, then (3) AI will be able to generate its own, fully synthetic data and be the best "player" on this planet, doing what AlphaZero did for chess (in the next section, I shall describe this process as *enveloping* (Floridi 2014a)). To quote Wiener:

> The best material model of a cat is another, or preferably the same, cat. (Rosenblueth and Wiener 1945, p. 316).

Ideally, the best data on which to train an AI are either the fully historical data or the fully synthetic data generated by the same rules that generated the historical data. In any board game, this happens by default. But insofar as any of these two steps (1)–(2) is difficult to achieve, the absence of rules or the presence of merely constraining rules is likely to be a limit. We do not have the actual cat, but only a more or less reliable model of it. Things can get more complicated once we realise that, in actual games, the constraining rules are simply conventionally imposed on a previously occurring activity, whereas in real life, when we observe some phenomena, e.g. the behaviour of a kind of tumour in a specific cohort of patients in some given circumstances, the genetic rules must be extracted from the actual "game" through scientific (and these days possibly AI-based) research. For example, we do not know, and perhaps we may never know, what the exact "rules" for the development of brain tumours are. We have some general principles and theories according to which we understand their development. So, at this stage (and it may well be a permanent stage), there is no way to "ludify" (transformation into a game in the sense specified above, I avoid the term 'gamifying' which has a different and well-established meaning) brain tumours into a "constitutive-rule game" (think of chess) such that an AI system, by playing according to the identified rules, can generate its own synthetic data about brain tumours that would be equivalent to the historical data we could collect, doing for brain tumours what AlphaZero has done for chess games. This is not necessarily a problem. On the contrary, Al, by relying on historical or hybrid data (e.g. brain scans) and learning from them, can still outperform experts, and expand its capabilities beyond the finite historical data sets provided (e.g. by discovering new patterns of correlations), or deliver accessible services where there is no expertise. It is already a great success if one can extract enough *constraining* rules to produce reliable data in silico. But without a reliable system of *constitutive rules*, some of the aforementioned advantages of synthetic data would not be available in full (the vagueness of this statement is due to the fact that we can still use hybrid data).

Ludification and the presence or absence of constraining/constitutive rules are not either-or, hard limits. Recall that hybrid data can help to develop synthetic data. What is likely to happen is that, in the future, it will become increasingly clear when high-quality databases of historical data may be absolutely necessary and unavoidable-when you need the actual cat, to paraphrase Wiener-and hence when we will have to deal with issues about availability, accessibility, legal compliance with legislation, and, in the case of personal data, privacy, consent, sensitivity, and other ethical questions. However, the trend towards the generation of as-synthetic-as-possible (synthesised, more or less hybrid, all the way to fully synthetic) data is likely to be one of Al's holy grails, so I expect the AI community to push very hard in that direction. Generating increasingly non-historical data, making the dial move as far as possible to the right in Fig. 22.2, will require a "ludification" of processes, and for this reason I also expect the AI community to be increasingly interested in the gaming industry, because it is there that the best expertise in "ludification" is probably to be found. And in terms of negative results, mathematical proofs about the impossibility of ludifying whole kinds or areas of processes or interactions should be most welcome in order to clarify where or how far an AlphaZero-like approach may never be achievable by AL.

22.4 Al's Future: From Difficult Problems to Complex Problems, and the Need for Enveloping

I have already mentioned that AI is best understood as a reservoir of agency that can be used to solve problems. AI achieves its problem-solving goals by detaching the ability to perform a task successfully from any need to be intelligent in doing so. The App on my mobile phone does not need to be intelligent to play chess better than I do. Whenever this detachment is feasible, some AI solution becomes possible in principle. This is why understanding the future of AI also means understanding the nature of problems where such a detachment may be technically feasible in theory and economically viable in practice. Now, many of the problems we try to solve through AI occur in the physical world, from driving to scanning labels in a supermarket, from cleaning flows or windows to cutting the grass in the garden. The reader may keep in mind AI as robotics in the rest of this section, but I am not discussing only robotics: smart applications and interfaces in the Internet of Things are also part of the analysis, for example. What I would like to suggest is that, for the purpose of understanding AI's development when dealing with physical environments, it is useful to map problems on the basis of what resources are needed to solve them, and hence how far AI can have such resources. I am referring to *computational resources*, and hence to degrees of *complexity;* and to *skill-related resources,* and hence to degrees of *difficulty*.

The degrees of complexity of a problem are well known and extensively studied in computational theory (Arora and Barak 2009; Sipser 2012). I shall not say much about this dimension but only remark that it is highly quantitative and that the mathematical tractability it provides is due to the availability of standard criteria of comparison, perhaps even idealised but clearly defined, such as the computational resources of a Turing Machine. If you have a "metre", then you can measure lengths. Similarly, if you adopt a Turing Machine as your starting point, then you can calculate how much time, in terms of steps, and how much space, in terms of memory or tape, a computational problem consumes to be solved. For the sake of simplicity-and keeping in mind that finely grained and sophisticated degrees of precision can be achieved, if needed, by using tools from complexity theory-let us agree to map the complexity of a problem (dealt with by AI in terms of space-time = memory steps required) from 0 *(simple)* to 1 *(complex).*

The degrees of difficulty of a problem, understood in terms of the skills required to solve it, from turning on and off a light to ironing shirts, need a bit more of a stipulation to be mapped here because usually, the relevant literature, e.g. in human motor development, does not focus on a taxonomy of problems based on resources needed, but on a taxonomy of the performance of the human agents assessed and their abilities or skills demonstrated in solving a problem or performing a task. It is also a more qualitative literature. In particular, there are many ways of assessing a performance and hence many ways of cataloguing skill-related problems, but one standard distinction is between gross and fine motor skills. Gross motor skills require the use of large muscle groups to perform tasks like walking or jumping, catching or kicking a ball. Fine motor skills require the use of smaller muscle groups, in the wrists, hands, fingers, and the feet and toes, to perform tasks like washing the dishes, writing, typing, using a tool, or playing an instrument. Despite the previous difficulties, you can see immediately that we are dealing with different degrees of difficulty. Again, for the sake of simplicity- and recalling that finely grained and sophisticated degrees of precision can be achieved, if needed, by using tools from developmental psychology-let us agree to map the difficulty of a problem (dealt with by AI in terms of skills required) from 0 *(easy)* to 1 *(difficult).*

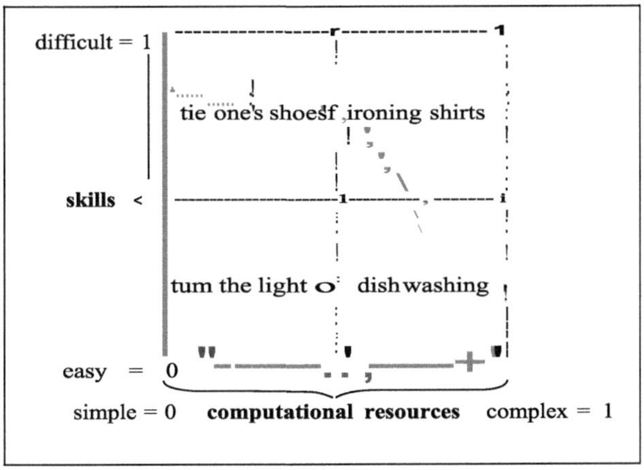

Fig. 22.3 Translating difficult tasks into complex tasks

We are now ready to map the two dimensions in Fig. 22.3, where I have added four examples.

Turning the light on is a problem whose solution has a very low degree of complexity (very few steps and states) and of difficulty (even a child can do it). However, tying one's own shoes requires advanced motor skills, and so does lacing them, thus it is low in complexity (simple), but it is very high in difficulty. As Adidas CEO Kasper Rorsted remarked in 2017:

> The biggest challenge the shoe industry has is how do you create a robot that puts the lace into the shoe. I'm not kidding. That's a complete manual process today. There is no technology for that.[7]

Dishwashing is the opposite: it may require a lot of steps and space, indeed increasingly more the more dishes need to be cleaned, but it is not difficult, even a philosopher like me can do it. And of course, top-right we find ironing shirts, which is both resource- consuming, like dishwashing, and demanding in terms of skills, so it is both complex and difficult, which is my excuse to try to avoid it. Using the previous examples of playing football and playing chess, football is simple but difficult, chess is easy (you can learn the rules in a few minutes) but very complex, this is why AI can win against anyone at chess, but a team of androids that wins the world cup is science fiction. Of course, things are often less clear-cut, as table tennis robots show.

The reader will notice that I placed a dotted arrow moving from low-complexity high-difficulty to high-complexity low-difficulty.[8] This seems to me the arrow that successful developments of AI will follow. Our artefacts, no matter how smart, are not really good at performing tasks and hence solving problems that require high

---

[7] https://qz.com/966882/robots-cant-lace-shoes-so-sneaker-production-cant-be-fully-automated-just-yet/

[8] I am not the first to make this point, see for example: https://www.campaignlive.eo.uk/article/hard-things-easy-easy-things-hard/1498154

degrees of skillfulness. However, they are fantastic at dealing with problems that require very challenging degrees of complexity. So, the future of successful AI probably lies not only in increasingly hybrid or synthetic data, as we saw, but also in translating difficult tasks into complex tasks.

How is this translation achieved? By transforming the environment within which AI operates into an AI-friendly environment. Such translation may increase the complexity of what the AI system needs to do enormously but, as long as it decreases the difficulty, it is something that can be progressively achieved more and more successfully. Some examples should suffice to illustrate the point, but first, let me introduce the concept of *enveloping*.

In industrial robotics, the three-dimensional space that defines the boundaries within which a robot can work successfully is defined as the robot's *envelope*. We do not build droids like Star Wars' C3PO to wash dishes in the sink exactly in the same way as we would. We envelop environments around simple robots to fit and exploit their limited capacities and still deliver the desired output. A dishwasher accomplishes its task because its environment-an openable, waterproof box-is structured ("enveloped") around its simple capacities. The more sophisticated these capacities are, the less enveloping is needed, but we are looking at trade-off, some kind of equilibrium. The same applies to Amazon's robotic shelves, for example. It is the whole warehouse that is designed to be robot-friendly. Ditto for robots that can cook[9] or flip hamburgers,[10] which already exist. Driverless cars will become a commodity the day we can successfully envelop the environment around them. This is why it is plausible that in an airport, which is a highly controlled and hence more easily "envelopable" environment, a shuttle could be an autonomous vehicle, but not the school bus that serves my village, given that the bus driver needs to be able to operate in extreme and difficult circumstances (countryside, snow, no signals, no satellite coverage etc.) that are most unlikely (mind, not impossible) to be enveloped. In 2016, Nike launched HyperAdapt 1.0, its automatic electronic self-lacing shoes, not by developing an AI that would tie them for you, but by re-inventing the concept of what it means to adapt shoes to feet: each shoe has a sensor, a battery, a motor, and a cable system that, together, can adjust fit following an algorithmic pressure equation.[11] Enveloping used to be either a stand-alone phenomenon (you buy the robot with the required envelop, like a dishwasher or a washing machine) or implemented within the walls of industrial buildings, carefully tailored around their artificial inhabitants. Nowadays, enveloping the environment into an AI-friendly info sphere has started to pervade all aspects of reality and is happening daily everywhere, in the house, in the office, and in the street. We have been enveloping the world around digital technologies for decades, invisibly and without fully realising it. The future of AI also lies in more enveloping, for example,

---

[9] http://www.moley.com/

[10] https://misorobotics.com/

[11] Strange things happen when the software does not work properly: https://www.bbc.eo.uk/news/business-47336684

in terms of 5G and the Internet of Things, but also insofar as we are all more and more connected and spend more and more time "onlife", and all our information is increasingly born digital. In this case too, some observations may be obvious. There may be problems, and hence relative tasks that solve them, that are not easily subject to enveloping. Yet here it is not a matter of mathematical proofs, but more of ingenuity, economic costs, and user or customer preferences. For example, a robot that iron shirts can be engineered. In 2012, a team at Carlos III University of Madrid, Spain, built TEO, a robot that weighs about 80 kg and is 1.8 m tall. TEO can climb stairs, open doors and, more recently, has been shown to be able to iron shirts (Estevez et al. 2017), although you have to put the item on the ironing board. The view, quite widespread, is that

> 'TEO is built to do what humans do as humans do it,' says team member Juan Victores at Carlos III University of Madrid. He and his colleagues want TEO to be able to tackle other domestic tasks, like helping out in the kitchen. Their ultimate goal is for TEO to be able to learn how to do a task just by watching people with no technical expertise carry it out. 'We will have robots like TEO in our homes. It's just a matter of who does it first,' says Victores.

And yet, I strongly doubt this is the future. It is a view that fails to appreciate the distinction between difficult and complex tasks and the enormous advantage of enveloping tasks to make them easy (very low difficulty), no matter how complex. Recall that we are building autonomous vehicles not by putting robots in the driving seat, but by rethinking the whole ecosystem of vehicles plus environments, that is, removing the driving seat altogether. So, if my analysis is correct, the future of AI is not full of TEO-like androids that mimic human behaviour, but is more likely represented by Effie,[12] Foldimate,[13] and other similar domestic automated machines that dry and iron clothes. They are not androids, like TEO, but box-like systems that may be quite sophisticated computationally. They look more like dishwasher and washing machines, with the difference that, in their enveloped environments, their input is wrinkled clothes and their output is ironed ones.

Perhaps similar machines will be expensive, perhaps they may not always work as well as one may wish, perhaps they may be embodied in ways we cannot imagine now, but you can see how the logic is the correct one: do not try to mimic humans through AI but exploit what machines, AI included, do best. *Difficulty* is the enemy of machines, *complexity* is their friend, so envelop the world around them, design new forms of embodiment to embed them successfully in their envelop, and at that point progressive refinements, market scale, and improvements will become perfectly possible.

[12] https://helloeffie.com/

[13] https://foldimate.com/

22.5 Conclusion: A Future of Design

The two futures I have outlined here are complementary and based on our current and foreseeable understanding of AI. There are unknown unknowns, of course, but all one can say about them is precisely this: they exist, and we have no idea about them. It is a bit like saying that we know there are questions we are not asking but cannot say what these questions are. The future of AI is full of unknown unknowns. What I have tried to do in this article is to look at the "seeds of time" that we have already sowed. I have concentrated on the nature of data and of problems because the former are what enable AI to work, and the latter provide the boundaries within which AI can work successfully. At this level of abstraction, two conclusions seem to be very plausible. We will seek to develop AI by using data that are as much as possible hybrid and preferably synthetic, through a process of ludification of interactions and tasks. In other words, the tendency will be to try to move away from purely historical data whenever possible. And we will do so by translating as much as possible difficult problems into complex problems, through the enveloping of realities around the skills of our artefacts. In short, we will seek to create hybrid or synthetic data to deal with complex problems, by ludifying tasks and interactions in enveloped environments. The more this is possible, the more successful AI will be, which leads me to two final comments.

Ludifying and enveloping are a matter of *designing,* or sometimes re-designing, the realities with which we deal (Floridi 2019). So, the foreseeable future of AI will depend on our design abilities and ingenuity. It will also depend on our ability to negotiate the resulting (and serious) ethical, legal, and social issues (ELSI), from new forms of privacy (predictive or group-based (Floridi 2014c)) to nudging and self-determination. The very idea that we are increasingly shaping our environments (analog or digital) to make them AI-friendly should make anyone reflect (Floridi 2013). Anticipating such issues, to facilitate positive ELSI and avoid or mitigate any negative ones, is the real value of any foresight analysis. It is interesting to try to understand what the paths of least resistance may be in the evolution of AI. But it would be quite sterile to try to predict "which grain will grow and which will not" and then to do nothing to ensure that the good grains grow, and the bad ones do not (Floridi 2014b). The future is not entirely open (because the past shapes it), but neither is it entirely determined, because the past can be steered in a different direction. This is why the challenge ahead will not be so much digital innovation per se, but the governance of the digital, AI included.

Acknowledgements I would like to thank all members of the Digital Ethics Lab, OII, University of Oxford, for many discussions about some of the topics covered in this article, and Nikita Aggarwal, Josh Cowls, Jessica Morley, David Sutcliffe, and Mariarosaria Taddeo for their hugely helpful comments on a last draft.

References

Arora, S., and B. Barak. 2009. *Computational complexity: A modem approach*. Cambridge: Cambridge University Press.

Campbell, M., A. Joseph Hoane Jr., and F.-H.J. Hsu. 2002. Deep blue. *Artificial Intelligence* 134 (1–2): 57–83.

Corea, Francesco. 2018. AI knowledge map: How to classify AI technologies, a sketch of a new AI technology landscape. Medium – Artificial intelligence. https://medium.com/@Francesco_AI/ai-knowledge-map-how-to-classify-ai-technologies-6c073b969020

Crawford, K., and J. Schultz. 2014. Big data and due process: Toward a framework to redress predictive privacy harms. *BCL Rev.* 55: 93.

De, F., J.R.L. Jeffrey, B. Romera-Paredes, S. Nikolov, N. Tomasev, S. Blackwell, H. Askham, X. Glorot, B. O'Donoghue, D. Visentin, G. van den Driessche, B. Lakshminarayanan, C. Meyer, F. Mackinder, S. Bouton, K. Ayoub, R. Chopra, D. King, A. Karthikesalingam, C.O. Hughes, R. Raine, J. Hughes, D.A. Sim, C. Egan, A. Tufail, H. Montgomery, D. Hassabis, G. Rees, T. Back, P.T. Khaw, M. Suleyman, J. Comebise, P.A. Keane, and O. Ronneberger. 2018. Clinically applicable deep learning for diagnosis and referral in retinal disease. *Nature Medicine* 24 (9): 1342–1350.

Estevez, David, Juan G. Victores, Raul Fernandez-Fernandez, and Carlos Balaguer. 2017. *Robotic ironing with 3D perception and force/torque feedback in household environments*. 2017 IEEE/RSJ international conference on Intelligent Robots and Systems (IROS).

Floridi, L. 2008a. The method of levels of abstraction. *Minds and Machines* 18 (3): 303–329.

———. 2008b. Understanding epistemic relevance. *Erkenntnis* 69 (1): 69–92.

———. 2010. *Information: A very short introduction*. Oxford: Oxford University Press.

———. 2012. Big data and their epistemological challenge. *Philosophy & Technology* 25 (4): 435–437.

———. 2013. *The ethics of information*. Oxford: Oxford University Press.

———. 2014a. *The fourth revolution – How the info sphere is reshaping human reality*. Oxford: Oxford University Press.

———. 2014b. Technoscience and ethics foresight. *Philosophy & Technology* 27 (4): 499–501.

———. 2014c. Open data, data protection, and group privacy. *Philosophy & Technology* 27 (1): 1–3.

———. 2017. Digital's cleaving power and its consequences. *Philosophy & Technology* 30 (2): 123–129.

———. 2018. What the maker's knowledge could be. *Synthese* 195 (1): 465–481.

———. 2019. *The logic of information*. Oxford: Oxford University Press.

Floridi, L., and J.W. Sanders. 2004. On the morality of artificial agents. *Minds and Machines* 14 (3): 349–379.

Floridi, L., M. Taddeo, and M. Turilli. 2009. Turing's imitation game: Still an impossible challenge for all machines and some judges-an evaluation of the 2008 Loehner contest. *Minds and Machines* 19 (1): 145–150.

Goodfellow, I., J. Pouget-Abadie, M. Mirza, B. Xu, D. Warde-Farley, S. Ozair, A. Courville, and Y. Bengio. 2014. Generative adversarial nets. *Advances in Neural Information Processing Systems*.

Howe, Bill, Julia Stoyanovich, Haoyue Ping, Bemease Herman, and Matt Gee. 2017. Synthetic data for social good. *arXiv preprint arXiv:l710.08874*.

Liang, Huiying, Brian Y. Tsui, Hao Ni, Carolina C.S. Valentim, Sally L. Baxter, Guangjian Liu, Wenjia Cai, Daniel S. Kennany, Xin Sun, Jiancong Chen, Liya He, Jie Zhu, Pin Tian, Hua Shao, Lianghong Zheng, Rui Hou, Sierra Hewett, Gen Li, Ping Liang, Xuan Zang, Zhiqi Zhang, Liyan Pan, Huimin Cai, Rujuan Ling, Shuhua Li, Yongwang Cui, Shusheng Tang, Hong Ye, Xiaoyan Huang, Waner He, Wenqing Liang, Qing Zhang, Jianmin Jiang, Wei Yu, Jianqun Gao, Wanxing Ou, Yingmin Deng, Qiaozhen Hou, Bei Wang, Cuichan Yao, Yan Liang, Shu Zhang, Yaou Duan, Runze Zhang, Sarah Gibson, Charlotte L. Zhang, Oulan Li, Edward

D. Zhang, Gabriel Karin, Nathan Nguyen, Xiaokang Wu, Cindy Wen, Jie Xu, Wenqin Xu, Bochu Wang, Winston Wang, Jing Li, Bianca Pizzato, Caroline Bao, Daoman Xiang, Wanting He, Suiqin He, Yugui Zhou, Weldon Haw, Michael Goldbaum, Adriana Tremoulet, Chun-Nan Hsu, Hannah Carter, Long Zhu, Kang Zhang, and Huimin Xia. 2019. Evaluation and accurate diagnoses of pediatric diseases using artificial intelligence. *Nature Medicine*.

McCarthy, J., M.L. Minsky, N. Rochester, and C.E. Shannon. 2006. A proposal for the Dartmouth summer research project on artificial intelligence. *AI Magazine* 27 (4): 12.

Rosenblueth, A., and N. Wiener. 1945. The role of models in science. *Philosophy of Science* 12 (4): 316–321.

Silver, D., T. Hubert, J. Schrittwieser, I. Antonoglou, M. Lai, A. Guez, M. Lanctot, L. Sifre, D. Kwnaran, T. Graepel, T. Lillicrap, K. Simonyan, and D. Hassabis. 2018. A general reinforcement learning algorithm that masters chess, shogi, and Go through self-play. *Science* 362 (6419): 1140–1144.

Sipser, M. 2012. *Introduction to the theory of computation*. 3rd ed. Boston: Cengage Learning.

Turing, A.M. 1950. Computing machinery and intelligence. *Mind* 59 (236): 433–460.

Watson, David S., Jenny Krutzinna, Ian N. Bruce, Christopher E.M. Griffiths, Iain B. Mcinnes, Michael R. Barnes, and Luciano Floridi. Forthcoming. Clinical applications of machine learning algorithms: Beyond the black box. *British Medical Journal*.

GPSR Compliance

The European Union's (EU) General Product Safety Regulation (GPSR) is a set of rules that requires consumer products to be safe and our obligations to ensure this.

If you have any concerns about our products, you can contact us on ProductSafety@springernature.com

In case Publisher is established outside the EU, the EU authorized representative is:

Springer Nature Customer Service Center GmbH
Europaplatz 3
69115 Heidelberg, Germany

Batch number: 10732076

Printed by Printforce, the Netherlands